THE TECHNIQUES OF
MODERN
STRUCTURAL
GEOLOGY
Volume 2: Folds and Fractures

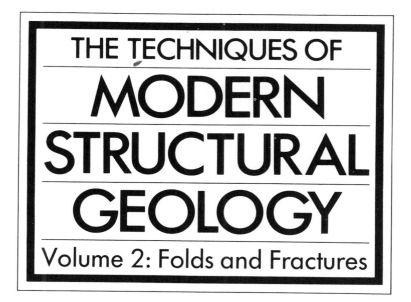

THE TECHNIQUES OF
MODERN
STRUCTURAL
GEOLOGY
Volume 2: Folds and Fractures

JOHN G. RAMSAY
MARTIN I. HUBER

1987

ACADEMIC PRESS
Harcourt Brace Jovamovich, Publishers

London   San Diego   New York   Berkeley
Boston   Sydney   Tokyo   Toronto

ACADEMIC PRESS LIMITED
24/28 Oval Road
London NW1

United States Edition published by
ACADEMIC PRESS INC.
San Diego, CA 92101

British Library Cataloguing in Publication Data
Ramsay, John G.
    The techniques of modern structural geology.
    Vol. 2, Folds and fractures
    I. Geology, Structural
    I. Title   II. Huber, Martin I.
    551.8   QE601

    ISBN 0-12-576902-4
    IBSN 0-12-576922-9 Pbk

Printed in Great Britain at the Alden Press, Oxford

Preface

The accumulation of small-scale structural data is a painstaking but fascinating business. It involves plotting not only dips and strikes of stratification, but also of schistosity, together with the plunges of linear structures, and the collection of oriented specimens for laboratory study. It is easy to be carried away by enthusiasm for the subject while in the field and to miss other equally important but less exciting geological observations. This means that the worker must submit himself to a strict discipline, because without corresponding attention to stratigraphy and its kindred subjects, a tectonic or historical synthesis of a region may be incorrect, or at least incomplete. The converse is also true; because without appreciation of the movements which the rocks have suffered, the stratigraphy itself may be incorrect or indecipherable.

Gilbert Wilson, 1961

. . . every structure in a rock is significant, none is unimportant, even if, at first sight it may seem irrelevant.

Ernst Cloos, 1946

Volume 2 of *The Techniques of Modern Structural Geology* has been written to complement Volume 1 so that the two books can form a basis for a two-semester university teaching programme in structural geology. The organization and format of Volume 2 are more or less on the same lines as those of Volume 1. Many of the basic concepts of strain, compatibility and the significance of rock fabrics set out in Volume 1 are extended so as to interlink with the wider geometrical and mechanical aspects of folding and fracturing processes seen in naturally deformed rocks. Each Session has been prepared to form the basis for discussion of some well-defined aspect of rock structure. Each begins with an introductory part setting out the groundwork of the topic and the reader is then confronted with questions of a practical nature which require constructive thought to find a solution. All these questions are based on real examples, mathematical principles of deformation geometry, or the results of actual laboratory experiments. The questions are then followed by an "answers and comments" section in which geometrical, numerical or mathematical aspects are resolved and a wide-ranging discussion of specific phenomena arising in the topic are presented. For the student to be able to advance to the following Session he should have studied all the basic questions and the relevant commentaries. For those students who have more time or are at a higher level of experience, a series of "starred" questions will provide a deepening of their understanding. From the way the book is constructed, we do not expect a student to systematically read it from cover to cover.

We have provided many photographic illustrations of the structures of naturally deformed rocks as seen in the field and in the laboratory over a wide range of scales from mountain side views, through the features of individual outcrops, and to the scale of microscopic views of thin sections. We are of the opinion that it is most important that the student acquires a sound impression of these natural phenomena. The examples we have chosen for our illustrations have been selected not only for their clarity but because they are different from the well-worn selections often found in text books. We did not want to rehash the classic examples yet again, important as these are, but, in consequence, we do expect the student to supplement this text with other books and journal publications.

We have not discussed the regional aspects of tectonics for reasons of philosophy as well as for reasons of space. In our opinion structural geology is just one facet of the subject of tectonics. Any adequate discussion of tectonics requires structural geology to be integrated with sedimentology, stratigraphy, igneous and metamorphic petrology, geochronology and many aspects of geophysics. Structural geology is a well-defined subject in its own right, and to do minimum justice to its range and content has required considerable text space. Today there is an unfortunate tendency of some of our earth science colleagues to put all their eggs into the plate tectonics basket and to consider phenomena on a smaller scale to be trivial details, irrelevant, or even obstructive to scientific progress. To our way of thinking this idea of "bigger is better" is intellectually and philosophically unacceptable.

Several colleagues and friends have helped us in providing material used in the practical exercises. Of our colleagues at the ETH Zürich, Ueli Briegel provided data on experimental rock deformation and Nazario Pavoni, of the Geophysics Institute, gave us data and advice on the first motion seismic analysis. Dorothee Dietrich provided two diagrams summarizing the regional patterns of shear vein arrays in the Helvetic nappes and Peter Huggenberger gave us permission to use the results of his palinspastic reconstructions of the Morcles Nappe. Friends outside Zürich also provided us with valuable input material. Special thanks go to Mark Cooper for giving us the original data for the line length balancing of the Hénaux quarry and to Dietmar Meier for providing data and much useful discussion on the significance of joint patterns in the Jura mountains. Photographic material showing thin section fabrics in sheared rocks was provided by Carol Simpson and Cees Passchier, a splendid example of folded cross bedding came from Jack Soper, and two results of laboratory experiments pertinent to our discussions of folding

came from the work of Peter Cobbold. Appendix F, on geological mapping, was inspired by a humourous but very instructive article which appeared in the first editions of the *Kingston Geological Review* written by Mike Fleuty.

Our special thanks go to our colleague Barbara Das Gupta for organizing our manuscript and for valuable advice, and also to Urs Gerber for providing black and white copies of our colour transparencies and for solving so expertly many other photographic problems. Warm and unsolicited thanks go to the staff of Academic Press, especially to Conrad Guettler for being so receptive to the rather unusual style and format of this book and for providing the expert professional help needed to produce these first two volumes.

Many teachers and instructors in the field of structural geology have expressed an interest in obtaining copies of the colour transparencies of the black and white photographs used to illustrate Volumes 1 and 2 of *The Techniques of Modern Structural Geology*. We have therefore prepared several partial and one complete collection of very high quality colour transparencies of this material. A brochure with further information can be obtained from Professor J. G. Ramsay, Geologisches Institut, ETH Zentrum, CH-8092 Zürich, Switzerland.

JOHN RAMSAY
MARTIN HUBER

Contents

Contents of Volume 1: Strain Analysis

These volumes are dedicated to the memory
of Gilbert Wilson

SESSION 15

Fold Morphology

After a general introduction to folding this session considers what special terms are required to describe the geometric features of folds: first those terms necessary to define features of a single folded surface, then those which describe the geometric relationships of adjacent folded layers. Special techniques to describe curvature variations based on harmonic—or Fourier—analyses are recommended and it is shown how this type of analysis is practical both for general descriptive purposes as well as for very exact analysis. The relationships of small wavelength folds in polyharmonic folds are described and practical ways of recording changing fold symmetry discussed.

INTRODUCTION

Folds are perhaps the most common tectonic structure developed in deformed rocks. They form in rocks containing planar features such as sedimentary bedding, lithological layering in metamorphic schists and gneisses, or planar anisotropic features, such as cleavage or schistosity, produced during an early deformational event. Folds are common because the growth rate of mechanical instabilities setting up sideways deflections during the shortening along a planar feature is generally rapid. Not all folds, however, are formed by such **buckling** processes. Folds can also be formed when non-uniform forces are applied across the layering; **bending** processes such as variable vertical subsidence of crustal layers, differential shearing in shear zones or by the sliding of rock masses over underlying floor irregularities can also give rise to folds.

The most striking features of folds are their great variations in geometric type, generally termed **fold style**, and in their size. A quick glance through the illustrations accompanying the next six sessions will give some idea of the variety of fold geometry. These variations in fold style are invariably linked with differences in the mechanical origin of the structures and especially with the rheological state of the rocks during the folding process. The correct interpretation of the significance of variations in fold style is important for the geologist wishing to relate geometric structure of deformed rocks to major tectonic events. Changing environmental conditions are typical of orogenic zones in both space and time, and differences in rock type, temperature, pressure and strain rate gives rise to a vast geometric spectrum of fold styles.

Correct evaluation and interpretation of fold geometry has especially important industrial applications. In the oil industry it is well known that folds and their associated structures can form oil traps, and an understanding of fold periodicity and geometry often plays a key role in the selection of drilling sites and in exploration and development

programmes. In the mineral industry and prediction of the extent of ore reserves in folded ore bodies is often of great importance in developing economically viable programmes on mine and regional scales.

In this session our first concern will be to discover what are the main geometric features of folded rocks. We need to find a number of essential descriptive terms with which we can unambiguously refer to particular parts of folds and to develop a primary classification of folds based on the geometry of these features. Accurate geometric descriptions and classifications are not just of academic interest. They are vital if we need to predict the continuity (or lack of continuity) of a fold both sideways and in depth, and they provide the key to comparing natural folds with folds produced in laboratory experiments and with the fold shapes predicted using mechanical theories of fold formation. We emphasize that these descriptions and classifications must be of practical application. A geologist often wants to be able to describe a fold quickly in the field; he does not always have the time or need to spend a long period of analytical effort before he is able to give a name to a particular fold shape. We have therefore focused attention on practical methods which give rapid, useful and accurate geometric descriptions, especially those methods which can be elaborated, should the problem demand it, to define precisely and in numerical terms the exact geometric features of folds.

Definitions: a single folded surface

A folded layer shows variations in curvature which are used to define important features of the fold. Curvature is a fundamental or invariant property of the surface, and the terms based on curvature variations define basic features of the folded layer. Curvature varies from particular maximum (positive) values through zero to minimum (negative) values.

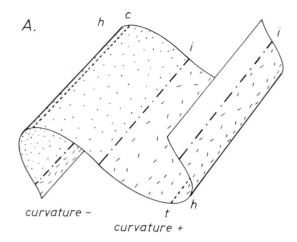

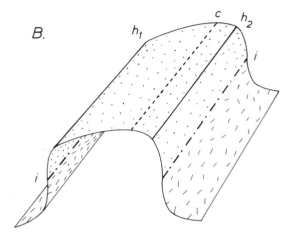

Figure 15.1. Main features shown by single hinge (A) and double hinge (B) folds. Hinge lines h, inflexion lines i, crest lines c, trough lines t. Antiformal domain with negative curvature shown with point ornament, synformal domains with positive curvature with dashed lines.

Lines joining the points of maximum curvature and those joining points of minimum curvature are known as **hinge lines**, and lines joining points of zero curvature are known as **inflexion lines** (Figure 15.1). Inflexion lines subdivide the folded surface into **fold domains**. Domains with negative curvature (upward closing) are known as **antiforms**, those with positive curvature (downward closing) as **synforms** (Figure 15.1). Although most folds can be described in this way, some close with neither upward nor downward

sense. Such sideways closing structures are known as **neutral folds** (Figure 15.2A), and where all surfaces in a neutral fold are vertically inclined the structure is termed a **vertical fold** (Figure 15.2B). The terms antiform and synform and neutral are based only on geometric criteria, and other terms are used to define the stratigraphic relationships, "younging" or polarity sense in folded sedimentary rocks. An **anticline** is a fold in which the oldest rocks are located in the core of the fold (or concave side of the fold surface), whereas a **syncline** is a fold in which the youngest rocks are located in the fold core. Although in most situations that will be encountered antiforms are generally also anticlines, and synforms generally synclines, this is not always so. Folds developed in rocks which have been previously inverted will show antiforms which are synclines. This double nomenclature is extremely useful in regions of superposed folding to define the exact geometric and stratigraphic relationships in the structure. Figure 15.3 shows a large sideways closing anticline which has been refolded by more upright structures. The superimposed antiform on the right of the diagram can be subdivided into two parts: the polarity directions indicated by arrows allow us to distinguish an upper **antiformal anticline** and a lower **antiformal syncline**. Another method that is sometimes used to incorporate the polarity of stratigraphic successions into descriptions of fold geometry was suggested by Shackleton (1958). He defined the **facing direction** in a fold in the sense of the polarity arrows in Figure 15.3. In this nomenclature the antiformal anticline would be termed an **upward facing anticline**, the antiformal syncline would be a **downward facing syncline**.

Within the domain of a fold lying between two inflexion lines a fold may have one (Figure 15.1A), two (Figure 15.1B) or more hinge lines. The surface region of the fold around the fold hinge line is known as a **hinge zone**, and the surface region of the fold between two adjacent hinge lines is generally known as a **fold limb**. Very precise definitions of hinge zone and fold limb have been suggested (Ramsay, 1967) based on a more detailed analysis of curvature variation, but these suggestions have not been generally used. The azimuth or direction of a hinge line is often termed the **axial trend** of the fold. A fold hinge may be horizontal or inclined, and the surface outcrop pattern of the folded layering in folds with inclined hinge lines show characteristic V or U shaped patterns which are important features used in the interpretation of fold geometry on geological maps. In general, fold hinge lines undulate, and these changes in orientation give rise to special structural forms. The feature whereby a hinge line plunging in one direction passes

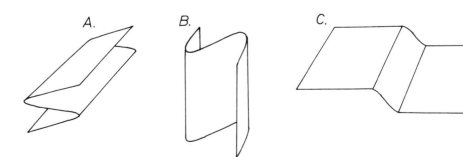

Figure 15.2. Sideways closing or neutral fold (A), vertical fold (B) and monocline (C).

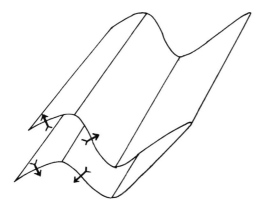

Figure 15.3. *Refolded neutral fold. The antiformal refold on the left can be subdivided into an upper antiformal anticline and a lower antiformal syncline. Arrows indicate stratigraphic polarity.*

through the horizontal to plunge in the opposite direction is described as a hinge **culmination** or **depression** depending upon whether the change of plunge is upbowed or downbowed respectively (Figure 15.4). Such changes of plunge set up **domes** (antiform with hinge line culmination), **basins** (synform with hinge line depression), **saddles** (antiform with depression), or **inverted saddles** (synform with culmination). In certain regions these changes of hinge line orientation may be very strong (greater than 90°), and the dome or basin morphology becomes extremely prominent. Such folds have been termed **eyed folds** (Ramsay, 1962—on account of the closed eye-like forms seen in map outcrop, Figure 15.5A) or **sheath folds** (Cobbold and Quinquis, 1980—on account of the overall three dimensional geometry, Figure 15.5B).

If at all locations on the fold surface a direction can be found which lies parallel to the fold hinge line, the fold is termed **cylindrical**, and the line is known as the **fold axis**. The cross section of a cylindrical fold need not be circular in form; the name implies only that the surface morphology is mathematically cylindrical in the sense that it could be generated by moving the axis in space parallel to itself. Although the geometric form of small sectors of many folds is often quite close to this model, most folds depart from true cylindricity and are known as **non-cylindrical** folds. Folds can never continue indefinitely along their hinge lines, and at some point they must end. In many instances the end of one fold (or fold pair) is connected with the start of a new fold (or fold pair). Folds which relay each other in this manner are termed **en-echelon folds** (Figures 15.6, 15.7). The terminations of individual en-echelon folds are always non-cylindrical. Sometimes they approach a form which is like that of a mathematically defined cone (a surface generated by a line moving througn a fixed point). Forms approaching **conical folds** can be found (Wilson, 1967) but generally the layer morphology is more complex than that of a truly mathematical conical shape.

On a fold surface there is generally found a line where the fold surface reaches its topographically highest position, and one where the layer is at its deepest. These lines are known as **crest line** and **trough line** respectively (Figure 15.1, lines *c* and *t*). In contrast to hinge lines these are not invariant features of the fold surface, but depend upon the orientation of the fold wave to the Earth's surface.

Fold dimensions

Folds are found over a wide range of scales and we often need to express the size in an appropriate way. It may be convenient in general descriptions to give a picture in average metric terms (e.g. kilometric, decametric, millimetric).

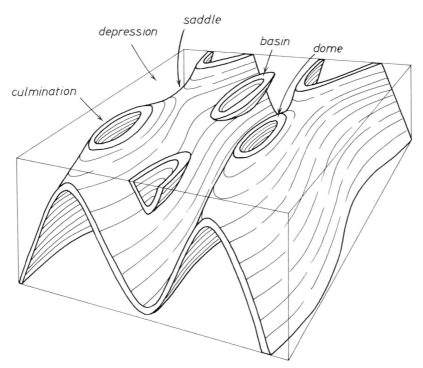

Figure 15.4. *Structural features arising from changes in the hinge line plunges of folds.*

Figure 15.5. (A) *eyed folds in section, Sokumfjell, Norway. (B): Sheath fold developed in a folded quartz vein, Isle de Groix, France. The fold axis changes direction through an angle of more than 130°.*

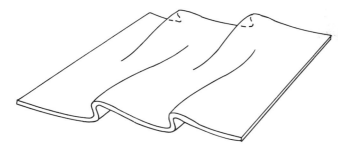

Figure 15.6. *Folds relaying each other in an en-echelon manner.*

For more exact descriptions we need to apply the mathematical concepts of wavelength and amplitude. If folded rocks showed the regular geometry of simple sinusoidal waves there would be no problem in doing this, but natural folds do not show such symmetry or periodic regularity. Each fold domain must be analysed separately, each fold having its own wavelength and amplitude. The general trend of the folded layer can be found by joining successive inflexion points, and this join defines the **median surface** of the folds. To determine the **wavelength W** of individual folds we measure the distance between the two inflexion lines defining the fold domain, and this distance is the half wavelength of the fold ($W/2$ in Figure 15.8A). For this domain we construct a line parallel to the inflexion line join so as to touch the extremity of the fold. The distance between these lines is the **fold amplitude A**.

When the median surface of a folded layer is constructed it may be found that this has a curving or folded form. The folded form of the median surface can be analysed in the same way as was done for the original folds and, in so doing, larger characteristic wavelengths and amplitudes of the folded surface can be established (Figure 15.8B). Fold wave trains with two or more orders of wavelengths and amplitudes are known as **polyharmonic folds**. We will study these in Session 20 and it will be shown that they generally arise where layers of different composition or thicknesses have been folded together. The folds of smallest wavelength are sometimes called **parasitic folds** on the larger wavelength structure. The terms **anticlinorium** and **synclinorium** describe polyharmonic folds where the larger scale folds are antiformal and synformal respectively.

Fold shape

One important feature of a fold is the degree of tightness. This is defined by determining the **interlimb angle**: the angle between tangents to the fold surface drawn through the inflexion lines (Figure 15.9). It is sometimes useful to be able to refer to this feature in words, and the following terms recommended by Fleuty (1964) will be found to provide a suitable range of names:

gentle	interlimb angle 180°–120°
open	120°–70°
close	70°–30°
tight	30°–0°
isoclinal	0°
elastica	negative interlimb angle

Figure 15.7. *Folds on a surface of Flysch siltstone showing complex relaying relationships, ultrahelvetic nappes at Frenières-sur-Bex, Switzerland.*

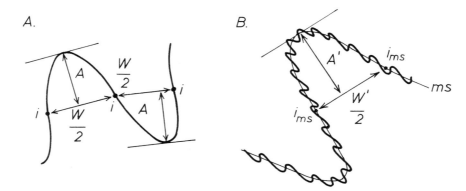

Figure 15.8. *Definitions of fold wavelength W (c.f. antiformal and synformal parts respectively of the folded surface) and fold amplitude A. B Shows how more than one wavelength may exist in a folded surface, and W' and A' define the higher order wavelength and amplitude in a polyharmonic fold where ms is the median surface of the lower order folds and i_{ms} the inflexion lines of this median surface.*

The appearance of a fold depends not only on the inter-limb angle but also on other features of curvature variation. These variations can be very wide and it is sometimes diffi-cult to describe in words the shapes one observes. Hinge zones may be rounded or angular, and limbs may be curved or straight. Folds with narrow angular hinge zones and straight limbs are known as **zig-zag**, **chevron** or **accor-dion** folds. Two major problems confront us here. The first is that, although verbal descriptions may be used to efficiently describe the geometry of certain styles of fold, one cannot devise an overall word usage to cover all possible forms. Second, for certain types of morphological analysis, one needs specific numerical parameters to exactly define the observed variations in curvature. A practical solution to this problem using a type of mathematical formulation known as harmonic or Fourier analysis has been suggested by Stabler (1968) and by Hudleston (1973a).

Fourier analysis of fold form

One basic method used in mathematics for analysing the curvature variations is to try and find a series of simple x–y functions, the sum of which gives a close fit to the more complex x–y function of the original curve. It is possible, for example, to fit a polynominal expression

$$f(xy) = a + bx + cx^2 + dx^3 \ldots \quad (15.1)$$

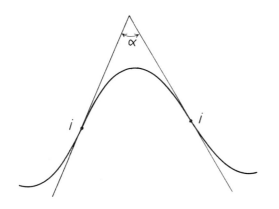

Figure 15.9. *The interlimb angle α of a fold.*

to a particular curve such that it can be expressed as a number of coefficients a, b, c, d, etc. to members of this series. In practice it is found that a large number of coef-ficients are necessary to express the form of even small sectors of folds.

The technique known as harmonic or Fourier analysis uses series of trigonometric functions to express the shape of any curve.

$$f(xy) = a_0 + a_1 \cos x + a_2 \cos 2x + a_3 \cos 3x \ldots$$
$$+ b_1 \sin x + b_2 \sin 2x + b_3 \sin 3x \ldots \quad (15.2)$$

The curve is specified by the various a and b parameters, known as Fourier coefficients, which refer to the amplitude of the various cosine and sine components (Figure 15.10). Polyharmonic waves of considerable complexity can be des-cribed by this series but generally a large number of Fourier coefficients have to be employed. If, however, we select only a small part of the complete wave, we can define the shape of this sector with very few coefficients. To do this with maximum simplicity we also need to select our reference directions so as to coincide with special features of the fold and to remove the scale factor so that the component wavelengths accord with that of the particular fold we wish to analyse.

To make the analysis we first select a sector of the complete layer between an inflexion line and a hinge line (Figure 15.11, i and h). The coordinate axis y is chosen so that it passes through the inflexion point i and is parallel to the axial surface of the fold (the **axial surface** is the surface joining the hinge line on the folded layer with the hinge line on an immediately adjacent folded layer—see p. 317). The x-axis of the coordinate system passes through i perpendicular to the y-axis so that the origin is the point i. With this coordinate frame all the a-type Fourier coefficients become zero; only the b-type Fourier coefficients can be represented in the function (cf. Figure 15.10). The fold sector i–h is now envisaged as being a quarter-wave sector of a completely regular symmetric wave form (dashed line in Figure 15.11). Because all the even numbered b-type coef-ficients (b_2, b_4, b_6, etc.) give rise to changes in shape of successive quarter wave sectors they must also be zero. Our fold sector can therefore be represented by a reduced Fourier series:

$$f(xy) = b_1 \sin x + b_3 \sin 3x + b_5 \sin 5x \ldots (15.3)$$

cosine functions   sine functions

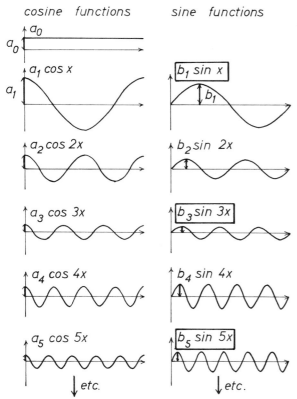

Figure 15.10. *Components of a full Fourier Series showing the significance of the coefficients a_0, a_1, . . . , b_1, b_2, . . . , etc.*

In practice it is found that the successive coefficients in this series become smaller and, in mathematical terminology, the series converges very rapidly. Only the first three coefficients provide significant input into the description of the wave shape, and the third of these takes on a very small value.

Figure 15.11 illustrates graphically the first (b_1), third (b_3) and fifth (b_5) harmonic components of the Fourier series defining the sector i–h of the original folded layer.

The values of these three components can be obtained by finding coordinate values of any three points (x_1, y_1), (x_2, y_2) and (x_3, y_3) on the original layer. From equation (15.3) we can write three equations:

$$y_1 = b_1 \sin x_1 + b_3 \sin 3x_1 + b_5 \sin 5x_1$$

$$y_2 = b_1 \sin x_2 + b_3 \sin 3x_2 + b_5 \sin 5x_2$$

$$y_3 = b_1 \sin x_3 + b_3 \sin 3x_3 + b_5 \sin 5x_3$$

The determinants of the following matrices are calculated:

$$D = \begin{vmatrix} \sin x_1 & \sin 3x_1 & \sin 5x_1 \\ \sin x_2 & \sin 3x_2 & \sin 5x_2 \\ \sin x_3 & \sin 3x_3 & \sin 5x_3 \end{vmatrix}$$

$$D_1 = \begin{vmatrix} y_1 & \sin 3x_1 & \sin 5x_1 \\ y_2 & \sin 3x_2 & \sin 5x_2 \\ y_3 & \sin 3x_3 & \sin 5x_3 \end{vmatrix}$$

$$D_2 = \begin{vmatrix} \sin x_1 & y_1 & \sin 5x_1 \\ \sin x_2 & y_2 & \sin 5x_2 \\ \sin x_3 & y_3 & \sin 5x_3 \end{vmatrix}$$

$$D_3 = \begin{vmatrix} \sin x_1 & \sin 3x_1 & y_1 \\ \sin x_2 & \sin 3x_2 & y_2 \\ \sin x_3 & \sin 3x_3 & y_3 \end{vmatrix}$$

(i.e. $D = \sin x_1 \sin 3x_2 \sin 5x_3 + \sin 3x_1 \sin 5x_2 \sin x_3 + \sin 5x_1 \sin x_2 \sin 3x_3 - \sin x_1 \sin 5x_2 \sin 3x_3 - \sin 3x_1 \sin x_2 \sin 5x_3 - \sin 5x_1 \sin 3x_2 \sin x_3$) and the solutions become

$$b_1 = D_1/D, \quad b_3 = D_2/D, \quad b_5 = D_3/D \qquad (15.4)$$

An especially convenient solution for the evaluation of these Fourier coefficients was suggested by Stabler. By dividing the quarter wavelength fold sector into three equal parts, the four determinants become very simple algebraic expressions and three ordinate measurements can be made to give the

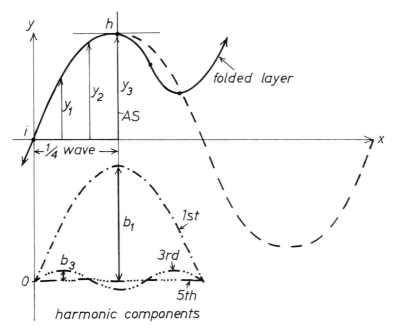

Figure 15.11. *Method of selecting coordinates for a quarter wave sector of a folded surface from the inflexion and hinge lines and the axial surface AS. The quarter wave section is envisaged as forming part of a symmetric full wave (dashed line). The significance of the Fourier coefficients b_1, b_3 and b_5 are shown in the lower part of the diagram as the harmonic components.*

values (Figure 15.11):

$$b_1 = (y_1 + \sqrt{3}y_2 + y_3)/3$$
$$b_3 = (2y_1 - y_3)/3 \qquad (15.5)$$
$$b_5 = (y_1 - \sqrt{3}y_2 + y_3)/3$$

The interaction of the first and third harmonics to provide the principal features of the fold shape is illustrated in Figure 15.12. The main amplitude is derived from the coefficient b_1, however this is modified slightly by the b_3 coefficient. Positive values of b_3 reduce the amplitude effect of b_1, while negative values increase it. The other significant effect of the b_3 coefficient is to modify the shape of the sinusoidal b_1 components. Positive values of b_3 broaden the hinge region and tend to produce box-fold shapes, whereas negative values tend to steepen and straighten the fold limbs and produce more chevron-like styles (Figure 15.12, cf. A and B).

Because the fifth harmonic component is very small and exerts very little geometric effect on the result, Hudleston suggested a very practical graphical method of plotting fold shapes by graphically representing b_1 as abscissa and b_3 as ordinate (Figure 15.13). Each fold shape has a characteristic b_3/b_1 ratio (e.g. chevron, sinusoidal, parabolic, semi-ellipse and box folds have b_3/b_1 values of -0.111, 0.000, 0.037, 0.165 and 0.333, respectively) and for each shape type the ratio of amplitude to wavelength increases with increase in the value of the coefficient b_1. Hudleston suggested that it might be practical to devise a shape classification scheme for single folded layers using shape type (indicated by letters A to F) versus amplitude (indicated by numbers 1 to 5). Folds may be classified according to their general style (e.g. D, parabolic), or may be given a definitive letter and number classification (e.g. D4) to denote more exactly the amplitude–wavelength relationships of a particular parabolic fold. This type of classification can be carried out very quickly by a technique he termed **visual harmonic analysis**. The 30 basic fold types he recommends for this classification are illustrated in Figure 15.14, the folds being drawn as if they were symmetric half waves. He recommends that this diagram be recorded on transparent paper so that the actual fold outcrop (or photograph) can be compared directly with the various shape types, the scale factor of the distance between hinge and inflexion point being adjusted by moving the transparent overlay towards or away from the observer.

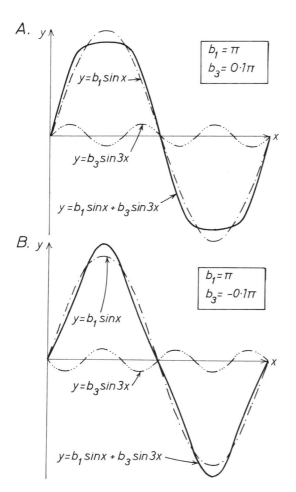

Figure 15.12. *Effect of the value of the coefficient b_3 on the fold shape. A shows how a positive value give box fold shapes and B how negative values give chevron-like styles.*

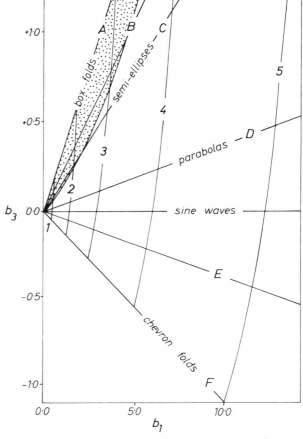

Figure 15.13. *Graphical method of plotting fold shape using Fourier coefficients (after Hudleston, 1973a). The stippled area is where double hinged folds occur. The letters A to F describe overall fold shape, and the numbers 1 to 5 relate to the amplitude number.*

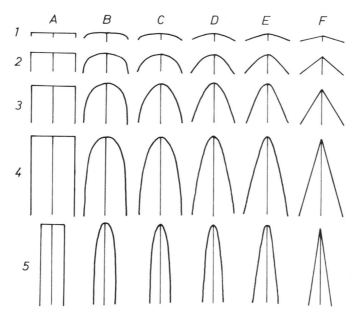

Figure 15.14. *Thirty basic folds with different shapes (*A *to* F*) and amplitude numbers (1 to 5) used for visual harmonic analysis (after Hudleston, 1973a).*

QUESTIONS

Wavelength–amplitude analysis

Question 15.1

Figure 15.15 shows an outcrop of folded metasedimentary rocks, amphibolite grade pelites and sandstones of the Moine Series of Mull, Northern Scotland. Several marker surfaces have been identified and lettered A to I. The joins of fold hinge lines in adjacent surfaces have been constructed (axial surfaces) as dashed lines, and the joins of inflexion lines in the surfaces are also indicated with dotted lines.

Using the nomenclature of fold wavelength and amplitude set out in Figure 15.8 determine and graphically represent the half wavelengths as abscissa and amplitudes as ordinate. Discuss the result. Can you suggest other ways of defining wavelength and amplitude relationships that might also be appropriate to this problem?

Visual harmonic analysis

Question 15.2

Using the visual harmonic analysis method of Hudleston determine which of the 30 basic types of folds can be found in Figure 15.15.

Fourier analysis

Question 15.3

In Figure 15.15 make an accurate determination of the Fourier coefficients b_1, b_3, and b_5 in the fold sectors c-c' and d-d' using the recommendations by Stabler (equations 15.5 and Figure 15.11). Note that the distances y_1, y_2 and y_3 must

be expressed in terms of units of length of the standardized quarter wavelength $(\pi/2)$.

Definitions: several folded surfaces

Several new definitions to describe fold geometry have to be introduced when we consider the relationships of adjacent folded surfaces. The inflexion lines on successive surfaces can be joined up to form surfaces known as **inflexion line surfaces**, and these completely define the various fold domains in three dimensions (Figure 15.16). An inflexion surface may end at some line in the rock. This implies that a fold in one layer becomes of progressively smaller amplitude in adjacent layers and that this amplitude continues to decrease until a point where the layers are unaffected by folds (Figure 15.17). In other instances adjacent inflexion surfaces converge to produce lens-shaped regions in three dimensions, and these "pods" completely delimit a particular fold domain (Figure 15.17).

Surfaces which connect the hinge lines on adjacent layers are **hinge line surfaces**, more usually called **axial surfaces** or, when planar, **axial planes** (Figure 15.16). A fold domain may show one or more axial surfaces. Folds with converging paired axial surfaces are generally termed **conjugate folds**. Conjugate folds with rounded hinge zones are generally known as **box folds**, whereas those with very sharp hinges are generally called conjugate **kink folds**. The axial surfaces of conjugate folds generally meet along a line somewhere in the structure. Beyond this line the fold may show only a single axial surface (Figure 15.18A) or the two axial surfaces may cross one another, with one surface being displaced across the other (Figure 15.18B). Rarely one finds fold domains with more than two axial surfaces; these are known as **polyclinal folds**.

The axial surfaces of single hinge folds may come to a stop in a similar way to that described with inflexion surfaces, or they may form closed pod-like forms which isolate a limb

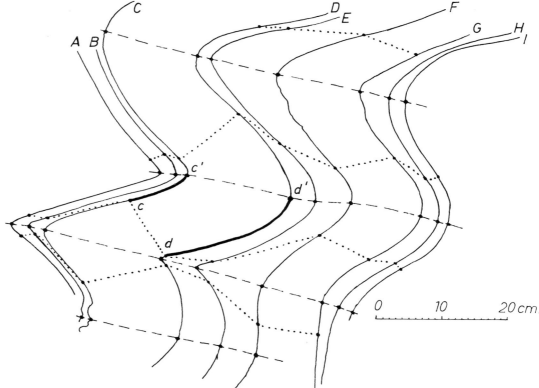

Figure 15.15. *Folded metasandstones and pelites from Mull, Scotland. See Questions 15.1, 15.2, 15.3.*

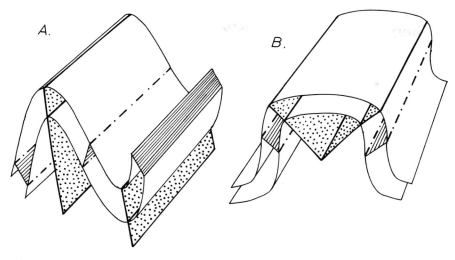

Figure 15.16. *Axial surfaces (dotted) and inflexion surfaces (lined) in* A, *single hinge and* B, *double hinge folds.*

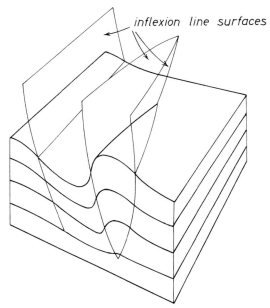

inflexion line surfaces

Figure 15.17. *Inflexion line surfaces defining fold domains in three dimensions. Where the two inflexion surfaces on the right-hand side of the block meet, the antiform ceases to exist as a geometric entity.*

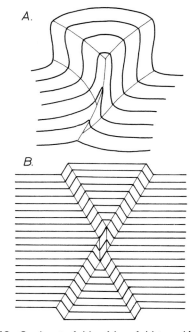

Figure 15.18. *Conjugate folds of box fold type* (A) *and kink fold type* (B).

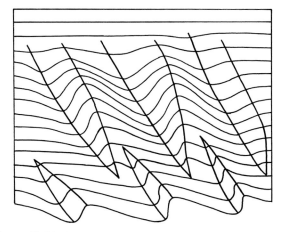

Figure 15.19. *Types of axial surface patterns in folded multilayers.*

region between two adjacent fold domains (Figure 15.19). The position of an axial surface of a fold on the ground surface is known as the fold **axial trace**. If a hinge line is horizontal, the strike of the axial trace is parallel to the trend of the hinge line, but if the axial surface is inclined and the fold hinge has a definite angle of plunge the **axial trace** and **axial trend** (hinge line azimuth) will not be parallel.

If the median surface of a fold and the axial surface are perpendicular and if the axial surface divides the fold domain into mirror symmetric quarter waves, then the fold is termed **symmetric**. If either of these conditions is not met, the fold is **asymmetric** (Figure 15.20). The relationships of the symmetry of parasitic folds in polyharmonic folds is very systematic: the limbs of the larger wavelength structure show asymmetric parasitic folds, whereas the hinge zone shows symmetric parasitic folds. Where the folds have inclined hinge lines the differing symmetry forms of the parasitic folds can be indicated by designating their shapes with symbols, M for symmetric folds, S or Z for asymmetric folds. These symbols express the *fold shape as it would appear on a horizontal surface* (Figures 15.21, 15.22). The relationships of these forms to position in the major structure and the methods for recording the symbols on a map should be clear from Figure 15.21B. The interpretation of these symbols on a map can be an extremely valuable tool for deciphering large scale fold geometry (see Appendix F). Where the fold hinge lines are horizontal these symbols cannot be used to indicate fold symmetry on a map, and some other method of describing the relationships of the axial surfaces of the parasitic folds to the median lines of the folded layers must be employed. In situations where the large scale structure has a normal and an overturned limb,

the asymmetry type of the parasitic limb folds can be expressed by indicating if the axial surface is steeper than the median surface ($A > M$, characteristic of normal fold limbs) or less steep than the median surface ($A < M$, characteristic of overturned fold limbs, Figure 15.23).

Fold orientation

The angle of dip of the axial plane and the angle of plunge of the fold hinge line are the features generally used to develop a terminology for fold orientation. Both features are generally easy to measure, and a terminology based on them is practical. The most complete set of descriptive names was proposed by Fleuty (1964) and is shown in Figure 15.24.

Axial trace construction

Question 15.4

Figure 15.25 shows a horizontal outcrop of banded hornblende–biotite–quartz–feldspar gneiss in the core region of one of the Pennine nappes of the Swiss Alps. Construct the axial traces of the folds, and indicate whether the traces are of antiformal or synformal folds. Why are the trends of the axial traces not parallel to the fold axis (axial trend)? Indicate the general dip directions of the axial surfaces. Discuss the relationships of the axial planes in the overall fold pattern.

Now proceed to the Answers section and return to Question 15.5* or to Session 16.

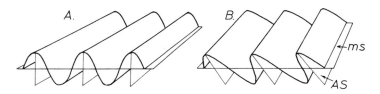

Figure 15.20. *Symmetric* (A) *and asymmetric* (B) *folds, where ms is the median surface of the fold train and AS the axial surfaces.*

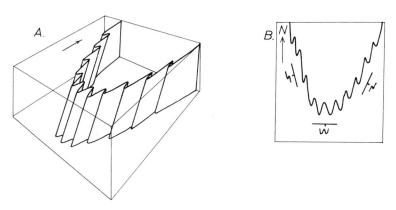

Figure 15.21. *Relationships between fold symmetry of the parasitic folds in a polyharmonic fold. B shows the map representation of the folds and the method of using symbols to communicate the type of fold symmetry.*

Figure 15.22. Z-shaped small scale folds in banded hornblende–biotite gneiss, Fusio, central Pennine Alps.

Figure 15.23. Folds in Cretaceous and Tertiary sediments on the west face of the Dent de Morcles, Helvetic nappes, Switzerland. The folds are facing towards the left of the picture, and the overall dip of the stratigraphic layering is to the right more steeply inclined than the fold axial surfaces. This characterizes the overturned part of a large anticlinal fold.

STARRED (★) QUESTION

Symmetry of small scale folds

Question 15.5★

Figure 15.26 is a map of a region of metamorphic rocks (the Moine Series of N.W. Scotland), metasandstones are unornamented and pelites are stippled. These rocks are cut by very coarse grained quartz–orthoclase–muscovite pegmatite sheets (black). The folds plunge to the southwest at angles of 50 to 60°. The shapes of the small folds, as seen on horizontal outcrop surfaces, are indicated with the symbols S, M, or Z according to their symmetry. Tectonic breccias are found at localities with X symbols.

Determine the traces of the major folds indicating whether they are synforms or antiforms and discuss the relationships of the pegmatite intrusions and the folds.

ANSWERS AND COMMENTS

Answer 15.1 Wavelength–amplitude analysis

Figure 15.27 shows the results of the measurements made on the folded metasediments of Figure 15.15. Both wavelength and amplitude vary over an order of magnitude and there is a general trend for amplitude to increase with wavelength. Some of this variation arises from inaccurate constructions;

for example, it is difficult to determine the position of an inflexion point where the fold limbs are fairly straight. There is also some small scale folding that superposes geometric "noise" on the main folds (e.g. surface F). However, most of the variation seen in Figure 15.27 is real, and the next procedure is to see if it is possible to subdivide the rather wide spectrum of data points in any way so as to determine the geological significance of the data spread.

A study of the geometry of the lines joining the inflexion points shows that in some layers these converge, whereas in others they diverge (convergence being used in the sense of following a line from an outer to an inner fold arc). In the photograph it is possible to relate these patterns to rock lithology; convergent lines occurring in the sandstone layers and divergent in the pelite. These recognizable geometric differences can be used to separate the folds into two groups. If a fold is situated on a part of a surface where the lines joining inflexion points generally converge towards that surface, the data points have been designated as open circles in Figure 15.27; if on a part where the lines converge away from that surface, the data are shown as filled circles. A study of the photograph will show that this division is generally connected to the inner arcs and outer arcs of the sandstone layers, respectively. The data are now grouped into two fields having small and large wavelength plus amplitude relationships respectively. The means of these two fields have been calculated (points with crosses in Figure 15.27). The inner sandstone fold arcs have an average wavelength $W = 18.7$ cm, and amplitude $A = 3.5$ cm,

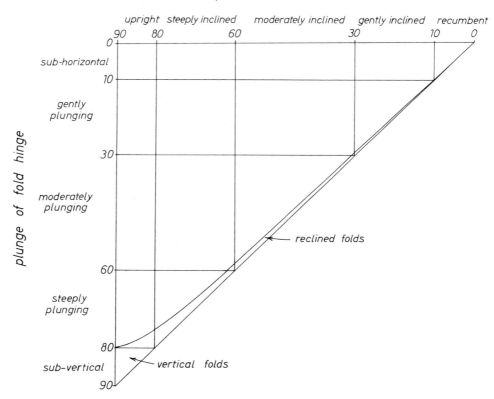

Figure 15.24. *Terms used to describe fold orientation (after Fleuty, 1964). Reclined folds have an hinge line pitch of 80° to 100° in the axial surface.*

Figure 15.25. Folded hornblende gneiss, Fusio, Lepontine Alps; see Question 15.4.

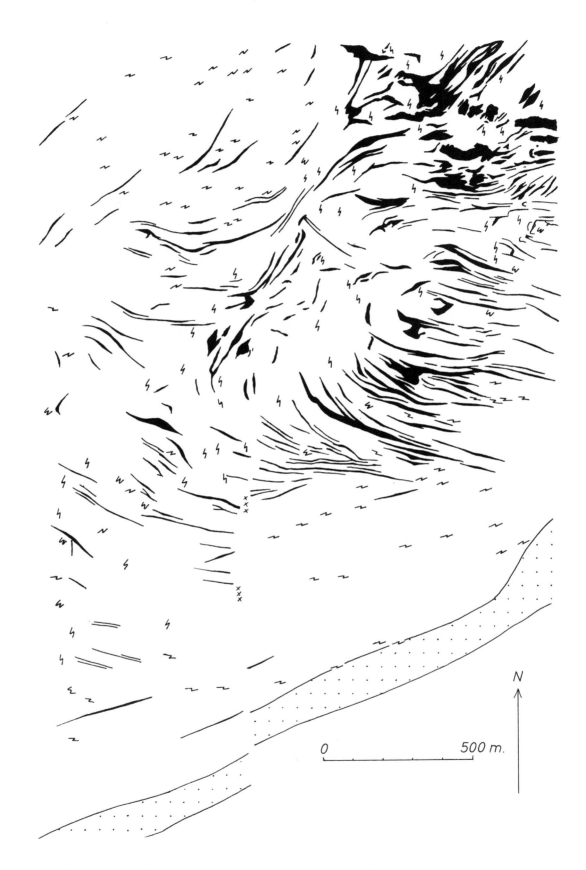

Figure 15.26. Map of the Moine rocks of Glenstrathfarrar, N.W. Scotland. See Question 15.5*.

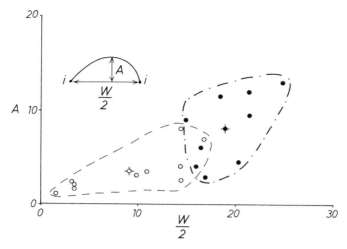

Figure 15.27. Wavelength–amplitude data from the folds of Figure 15.15.

whereas the outer arcs show both parameters of about twice this size ($W = 38 \cdot 1$ cm, $A = 8 \cdot 0$ cm). These averages have been illustrated schematically in Figure 15.28. It is worth comparing this with the real fold sizes of Figure 15.15: no actual contact is exactly like any we have deduced because our picture is an average view. We hope that you will agree that what, at first sight, might have looked to be rather a dull exercise has revealed something of considerable geometric and therefore geologic interest in these folds. The systematic differences in size of the folds along the pelite–sandstone contacts is of great importance, and it relates to the special growth properties of the mechanical instabilities along a competent–incompetent rock interface when it is com-

pressed. We will look into the significance of this geometry in more detail in Session 19.

In view of the special paired nature of alternate folds, another way of determining wavelength–amplitude relationships here would be to take adjacent fold pairs, determining the total wavelength $\bar{W}$ of a pair between every second inflexion point along the folded surface, and the amplitude sum ($2\bar{A}$) of the pair. Measuring the folds in this way (Figure 15.29) gives valuable information about average fold size in this outcrop, the measurements from combined folds being of much more restricted variation than those from the individual components.

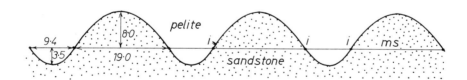

Figure 15.28. Schematic representation of fold sizes along a pelite–sandstone interface derived from Figure 15.27.

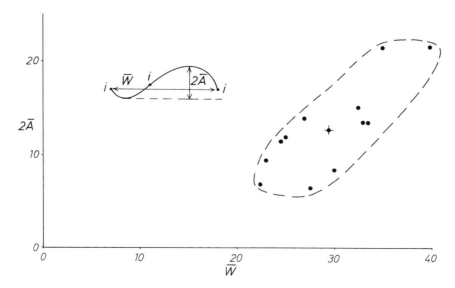

Figure 15.29. Wavelength–amplitude data from Figure 15.15 using adjacent fold pairs.

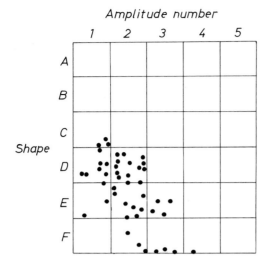

Figure 15.30. *Visual analysis of the folds of Figure 15.15 using the scheme of Figure 15.14.*

Visual harmonic analysis

Answer 15.2

Figure 15.30 illustrates the results of a visual analysis of the folds of Figure 15.15. The fold shapes range from semi-elliptical to chevron and there is a fair range of Fourier amplitude. A complete Fourier analysis of all the folds has been carried out and the b_1/b_3 values plotted graphically in Figure 15.31. Folds which lie on the inner arcs of sandstone layers are plotted as open circles, those on outer arcs as filled circles, and those of other relationships (e.g. inside sandstone or pelite layers) as small points. It will be seen from this analysis that the wide range of fold styles is, in part, related to the location of the folds. Those folds in the outer arcs of the sandstone layers show more rounded forms than those situated in the inner arcs, with b_3 values averaging to give an overall sinusoidal aspect. Of those in the inner arcs, all except one have negative b_3 values and have forms lying between sinusoidal and the sharp-crested, straight-limbed chevron types.

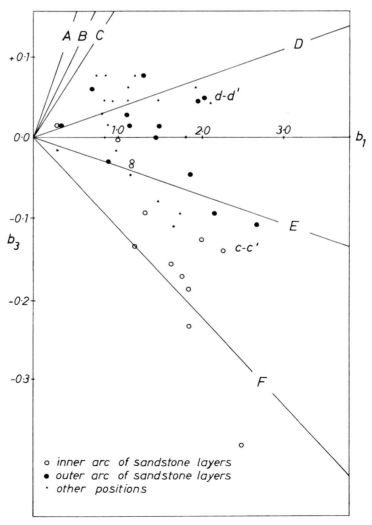

o   inner arc of sandstone layers
●   outer arc of sandstone layers
·   other positions

Figure 15.31. *Complete representation of the Fourier coefficients b_1 and b_3 of the folds of Figure 15.15.*

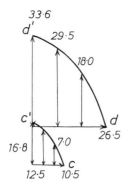

Figure 15.32. Measurements necessary on the fold arcs c–c' and d–d' to compute their Fourier coefficients.

It should be clear from the mathematical procedures used in the Stabler–Hudleston method for shape analysis that the term amplitude used in this analysis is not equivalent to the amplitude A defined by the absolute measured distance between the fold hinge and the line joining inflexion points that we have used earlier. The amplitude term of the Fourier method relates the relative height of the fold above the quarter wave base line, these measurements being described in a scale-independent form. To clarify the difference in meaning of the amplitude term of the Fourier analysis with that used earlier it might be best to refer to the Fourier amplitude as **relative amplitude**, or **amplitude number**.

Calculation of Fourier coefficients

Answer 15.3

The measurements made on the two sectors of folded surfaces are given in diagram millimetres in Figure 15.32. To carry out the calculations the values of y_1, y_2 and y_3 must be put into units where the quarter wave base (x-axis of the calculation) is of length $\pi/2$ units. For the fold sector c–c':

$$y_1 = 7{\cdot}0\pi/2 \times 10{\cdot}5 = 1{\cdot}05$$

$$y_2 = 12{\cdot}5\pi/2 \times 10{\cdot}5 = 1{\cdot}87$$

$$y_3 = 16{\cdot}8\pi/2 \times 10{\cdot}5 = 2{\cdot}51$$

Using equations (15.5) this gives Fourier coefficients $b_1 = 2{\cdot}27$, $b_3 = -0{\cdot}14$, $b_5 = 0{\cdot}11$. For the fold sector d–d'

$$y_1 = 18{\cdot}0\pi/2 \times 26{\cdot}5 = 1{\cdot}07$$

$$y_2 = 29{\cdot}5\pi/2 \times 26{\cdot}5 = 1{\cdot}75$$

$$y_3 = 33{\cdot}6\pi/2 \times 26{\cdot}5 = 1{\cdot}99$$

giving Fourier coefficients $b_1 = 2{\cdot}03$, $b_3 = 0{\cdot}05$, $b_5 = 0{\cdot}01$. The fold c–c' lying on the inner arc of a sandstone layer has a negative value of b_3 whereas d–d' on the outer arc has a positive value. This reflects the fundamental difference between the inner and outer arcs in that the inner arc fold is sharper crested than a sine wave, whereas the outer arc fold is more rounded than a sine curve. The positions of these two folds in a Hudleston b_1/b_3 graph are shown in Figure 15.31.

The determination of Fourier coefficients for the descrip-

tion of fold shapes is extremely practical and results are obtained quickly. The method has great potential for revealing features of naturally formed folds which are not immediately apparent, and Hudleston (1973a, c) has shown how effective it can be for comparing fold shapes from locality to locality. It also offers an extremely powerful tool for analysing the changes in shape of progressively evolving folds formed during laboratory experiments, and for comparing the folds arising from experiments with natural fold shapes (Hudleston, 1973b).

Axial trace construction

Answer 15.4

Figure 15.33 gives the locations of the axial traces of the folds of Figure 15.25. From the direction of the fold axis and the closure sense of the layers in the hinge areas it is a simple matter to decide the antiformal or synformal nature of individual folds. The general dip of the axial surfaces deduced from the fold axis azimuth must be towards the top of the figure, and triangle symbols have been used to convey this information on to the map. Several of the axial traces are curved. As the outcrop surface is planar, if the fold plunge is constant the curvature implies that the angle of dip of the axial surfaces must change.

The terminations of the axial traces are of several types. At locality A the axial traces converge and meet at a point (cf. Figure 15.17) and this fold pair may be relaying the next closed pair in an en-echelon manner (cf. the lower part of Figure 15.19). The axial traces at locality B close at both ends and have the form of a completely enclosed fold "pod". At locality C the axial surfaces come to an end without converging, the fold between them gradually losing amplitude to the left until it no longer exists (cf. the upper part of Figure 15.19).

We have seen that inflexion planes separate adjacent antiformal and synformal domains from each other whereas axial surfaces effectively split each fold domain into halves. Why have we not constructed the inflexion planes in this system: would it not have been more logical to have mapped the planes separating fold domains than construct the axial traces? The reason is a practical one. The inflexion lines, being positions of zero curvature, are generally much more difficult to locate accurately than the axial traces coinciding with positions of maximum or minimum curvature. This is especially true in folds which have long, rather straight limbs, such as we are confronted with in this problem. Although the axial traces of folds are generally easy to define it should be pointed out that there may be special problems of location if the two dimension section is markedly oblique to the fold axis because the maximum curvature of a surface in this section does not always coincide with the maximum curvature in three dimensions which defines the true position of the axial surface. We will return to this particular problem in Session 18.

Although the physical scale of Question 15.4 was small, many of the features seen in this outcrop occur on a regional scale. Figure 15.34 is a map of one of the Helvetic nappes of western Switzerland, the Morcles nappe, and shows the position of the axial traces of large folds in the Mesozoic and Tertiary sediments. The sectors of steeply inclined or overturned rocks between pairs of anticline–syncline axial traces

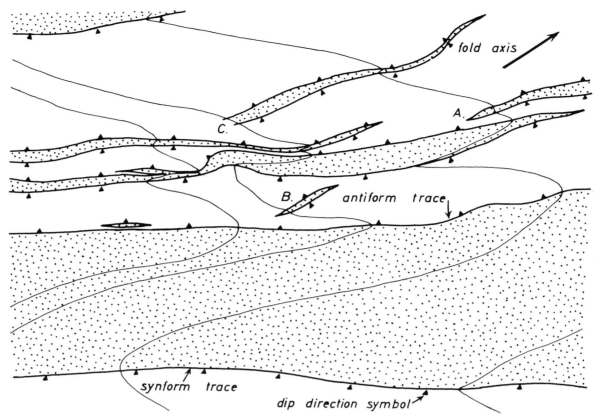

Figure 15.33. Structural analysis of the folds of Figure 15.25.

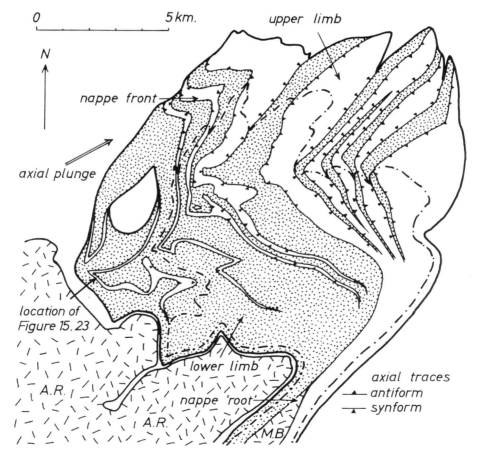

Figure 15.34. Map of the Morcles nappe, one of the Helvetic nappes of W. Switzerland. Underlying basement of the Aiguilles Rouges (A.R.) and Mont Blanc (M.B.) massifs shown with diagonal dash ornaments, and location of the Jurassic–Cretaceous contact in the nappe indicated as a dash–dot line. The folds in the nappe plunge from 5° to 30° to N 60° E, and a profile of the nappe is shown in Figure 11.10 of Volume 1.

are shown with a stipple in the same way as was done in Figure 15.33. To appreciate the scale of these folds we have located the folds seen on the southwest face of the Dent de Morcles shown in Figure 15.23. The cliff face is about 300 m high and 700 m in horizontal extent. The syncline (above)–anticline (below) fold pair in the photograph is shown on the map as two axial traces which isolate a strip of rocks in normal orientation from inverted rocks above and below. The stipple pattern on the map illustrates very clearly that the overall geometry of the nappe is consistent with that of a large anticline closing to the northwest with predominantly normally oriented Mesozoic and Tertiary sediments in the eastern part of the nappe outcrop, and predominantly over-turned sediments on the western side (see NW–SE profile in Figure 11.10, Vol. 1.). Because of topographic variations of up to 2500 m in the region, and because the axial surfaces are generally dipping at low to moderate angles, some of the curves of the axial traces seen on the map are the result of complex intersections of axial surfaces with topography. The method of constructing axial traces developed in this region has been particularly helpful in assisting the correct correlation of folds from locality to locality. Perhaps the most important mental stimulant it has provided is that the field geologist cannot "lose" or "forget about" any individual fold. He is forced into deciding where and how an axial trace ends. So useful are the results of this type of analysis that we are of the opinion that the mapping of axial traces should be adopted as a standard part of the field mapping techniques of structural geology.

Symmetry of small scale folds

Answer 15.5★

The principal strike lines and positions of the fold axial traces are shown in Figure 15.35. The axial traces have been located either where the small scale folds are symmetric and M-shaped, or in regions which separate S-shaped and Z-shaped asymmetric folds. A north–south strike fault exists in the south of the region. It has displaced the boundaries of the horizon of pelitic rock. Its northward continuation has been drawn along the breccia zone. The trace of the major antiformal fold appears to show a shift across the fault, likewise the southern part of the complex of pegmatite sheets. The pegmatite sheets generally cut across the bedding and the fold axial traces, although in a few localities the sheet orientation may have been guided to some extent by the bedding surfaces. In the field the pegmatites have very sharp cross-cutting contacts and are undoubtedly intrusive igneous rocks. No granite intrusions are known in this region, but the pegmatites could represent sheet intrusions into the roof of a granite which might exist at depth.

The structural sequence that can be deduced from this map is: (1) sedimentation; (2) folding (and metamorphism?); (3) intrusion of pegmatite dykes; (4) faulting.

KEYWORDS AND DEFINITIONS

There are too many keywords and definitions in this session to justify a repetition here. We list below the terms every student should known and where descriptions may be found in the previous text.

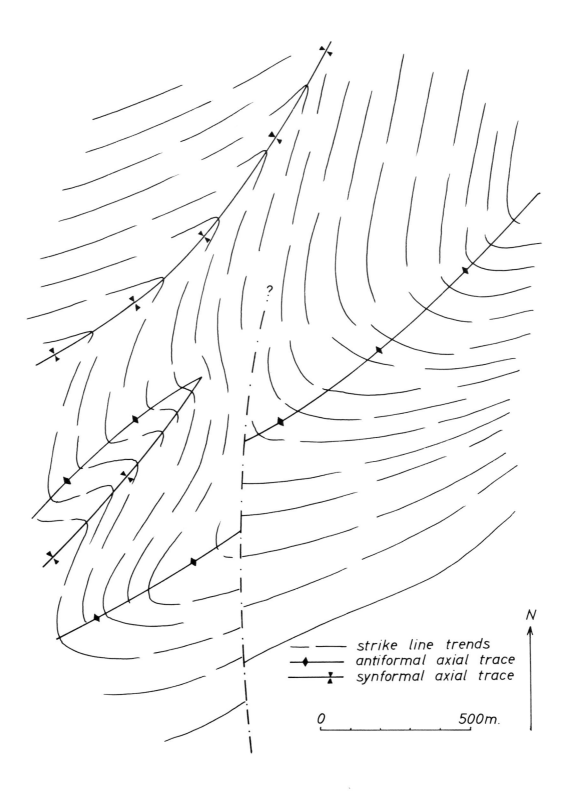

strike line trends
antiformal axial trace
synformal axial trace

N

0   500m.

Figure 15.35. *General trends to bedding planes and location of the axial traces from Figure 15.26.*

KEY REFERENCES

Fleuty, M. J. (1964). The description of folds. *Proc. Geol. Assoc. London* **75**, 461–492.

Hudleston, P. J. (1973a). Fold morphology and some geometrical implications of theories of fold development. *Tectonophysics* **16**, 1–46. (1973b). An analysis of "single layer" folds developed experimentally in viscous media. *Tectonophysics* **16**, 189–214. (1973c). The analysis and interpretation of minor folds developed in the Moine rocks of Monar, Scotland. *Tectonophysics* **17**, 89–132.

Ramsay, J. G. (1967). "Folding and Fracturing of Rocks", 568 pp. McGraw Hill, New York.

Shackleton, R. M. (1958). Downward-facing structures of the Highland Border. *Jour. Geol. Soc. London* **113**, 361–392.

Stabler, C. L. (1968). Simplified Fourier analysis of fold shapes. *Tectonophysics* **6**, 343–350.

Wilson, G. (1967). The Geometry of cylindrical and conical folds. *Proc. Geol. Assoc. London* **78**, 179–210.

A pioneering paper which analysed the rather confused fold nomenclature existing at the time it was written and which made many useful suggestions for putting in order the nomenclatorial chaos. This paper is very clearly and logically written and contains an extensive and useful bibliography.

Parts of these three papers would form valuable background reading to the Fourier analysis methods used in this session and would give a suitable lead in to ideas and methods we will meet in Sessions 17, 19 and 20. These papers are models of clear exposition and are very well illustrated.

Pages 345–360 set out much of the basic nomenclature used here, together with a few additional suggestions for defining exactly the extent of limbs and hinge zone in a fold.

This paper was one of the first to put forward the concepts of the facing direction of a fold. It illustrates very beautifully, with examples from the Caledonides of Scotland, how field observations of sedimentation structures can be combined with fold geometry and small scale tectonic structures to evaluate the major tectonics of a region.

Although various attempts previous to this paper had been made to apply the method of Fourier analysis to analyse the shape of folded surfaces they never caught the imagination of practising geologists. This paper showed how three appropriately located points could be chosen on a quarter wave sector to define fold shape very simply. Note that a mistake in the formula for the second Fourier coefficient was corrected in *Tectonophysics* (1969) **7**, 356.

This paper describes the geometrical features of folds and discusses in some detail the problems associated with the formation of conical folds in a multi-layered rock sequence. It includes a discussion of the constructions of fold profiles which would make a good introduction into Session 17.

SESSION 16

Fold Orientations: Projection Techniques

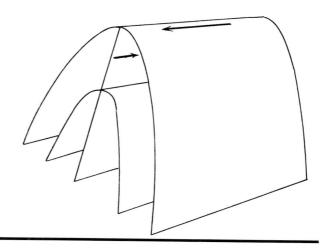

The methods by which field data on the orientations of planar and linear structures can be used to determine the principal geometric features of large scale folds are established. In cylindrical or sub-cylindrical folds the π-diagram and the β-diagram can be used to determine the fold axis, but with some types of data these methods may have limitations. Methods are given for determining the orientations of the axial surface, axial trace, crest surface and crest trace by combining information on angular relationships, derived from projections, with map analysis. The geometric features of non-cylindrical folds are briefly discussed.

INTRODUCTION

During the course of a field investigation of the structure of an area the geologist records on his field map the orientations of the various structural features he sees in individual outcrops. Some of this information will be recorded in his field note book, but the information on orientations of structures should be recorded, as accurately as possible and with the correct geographical orientation, on a map. The practical techniques for doing this are set out in Appendix F.

The principal orientation data acquired during the field study consist of measurements of several types of planar and linear structure:

1. Lithological layering, bedding planes in sediments, compositional banding in metamorphic rocks and gneisses, primary layering in igneous bodies.
2. Planar fabrics produced by the preferred alignments of minerals, cleavage and schistosity perhaps with attendant stretching lineation fabrics parallel to the X-axis of the finite strain ellipsoid.
3. Geometric elements of small scale folds, their axial planes and hinge lines.
4. Shear zones, faults and joints, and linear striations giving information on the relative movements (if any) between the fracture walls.

This information provides the framework on which much subsequent analysis of structural geometry is built. In some spectacularly exposed terrain, such as is found in high mountain ranges (e.g. Figure 15.23), many large scale structures can be directly observed in the field. However, such terrains are exceptional, and because of limited rock exposure or, even if the region is well exposed, because the topography is subdued, immediate recognition of the larger structures may be impossible. Generally a field map provides a method of recording data from outcrop to outcrop, and enables the regional picture to become gradually

revealed as the jigsaw-like fragments become assembled. In such a study the importance of measuring as many orientations of the linear and planar features of deformed rocks cannot be emphasized too strongly. The reasons are as follows. First, the more measured data we have, the more accurate will be the final geometric analysis and the deeper will be our understanding of the overall structure. Second, we have found from experience that the actual process of making measurements focuses attention on structural features that might otherwise go unnoticed or be overlooked, and these features may later be found critical in assessing the validity of some conclusions. The tectonic interpretations of some geologists who try too rapidly in the field to evaluate the regional structure often turn out to be crude and oversimplified in a final analysis. We have found that even in superficially simple terrain, for example in gently folded strata, geometric "catch questions" often arise, while in the structurally complex metamorphic terrains, where structural geometry is generally the result of complex polyphase deformation, the correct tectonic explanation will *only* emerge from a careful study of orientation data. In economic studies, where the formation of an accurate picture of structural geometry is often the key to future mining or oil drilling operations, the geologist who has accumulated the most orientation data will generally be in the best position to make the most reliable predictions.

In the sections which follow we will discuss the methods which are especially useful in analysing the geometric features of folded rocks, and we will leave the analysis of fracture geometry until Sessions 24, 25, 26 and 27.

Before reading the discussion and undertaking the questions below it might be a good idea to revise the general methods for plotting projection surfaces and their poles, and lines (Session 9, Vol. 1, pp. 154–155) and the methods for measuring angles between planar and linear elements (pp. 156–159, Vol. 1).

Fold axis

Folds may be **cylindrical** or **non-cylindrical** depending upon whether or not the folded surface can be generated by the parallel movement of a line (the fold axis) in space. Many natural folds show surfaces which are either very close to that of the cylindrical model, or which can be subdivided into smaller sectors which are effectively cylindrical. In a cylindrical fold it is possible to determine the axial direction by analysing the angular relationships of various planar components comprising the fold, namely lithological layering, planar deformation fabrics such as cleavage and the axial surfaces of congruent small scale parasitic folds. Two basic graphical procedures are used, both of which use projection techniques. These are known as the π-diagram and the β-diagram.

π-diagrams

This method of data anlysis uses a stereographic or (better) an equal area projection to plot the perpendiculars (π-poles) to the various folded surfaces making up a large fold. If the fold is cylindrical then each π-pole is oriented at an angle of $90°$ to the fold axis (Figure 16.1), and adjacent π-poles will be located on a partial great circle locus known as the π-circle. The perpendicular to this great circle is known as the π-axis, and it lies parallel to the fold axis (Figure 16.1F). In practice, the π-poles do not lie exactly on a unique great circle but fall in a zone around some fixed great circle. The reasons for this spread are: first, field measurements always contain some degree of error ($\pm 2°$) and will therefore show a slight scatter even if the fold is perfectly cylindrical. Second, folds are rarely perfectly cylindrical, so the model we are using will never provide an exact fit to the data. Generally, a great circle or π-circle is selected which provides the best "eyeball-in" fit of the data points. There are methods whereby a best fit great circle may be chosen using statistical criteria in much the same way that statistical methods are available to fit a straight line to data points on a two-dimensional graph (Ramsay, 1967; Mancktelow, 1981). In our opinion the accuracy contained in such methods is only justified if it is imperative to exploit the full potential of

very precise primary data. The nature of much geological data does not warrant the use of such techniques, and, in most "normal" structural analyses, such high precision is not demanded. Natural folds are unlikely to be perfectly cylindrical and it would seem a good idea to develop some practical terminology and simple testing procedure to define whether a fold is cylindrical or not. Such an attempt has not previously been made (see Fleuty, 1964; p. 465). We suggest that if more than 90% of the π-poles fall within an angle of $10°$ from the constructed π-circle the fold should be termed **cylindrical** for the purposes of geological description. If less than 90% lie in this zone, and more than 90% of data points fall within $20°$ of the π-circle the fold is then termed **sub-cylindrical**. Folds with data outside this limit are termed **non-cylindrical**. A simple scheme such as this is very easy to apply, because it is a comparatively rapid process to construct small circles at $10°$ and $20°$ from the best fit great circle and carry out the necessary counting procedure.

In some folds planar fabrics arising from deformation during folding may be arranged in a geometrically systematic way (axial plane cleavage, convergent and divergent cleavage fans: see Session 10, Vol. 1, p. 181) with respect to the fold axis. In such instances the poles of these secondary surfaces should also plot in the same π-circle zone as do the main folded layers (Figure 16.1, π_{Cl}). In situations where **transected cleavage** planes are skewed across the fold (Vol. 1, Figure 12.17) the cleavage poles will show no simple geometric relationship to the main structure.

The advantages of the π-diagram method above all others is the simplicity by which large amounts of data may be incorporated into the analysis. Errors inherent in the primary measurements therefore tend to cancel each other out, with the result that the mean position of the π-circle, and the fold axis derived from it, can be determined with the maximum accuracy. A feature of most π-diagrams is that the π-poles are not dispersed along a complete $180°$ sector of the great circle zone, implying that there are constraints on the possible orientations of the surfaces in the cylindrical fold. By measuring the value of this restrictive spread it is possible to determine the interlimb angle of the fold. The π-poles are

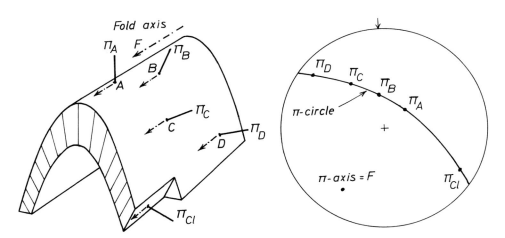

Figure 16.1. Geometric features of a cylindrical fold with normals to layering π_A, π_B, π_C, π_D draw at observation points A to D. The projection on the right shows the way these π-poles define the π-circle, and how the fold axis F is the normal to the π-circle plane.

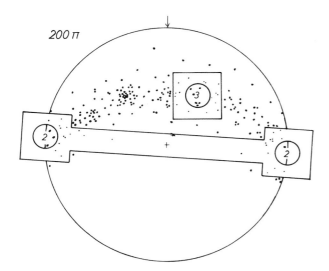

200 π

Figure 16.2. *Method of contouring a pole diagram using 1% contouring counters.*

often concentrated within the part of the zone they occupy. This may be because the sampling locations within different parts of the structure were uneven, but it might relate to real variations in frequency of certain planar orientations in the structure. High concentrations of π-poles occur when a fold has well defined planar limbs and a sharply demarcated hinge zone. By locating the centres of maximum concentration of π-poles we can obtain a best fit orientation for the fold limbs. This procedure is normally done by using a data **contouring technique.** Contouring can only be carried out with data plotted on an equal area projection. A card with a circular counting hole of diameter one tenth that of the projection is made so that the area of the hole is one per cent of that of the total diagram. This counting circle is moved across the data points and the data concentration percentage written in the centre of successive counting regions (Figure 16.2). If, for example, 200 data points have been plotted, and if six points occur within the region of the

counting circle the concentration is $(6 \times 100)/200 = 3\%$. A special double circle is used at the periphery of the projection so that a complete one per cent counter incorporates both sides of the projection (Figure 16.2). Several types of more elaborate contouring methods have been suggested, some employing counting regions which come progressively more ellipse shaped towards the edge of the projection, as required by the geometry of the projection. In our experience the extra time involved is quite large, and the increase in accuracy is practically negligible. When the final contoured diagram is produced it should always specify the number of data points used in its construction and the values of the contour lines.

β-diagrams

The theoretical concept used to develop this type of graphical method is essentially the same as that for the π-diagram, but the application is somewhat different. In a cylindrical fold, each surface contains the fold axis. If we intersect any two measured surfaces in the fold (Figures 16.3A and B) their intersection should define (unless they are absolutely parallel) the line common to both. This constructed line is known as a β-axis and should be parallel in space to the fold axis. In a similar way, the intersection of a congruent cleavage in the fold with part of the folded surface also gives a β-axis (Figures 16.3, B and Cl). The projection clearly offers a practical way to determine such β-axis intersections. If the fold was perfectly cylindrical and if all measurements were made without error all the great circles representing measured surfaces in the structure would pass through the same β-axis. Because neither of these conditions holds in real folds or with real data, each pair of planes produces a slightly differently oriented β-axis. In practice the construction of a β-diagram with many data leads to a scattered group of β-axes, the centre of concentration of this group giving the best fit for the average β-axis and therefore for the fold axis. The number of accumulating β-axes increases very rapidly with the number of primary data, and very few β-axes fall together in one place. Two data give one intersection, the third datum intersects the previous two to give two more

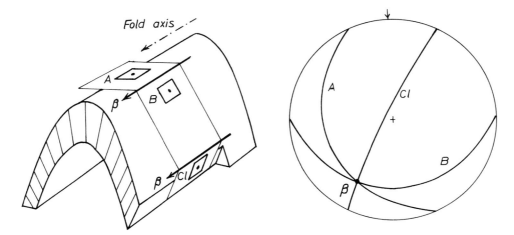

Figure 16.3. *Method of determining a β-axis from the intersection of planar features A and B in a cylindrical fold. The orientation of a cleavage surface Cl in the congruent convergent cleavage fan can also be used to obtain a β-axis.*

β-axes. The number N of β-axes produced from n primary data (Figure 16.4) consists of an arithmetic series

$$N_\beta = 1 + 2 + 3 + \cdots + (n - 1) = n(n - 1)/2$$
$$(16.1)$$

One of the major drawbacks of the β-diagram technique can be deduced from this formula. For example, with $n = 500$ the total number of β-axes is 124 750! This number of primary data would be quite easy to incorporate into a single π-diagram, but the technical difficulty of producing a β-axis graph with 124 750 accurately identified β-axes is clearly insuperable. If this method is employed it is recommended that diagrams are made using only about 15 data great circles, which contain 105 β-axes. The method is a good example of one that, in principle, is sound but which, in practice, poses so many difficult constructional problems that it cannot be enthusiastically recommended.

Axial surface

Because the axial surface of a fold is defined by the locations of hinge lines in successive folded layers, it has no simple relationship to the orientations of individual parts of the folded surface. Although it is not possible to construct the position of the axial surface from the data of a π-diagram, under special circumstances it may be possible to combine information from such a diagram with further information derived from a map to make a calculation of the axial surface orientation. In situations like that shown in Figure 16.5A, information derived from folded surfaces cannot be used to compute the orientation of the axial surface; however, in Figure 16.5B a solution can be made. This necessitates constructing the axial trace AT (trace of the axial surface on the ground surface) on the map, and then finding the great circle joining AT and the plunging fold axis F (derived from a π-diagram): the surface represented by this great circle is the axial surface. The tighter the fold (to obtain maximum resolution of the direction AT) and the steeper the plunge of the fold (to obtain maximum separation of AT and F in the projection) the more accurate is the resulting computation. Such a construction is impossible in folds like that shown in Figure 16.5A, because AT and F are effectively parallel. It should be emphasized that the axial plane of a fold does not necessarily bisect the fold limbs. If the axial plane is closer to one limb than to the other, then that limb is preferentially thinned.

Axial trace

The axial trace determined on a map is often incorrectly located because it is sometimes difficult to determine the maximum layer curvature in the profile section from an oblique section through the fold (Session 18, Question 18.2). It may be necessary to locate the approximate direction of the axial trace of the fold, and correct this position at a later construction stage. If the orientation of the axial surface is known, the axial trace is located by joining points where the layers are oriented perpendicular to the axial plane (see the positions of the great circles of planes P in Figures 16.5A and B). Where the axial plane and axis of the fold are both inclined, the axial trace is not parallel to the axial trend of the fold axis.

Crest plane trace

The crest plane joins the positions of lines occupying the topographic crest of successive layers (Figure 16.5, CP). These crestal lines are found in layers which have the lowest angle of dip in the fold. In cylindrical folds these layers have their strike lines at right angles to the direction of plunge of the fold axis (see positions of planes Q in Figure 16.5).

QUESTIONS

Question 16.1

Figure 16.6 shows a field map with data collected from a region of strongly folded metamorphic rocks in the Caledonian mountain chain of N.W. Scotland (Beinn nan Caorach, Inverness-shire). The terrain is generally well exposed and the limits of the actual rock outcrops are shown in the figure. It is undulating, but the height variation nowhere exceeds 150 m. The outcrops consist of metamorphic sandy sediments (Moine Series) shown with stippled ornament and banded hornblendic gneisses (Lewisian Series) left unornamented. These are cut by sub-vertical dykes of Vogesite lamprophyre, shown black. The metamorphic rocks show a well developed banding; this either represents traces of the original bedding planes in the metasediment or layering in the gneisses. The orientations of these structures are shown with a strike line and dip tick

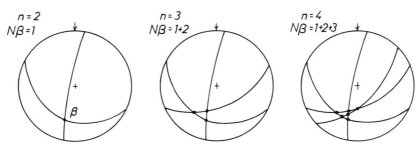

Figure 16.4. Progressive increase in number of β-axes in a fold which is not perfectly cylindrical.

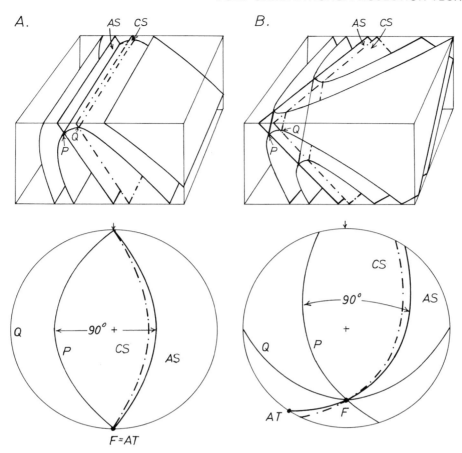

Figure 16.5. *Relationships of axial surface (AS) and crest surface (CS) in cylindrical folds, A with sub-horizontal fold axis and B with plunging fold axis (F). The axial trace of each fold is indicated AT in the projections. The surfaces P and Q give the orientations of the folded layer at the axial surface and crest surface respectively.*

and the two numbers by the side of each symbol give the strike direction and angle of dip, respectively. Small scale folds are sometimes present and their axial planes and hinge lines have been recorded at a few localities. The separations of the main lithological units were sketched in the field from the outcrop distribution and are shown as dashed lines.

The problem here is to analyse as accurately as possible the geometric form of the major structure and to determine the relationships of the small scale structures to the larger features.

From an inspection of the map and a visual assessment of the various orientations, interpret the main geometric features of the structure. To do this it is probably best to construct a simplified strike line trend map indicating the average strikes and dips of the layers. Describe the main features of the hinge zone, fold limbs and probable orientation of the axial surface. What is the significance of the steeply inclined or vertical layers in the structure? Can they be used to determine the general orientation of the fold hinge?

Now check the Answers and Comments section and return to Question 16.2.

Question 16.2

Plot the poles to layering planes on to an equal area projection

(π-diagram). Is the folding cylindrical? Determine the orientation of the fold axis. Note any regions where the data depart from that of a cylindrical fold. Discuss the significance of the unequal distribution of π-poles in the projection.

Question 16.3

Determine the general strike of the axial plane (i.e. axial trace) on the map. Combine these data with those of the π-diagram and calculate the orientation of the axial plane. What are the angles between the axial plane and the fold limb, and what is the likely significance in terms of limb thickness?

Question 16.4

What is the orientation of the folded surfaces along the axial trace of the fold? Locate the axial trace on the map. What is the orientation of the folded surfaces where they transect the crest plane? Locate the crest plane trace on the map.

Question 16.5

What are the relationships of axial planes and axes of small folds to the major structure?

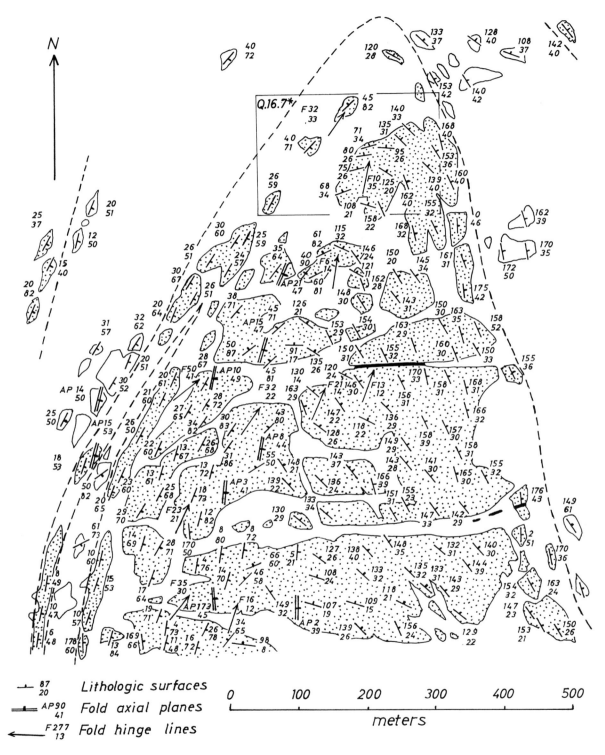

Figure 16.6. *Structural field map of part of N.W. Scotland giving data for Questions 16.1 to 16.7★.*

Question 16.6

What is the likely geological relationship between the lamprophyre dykes and the fold structure?

Now go the the Answer and Comments section, then proceed to Session 17 or to the starred question below.

STARRED (★) QUESTION

Question 16.7★

A small rectangular area has been located in the northern part of Figure 16.6. Make a β-diagram from these data and discuss the significance of the locations of the β-axes. What special problems did you encounter in constructing the β-diagram? If you have time, make a β-diagram from a small area on one of the fold limbs. Discuss the main disadvantages of the β-diagram method of analysis over the π-diagram method.

ANSWERS AND COMMENTS

Answer 16.1

Before any detailed geometric analysis of a structure using projection techniques is made it is most important that as many geometric facts as possible are accumulated from an inspection of the map. This procedure is critical in regions where the folding is non-cylindrical or where more or less cylindrical folds undergo changes in orientation from area to area in a region as a whole. The whole basis of analysis using projection techniques depends on having the sub-area as nearly structurally homogeneous as possible. If data from areas with differing axial directions are combined in the same projection, a conflicting muddle of data points results which cannot be easily interpreted.

The main features of the map are simplified in the strike-line map of Figure 16.7. The lithological contacts and strike lines show a northward closing V-form. From an inspection of the dips of these data this form represents an *antiform plunging to the NE*. The structure has well defined limbs which dip to the ENE and ESE respectively, the eastern limb (L_E) dipping less steeply than the western limb (L_W). The west limb appears to be "overturned" relative to the "normal" eastern limb, the beds passing through vertical orientations along a zone indicated in Figure 16.7 with stippled ornament. In this zone the strike lines strike from 20° to 40°. If the fold is cylindrical the *strike of vertical layers is parallel to the azimuth of the fold axis*. To obtain further information on the angle of plunge of the fold we seek layers with a strike direction at right angles to that of the vertical beds. The angles of dip of such layers will be the lowest possible angles of any folded layers in the structure. In a cylindrical fold the *angle of dip of layers which strike perpendicular to nearby vertical layers is the same as the angle of plunge of the fold*. In our example strike directions with values of from 110° to 130° degrees have dips varying between 11° and 32° with an average in the mid-twenties, and these values provide the mean for the major fold axis plunge. As far as can be determined there are no marked systematic changes of these features through the area so the folds might be cylindrical. Close inspection of the map data shows that there are variations of the dips of layers with the same strike, and changes in strike direction of layers with the same dip. These variations could be due to imperfections in the model of perfect cylindricity or, because the data are real, they may contain observational errors. To proceed further we must test the appropriateness of the cylindrical fold model using projection techniques.

By placing a ruler along the sharply defined hinge zone of the map the general trend of the axial trace can be determined. It has a strike of 12°. This strike direction has a lower numerical value than that of the azimuth of the fold axis (we deduced 20° to 40°). Because, by definition, the hinge line

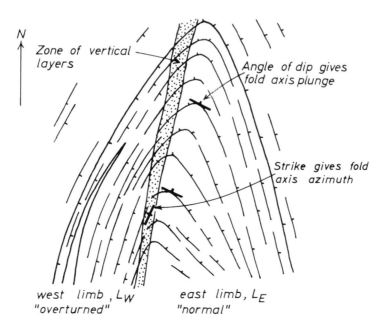

Figure 16.7. *Strike line sketch diagram of Fig. 16.6.*

must lie in the axial plane, this difference in trend implies that the axial plane dips to the ESE. In view of the deductions we made about "normal" and "overturned" limbs this is to be expected, but in fact the two methods of deducing the axial plane orientations are quite independent of each other. For example, both fold limbs might have been normal and symmetric about the ground surface, but the axial surface could still have been inclined.

Before we leave this general discussion of fold geometry it is worth examining some special problems which can arise from interpretation of layer orientations which are sometimes a puzzle to the beginner. Regions of low dips generally show much greater variation in strike direction than those of high dips. Does this variation imply large or unsystematic variations in layer orientation? Generally it does not, and we can see the reason for such variation in the diagrams of Figure 16.8. This shows three regions in a projection in each of which are situated seven points arranged at equal angular intercepts of 10° to each other. In each group six points are symmetrically arranged around a centrally located "mean" point. These groups could represent identical variations of a group of layering poles about a mean pole. Consider now the variations of dip and strike of each group of seven π-poles. The planes which are steeply inclined (group A) show variations in dip and strike which do not differ appreciably from the angular variation in the original pole data. In contrast those planes which are inclined at low angles (group C) show large variations of strike direction exceeding 40° on either side of the mean strike, but the angles of dip are containined within the initial ± 10° variation. This simple geometric fact is of great importance when we come to discuss the significance of map data. Regions with wildly varying strike directions do not necessarily imply extreme angular changes in the layering, providing that the dip angles are low. The strikes of layers in zones of steeply inclined layering often show rather constant orientations. If they do not then variable strikes indicate either strongly varying fold hinge lines (non-cylindrical overall geometry) or, if the folds are cylindrical, the fold axes must be steeply inclined.

Answer 16.2

A π-diagram of the data is shown in Figure 16.9A. The 190 data points do not lie on a simple great circle but scatter in a broad zone with approximate great circle affinities. Such a distribution is very typical of most real data plots. The variations in the location of the points are not due to errors in measurement. The angular departure of some π-poles from the line of the best fit great circle drawn through the data sometimes exceeds 20°, a figure which greatly exceeds any measurement area. The measurements made here were recorded to an accuracy of ± 2°, so the variation in π-pole position occurs because the fold is not exactly cylindrical. Figure 16.9C indicates the small circles of 10° and 20° of deviation of poles from the mean. In these data 20% fall outside the 10° small circle and 2·5% outside the 20° small circle. Using the criteria we recommended on p. 334, this fold is *sub-cylindrical*.

If you were watching your π-poles progressively accumulate you may have noticed the regions which supplied the strongest non-cylindrical effects into the plot. However, people are often so involved with the technicalities of making the construction that they do not always observe such features. That is a pity, because the development of an intelligent interest when plotting data not only takes the tedium from making the compilation but it also provides information which assists a more efficient subdivision of the total mapped area into appropriate sub-areas as well as indicating all sorts of subtle features about the fold geometry. In the region of Figure 16.6 the most severe departures from the best fit great circle occur on the western side of the antiform, especially along the contact between metasandstone and gneiss. In fact, this is a region which shows the effect of two periods of folding, and such a deformation history does not always produce cylindrical folds.

Notice the appearance of the scatter distribution of π-poles in Figure 16.9A: sometimes adjacent points are tightly grouped, perhaps even with occasional points falling at the same position, and sometimes they are separated. There is never a regular distribution of π-poles. It might be worthwhile to look at some published projection data to see how the

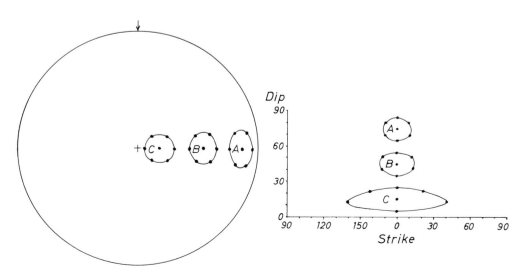

Figure 16.8. *Illustration of spread of strike trends in layers of low dip.*

point distributions do or do not compare. A few workers take dip and strike measurements only to the nearest 5°, being of the opinion that no field geologist can measure to the nearest degree. Although this is correct with reference to isolated observations, it is bad policy when many measurements are made and where statistical averages begin to produce significant weight to the results. Another technical problem with the 5° measurement approach is that the points in a projection become artificially grouped into locations at the corners of 5° sectors of the net and, if a reasonably large number of data has been collected, many will fall in exactly the same point on the net.

The data of Figure 16.9A show two main concentrations within the great circle zone. From a study of the map of the area the data do seem to be drawn uniformly from all parts of the fold in the metasandstone horizons, but are neither so abundant nor so uniformly distributed in the more easily weathered hornblende gneiss. From the viewpoint of the significance of the two π-pole maxima, it is valid to relate these concentrations to geometric features present in the fold. By contouring the data (Figure 16.9B) the two maxima can be pinpointed. From the discussion of Question 16.1 it is clear that they represent the fold limbs, and the best fits can consequently be obtained to give orientations L_w 21/64E

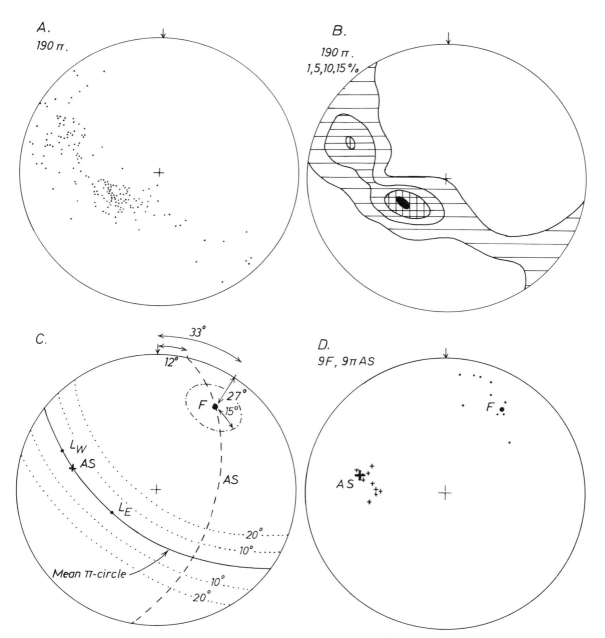

Figure 16.9. Data from Figure 16.6 plotted in equal area projections: A, poles to lithological layering; B, contoured version of A. C gives the geometric analysis—L_E and L_W are the poles to the two fold limbs, F the fold axis, AS the axial surface. The dotted lines around the mean π-circle are those used to assess whether the fold is cylindrical or sub-cylindrical. The 15° small circle around F gives the range of fold axis position satisfying 90% of the π-poles. D shows plots of fold axes (dots) and axial surfaces of small folds (crosses) and their relationships to the features of the large scale structure.

and L_E 152/30 NE. The paucity of π-poles in the SE quadrant of the diagram relates to the few readings taken across the hinge zone because of the sharp crested nature of the fold. To solve specific problems concerning the orientation of the fold axis it might be valuable to take a higher proportion of readings in this zone, because more data points from this region would enable the π-circle to be located with greater accuracy. Such special data selection would invalidate the general conclusions we have drawn from π-pole density based on random and fairly uniform choice of data stations.

In Figure 16.9C the fold axis F was determined by constructing the π-circle through the centres of concentrations of the two fold limbs. This great circle fits the rest of the data quite well, dividing the π-pole into more or less equal distributions on either side of the line. The greatest departure of poles from this circle occurs in the NW quadrant (layers steeply overturned to dip to the SE and located in the NW

limb of the structure). The pole to the best fit π-circle gives a value for the average plunge of the axis of the sub-cylindrical fold of 27° to 33°. In order to indicate the variations possible in a sub-cylindrical fold it might be an idea to construct the 90% data line for this fold as a small circle around the fold axis. If you count out the data points lying within this range you will find that this will have a 15° separation from F.

Answer 16.3

The axial plane of the fold has a pole which lies somewhere in the region of low π-pole density between the two π-pole concentrations representing the fold limbs. The axial trace strike (12°) is combined with the fold axis (F) to fix the position of the axial plane (12/55 SE). In Figure 16.9C it is seen that the pole (AS) lies closer to the pole to the western

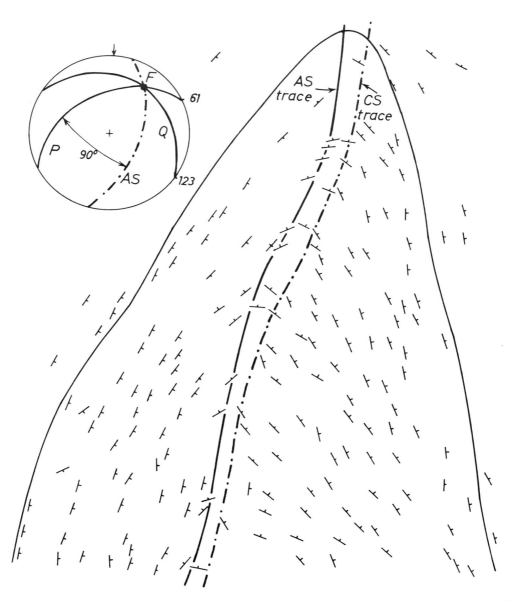

Figure 16.10. *Locations of the axial surface trace and crest plane trace in the fold. The projection shows the surfaces P and Q used to locate these features.*

fold limb than to the eastern limb. This implies that the layers on the western limb have suffered more thinning than those in the east, and we will verify this conclusion in Session 18 when we construct the true profile across this fold.

Answer 16.4

The orientation of the layers where they pass through the hinge line and the axial surface can be found by determining the surface which passes through the fold axis perpendicular to the axial surface. In Figure 16.10 this is plane *P* with orientation 61/47 NW. The axial trace of the fold has been found by joining localities which have (or which are likely to have from an analysis of nearby data) this orientation (Figure 16.10, *AS* trace).

Surfaces along the crestal plane strike perpendicular to the trend of the fold axis (Figure 16.10, plane *Q*, 123/27 NE), and the crestal plane trace has been constructed from the location of such surfaces (*CS* trace). Note that, because both axial and crestal surfaces are inclined, their positions do not coincide, and both traces lie to the east of the zone of vertical layer strikes (cf. Figures 16.7 and 16.10). Because the fold is not perfectly cylindrical, it is not always possible to find surfaces with the exact orientations of *P* and *Q* given above, and both traces have been constructed with a certain degree of compromise.

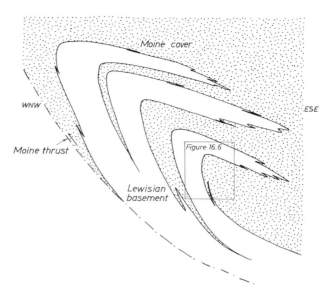

Figure 16.11. *Schematic cross section across the Caledonian front of N.W. Scotland indicating the regional situation of the map of Figure 16.6. The Lewisian basement occupies the cores of refolded anticlinal folds. The fold of Figure 16.6 is an antiformal refold of these earlier formed structures.*

Answer 16.5

The poles to the axial planes and hinge lines group around the calculated positions for the pole of the mean axial surface and fold axis respectively (Figure 16.9D). It is therefore possible to conclude that they are congruent to the main structure, and probably represent parasitic folds developed during the formation of the larger fold. The scatter of hinge lines of smaller folds is wider than that of the 15° small circle drawn around *F*, and it seems possible that some of the small folds might have been developed during a non-coaxial progressive deformation sequence (cf. Figure 12.13, Vol. 1).

Answer 16.6

The steeply inclined lamprophyre dykes cut across the fold structure and show no geometrical coincidence with any of the features of the fold. They were intruded much later than the deformations which produced the fold, and their intrusion is related to the opening of extension fissures during the late stage uplift phase of the Caledonian orogeny.

Before we leave this geometrical study of the Beinn nan Caorach fold it is perhaps worth commenting on the regional significance of the structure in the Scottish Caledonides. The two main rock types represent a series of late Precambrian cover sediments (Moine) and early Precambrian basement gneisses (Lewisian). In this antiform the stratigraphically youngest rocks occupy the core. The antiform is therefore a *downward facing syncline*. The contact between the basement and cover might be either a modified unconformity or a pre-folding fault surface. No evidence for fault derived rocks (cataclasites or mylonites) were found along this contact. Observations in other parts of the general area show that locally an angular unconformity is preserved along the contact, with a basal conglomerate in the Moine sediments, and also current bedding is present in the Moine sand-

stones indicative of sediment polarity facing away from the Lewisian–Moine contact. The contact between the two groups does appear to be best interpreted as a modified unconformity. The angular discordance has become almost obliterated, and we will see how such a relationship can come about in Session 22.

The antiform we have analysed is in fact a fold which has been superimposed on originally eastward facing recumbent folds (Figure 16.11). The anomalous stratigraphic relationships have arisen because the rocks have been previously overturned by an earlier folding event.

Answer 16.7★

The *β*-diagram for this problem is shown in Figure 16.12. The *β*-axes are strongly dispersed, although there is a general concentration in the NE quadrant of the projection. It is not easy to determine the fold axial direction from this diagram. The *β*-axis scatter arises partly from non-cylindrical features to the fold and partly from field measurement errors. Sharply defined *β*-axes are produced where two planes cross at a high angle, but where the two crossing surfaces are closely oriented location of their *β*-intersection is not easy. Any small measurement error contained in either of the two planes which lie at a small angle produces a very large shift in the resulting *β*-axis. If a *β*-diagram is made from a part of the fold where changes in layer orientation are not high (e.g. a fold limb), the resulting *β*-axes show little or no point concentration, but spread in a great circle zone along the great circle representing the average datum plane.

Although *β*-diagrams have been and sometimes continue to be used as a method of fold axis determination, they are markedly inferior in this respect to *π*-diagrams. The amount of data that can be incorporated into an analysis is in

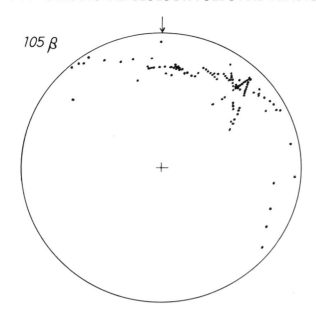

105 β

Figure 16.12. *β-diagram prepared from the small area within Figure 16.6.*

practice very limited, and if folds have surfaces of sub-parallel aspect many impressively (and also spurious) concentrated β-axes arise.

CONCLUDING REMARKS

It should be clear from the exercises carried out in this session that many measurements are required if the types of analysis recommended here are to be effective. We have found that using these techniques field mapping greatly improves (Appendix F): more data of higher accuracy result, and this leads to a deeper understanding of structural geometry. The technique of field mapping, connected in some people's minds with the somewhat tedious and imprecise tracing of geological boundaries, takes on a new interest by enabling new and exciting geometric facts to be appreciated.

The general use of projection techniques as a working tool for analysis has been emphasized. However, it must also be stressed that projections do not always make the best illustrative material for papers and reports. Publications relying too heavily on collections of projections or, worse still, discussion of such collections, are guaranteed to put off many readers, expert and non-expert alike. Projections may provide the clearest way possible to demonstrate certain aspects of three-dimensional orientation. If this is the case, use them. However, it is often better to abstract relevant information from the projections and represent this information in synoptic diagrams, trend maps, etc. which can be easily assimilated visually.

The region used for the practical work of this session was quite small, and was chosen for its geometric homogeneity. The effective application of projection methods to larger areas requires careful analysis of maps well provided with numerical data (cf. Question 16.1) so that a large region may be divided into more or less homogeneous sub-areas. This session has concentrated on cylindrical and sub-cylindrical fold models because these do seem to predominate in geology. Other fold models are theoretically possible and the geometric forms of circular-, elliptical-, and irregular-cones are worth investigating. However, most published accounts of fold geometry give a strong impression that the sub-cylindrical fold with axial direction changing from sub-area to sub-area does provide the best model for the most abundant folds found in naturally deformed rocks.

KEYWORDS

Perfect cylindrical fold	A fold whose π-poles lie precisely on a great circle.
Cylindrical fold	For geological purposes the definition above is too restricted. A fold with 90% of the layering π-poles lying within 10° of the mean π-circle.
Sub-cylindrical fold	A fold with 90% of the layering π-poles lying within 20° of the mean π-circle.
Non-cylindrical fold	A fold with more than 10% of π-poles lying outside a 20° zone around the mean π-circle.

KEY REFERENCES

Cheeney, R. F. (1983). "Statistical methods in geology", 169 pp. Allen and Unwin, London and Boston.

Chapter 9 of the well written introduction to statistical methods concerns itself directly with the analysis of three-dimensional orientation data and discusses the mathematical procedures that can be used in conjunction with projection techniques to determine various geometric features of folds.

Cruden, D. M. and Charlesworth, H. A. K. (1972). Observations on the numerical determination of axes of cylindrical folds. *Geol. Soc. Am. Bull.* **83**, 2019–2024.

This paper points out errors in previously published methods, and the authors put forward a new method for calculating the orientation of fold axes and evaluating the extent of any conical geometric components in the fold surface.

Mancktelow, N. (1981). A least squares method for determining the best fit point maximum, great circle and small circle to non-directional orientation data. *Math. Geol.* **13**, 507–521.

Ramsay, J. G. (1964). The uses and limitations of Beta-diagrams and Pi-diagrams in the geometrical analysis of folds. *J. Geol. Soc. Lond.* **120**, 435–454.

Turner, F. J. and Weiss, L. E. (1963). "Structural analysis of metamorphic tectonites", 545 pp., McGraw-Hill, New York and London.

Methods are described using a minimization of the angular residuals of orientation data from some chosen best fit model to find particular geometric fits for the determination of various features of cylindrical and conical folds. Fortran programming of the data provides what is probably the best practical method for evaluating these geometric features.

This discusses the problems that arise using projection techniques with particular emphasis on the errors that can arise when the methods are applied to certain types of fold geometry.

A very thorough discussion of rock fabrics on all scales is presented in this book with special emphasis on projection techniques. Particularly good background reading to this session, which would form a useful bridge to our studies of the geometry of superposed folds (Session 22), will be found in Chapters 3, 4 and 5.

SESSION 17

FOLD CLASSIFICATION

A brief introduction to some historical aspects of fold classification and a discussion of the geometric features of the parallel and similar fold models leads to more general methods of describing thickness variations in folded layers. Descriptive fold classification schemes using t_α/α graphs (layer thickness variation with dip) and dip isogons enable fold geometry to be specified exactly. The two schemes are interrelated and their merits and technical limitations are discussed. It is shown how exact equations can be set up to specify thickness variations in certain simple fold models.

INTRODUCTION

In Session 15 we saw that the shape of a single folded layer can be specified using the Fourier series method. Although some of what we understand under the heading of "fold style" relates to the shapes of single surfaces in a fold, most geologists tend to think of style as the way in which the two surfaces that enclose a layer interrelate, and especially the manner in which the thickness of a layer varies around a fold. In theory, we could describe these thickness variations quite precisely by specifying the two Fourier series which describe the shapes of the upper and lower surfaces and by relating these by some appropriate scaling and separation factors. However, such a technique would be tedious to develop in a practical way and, more important, it would be unlikely that the sets of numbers emerging from such a technique could be visualised in a way that would allow us to relate them simply to the geometry we observed.

Historically, the problem of describing layer shape developed around two geometric models, the **parallel fold** and the **similar fold** (Van Hise, 1896). In a parallel fold the thickness t measured orthogonally across the layer boundaries is constant throughout the fold (Figure 17.1A). Occasionally fold domains with this geometry show individual surfaces with almost constant curvature and such folds with circular boundary layers have been termed **concentric folds**. However, the layers within a parallel fold need not form concentric arcs around some specific centre (Figure 17.1, cf. A and B). Parallel folds are generally found in especially competent layers of sedimentary sequences (massive limestone, quartzite) which have been folded at high levels in the crust where competence contrasts are marked (hence the term **competent fold** given by Busk). **Similar folds**, in contrast to parallel folds, show considerable variations of layer thickness, and always a thinning on the fold limbs relative to that seen at the hinge zone (Figure 17.1C). The term **incompetent fold** has been used for this style, but

we recommend that both this term and that of competent fold be discontinued because neither is mechanically correct. In a fold with true similar geometry the shapes of the bounding surfaces are identical, and this means that the layer thickness T measured anywhere in the fold in a direction parallel to the axial surface is constant (cf. a parallel fold where T shows an increase from fold hinge to fold limb). The name "similar" does not provide a good geometric description ("identical" or "congruent" would be better adjectives —e.g. contrast congruent and similar triangles in Euclidean geometry), but the name is so firmly established in publications it would be unwise to attempt to change it.

These two models held the main attention of structural geologists for many years, and it was not until 1958 that de Sitter pointed out that natural folds often showed geometric forms with characteristics intermediate between these two models. He further showed how a fold which started its development with parallel geometry could be modified by a flattening process so as to come to lie in an intermediate model position. Ramsay (1962) showed how such intermediate fold styles might be accurately categorized using orthogonal thickness (t) and axial surface thickness (T) measurements, and suggested some ways whereby these measurements might be used to determine the extent of the flattening components. In 1967 these ideas were further extended to develop a more comprehensive descriptive classification based on recording thickness variations. One special point of interest of this new classification was that it showed that the parallel and similar folds were not the true geometric end members of a morphological spectrum, and that geometric possibilities existed outside the range of these special models. The realization that all of these geometric types exist in nature, and the simple application of the classification for practical purposes, have led to its adoption as a standard technique for describing folds.

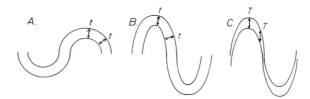

Figure 17.1 *Geometric features of:* A, *concentric parallel folds;* B, *non-concentric parallel folds; and* C, *similar folds. Orthogonal layer thickness, t; layer thickness parallel to axial plane, T.*

Fold classification using layer thickness variations (Ramsay, 1967)

This method is based on a description of the changes in thickness with angle of dip, these measurements being expressed as a proportion of the layer thickness at the fold hinge. As with Fourier analysis methods for describing single surfaces, a folded layer is analysed in quarter wave sectors. The procedure is as follows:

1. Construct a true profile section of the fold perpendicular to the hinge line.
2. Determine the hinge lines h_A, h_B and inflexion lines i_A, i_B on the bounding surfaces A and B of the layer. Construct the axial plane by joining h_A and h_B; this will be the reference direction for dip directions and thickness variations (Figure 17.2A). Construct the tangents to the fold at h_A and h_B; these provide the zero dip reference direction. Measure the orthogonal thickness between the tangents; this provides the standard hinge thickness reference t_0.
3. Select an angle of dip α, and construct tangents to the fold surfaces with this dip. Measure the orthogonal distance between these tangents t_α. Express t_α as a proportion of t_0, i.e. $t'_\alpha = t_\alpha/t_0$. The t_α measurements are simply and rapidly made with a transparent overlay marked out with a series of parallel lines of known spacing (Figure 17.2B).
4. Plot t'_α as ordinate against α as abscissa.
5. Repeat for other values of α and construct a continuous t'_α/α curve for the quarter wave fold sector.

The range of possible layer shapes is best analysed by starting from the well established parallel and similar fold models. The parallel fold model shows constant orthogonal thickness throughout the fold and shows $t'_\alpha = 1$ for all values of α (Figure 17.3). In a similar fold the distances between tangents drawn to the fold surfaces measured in a direction parallel to the axial plane are constant for all values of α.

$$T_\alpha = T_0 = t_0 \qquad (17.1)$$

from trigonometric relationships of Figure 17.2C

$$t_\alpha = T_\alpha \cos \alpha \qquad (17.2)$$

and combining (17.1) and (17.2) we find that for similar folds

$$t'_\alpha = \cos \alpha \qquad (17.3)$$

The two lines $t'_\alpha = 1$ and $t'_\alpha = \cos \alpha$ enable the fold shape field to be subdivided into compartments (Figure 17.3). In each of these compartments folds with the t'_α values can always exist. Between the parallel and similar fold lines are those folds first recognized by de Sitter. Above the parallel fold line and beneath the similar fold line are those types first emphasized by Ramsay. Folds which occur in these three field areas have never been designated with simple names; they have been given numbers and letters. Folds of Class 1A have orthogonal thickness on the limbs which exceed that at the hinge. Those of Classes 1C, 2 and 3 all show limb thinning, with Class 1C and 3 showing respectively less and more thinning than that of the true similar fold (Class 2). Figure 17.3 should clarify the differences in shapes between the main classes. All of the five folds shown here have been developed with an inner arc with constant geometric form. It should be emphasized that not only are all types geometrically possible, but all types do occur commonly in naturally deformed rocks. The significance of the number and letter

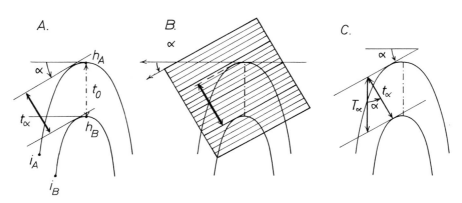

Figure 17.2. *Method of determining the orthogonal layer thickness* t_α *at angle of dip* α. *In* A, h_A, h_B, i_A *and* i_B *are the hinge lines and inflexion points defining the quarter wavelength domain, respectively.* B *illustrates an overlay method for measuring* t_α. C *shows the relationships between orthogonal thickness* t_α *and thickness* T_α *measured parallel to the axial plane.*

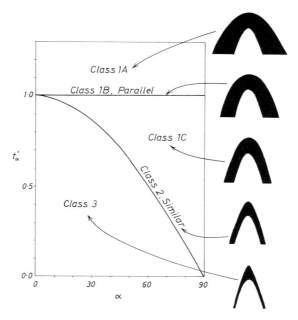

Figure 17.3. *Graphical plot of standardized orthogonal thickness t'_α plotted against angle of dip α and the main types of fold classes.*

reference scheme will become apparent later in this session when we will discuss the significance of thickness variations in terms of layer boundary curvature.

QUESTIONS

Layer thickness variations

Question 17.1

Figure 17.4A shows a thin section of a strongly folded Palaeozoic sediment consisting of alternating fine silts (pale) and clay-slates (dark). A part of this section has been enlarged in Figure 17.4B. The main features of the lithological layering have been represented, the silt bands are stippled, the argillaceous parts left unornamented. Hinge lines and inflexion lines have been identified in some of the folds and six fold sectors (1 to 6) have been indicated.

Using the thickness variation method described above construct t'_α/α graphs for each of the six sectors. Compare sectors 1 and 2 from a silt layer with 3 and 4 from a clay-slate layer. What are the main differences in fold style? In the thin section certain isolated silt layers show some of the characteristic features of ptygmatic structures (see p. 12) suggesting that the silt was more competent than the clay-slate. Suggest a correlation between fold class and layer competence.

Compare the fold styles in sectors 1, 2, 5 and 6 in the same layer and suggest reasons for the variations.

Check your results in the Answers and Comments section and return to the section below.

Fold classification using isogons

This method of fold classification is based on the construction of **dip isogons**: lines joining points of equal dip on either side of the folded layer. Dip isogons can be constructed very

rapidly and this graphical method leads to a very quick determination of the main fold class of the layer.

First, some convenient datum direction in a fold profile is chosen. That most frequently selected is a zero datum in the direction parallel to the layer surfaces at the hinge line, but in principle any direction may be chosen without invalidating the essential results of the classification of layer shapes based on the isogon pattern. This is useful because it means that we do not necessarily have to analyse successive quarter wave sectors in a layer showing several folds. Once the zero datum has been selected tangents are drawn to the two sides of the layer to make angles of α with the datum (Figure 17.5). The two points having this dip are then joined with a straight line—the α-*dip isogon*. Other isogons are constructed through the fold for other layer inclinations usually at periodic angular intervals. Adjacent isogons can be *parallel*, or they can *converge* or *diverge* (using these terms always in the sense of tracing isogons *from the outer to the inner arc* of the fold). The relationships shown by adjacent isogons provide information about the average curvature of the folded surfaces between the two isogons that are being considered. If the isogons are parallel, the average curvatures of the two surfaces are equal. Convergent isogons imply that the inner arc curvature exceeds that of the outer arc, whereas divergent isogons mean that the outer arc curvatures exceed that of the inner arc. Using the overall isogon pattern we find that there are three main classes which can be specified and that one of these may be subdivided into three sub-classes. These classes are exactly the same as those we met with the scheme based on layer thickness variations. They are shown in Figure 17.6 and can be defined:

Class 1   Convergent isogons
    sub-class 1A   strongly convergent
    sub-class 1B   **parallel fold** with isogons
                      perpendicular to layering
    sub-class 1C   weakly convergent.

Figure 17.4. *Thin section of a folded sediment. See Question 17.1.*

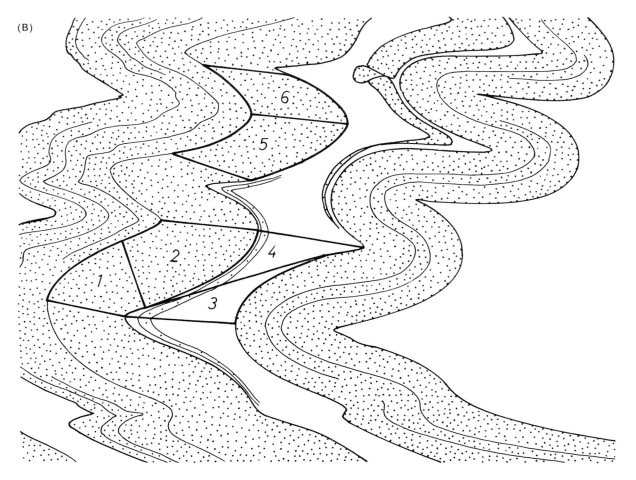

(B)

Class 2   Parallel isogons, similar fold
Class 3   Divergent isogons

It is unfortunate that the term parallel can be used in two differing senses (a parallel fold and a fold with parallel isogons, which is a similar fold), but the term parallel fold is so well established it would be unwise to change its well known meaning.

Dip isogons

Question 17.2

Figure 17.7 shows a ptygmatic vein of quartz–feldspar pegmatite in a matrix of granite gneiss from the Chindamora batholith of Zimbabwe. The main vein was originally intruded along the gneissic layering and was subsequently compressed to take up a folded form. The smaller vein in the lower part of the photograph was injected, in part discordantly, across the layering. In the sketch some of the slight irregularities along the vein–gneiss contact caused by the very coarse grain structure of the pegmatite have been smoothed to the average fold form.

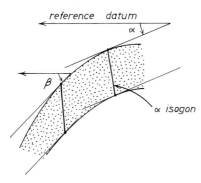

Figure 17.5 *Method of constructing dip isogons of value α and β.*

Construct dip isogons at 20° dip intervals. This can be done using a parallel rule to determine points of equal dip, by using a transparent measuring overlay (Figure 17.2B) or perhaps the quickest method is by placing the sketch over a piece of graph paper on a light table and making successive 20° rotations of the overlay over the graph lines.

Discuss the main geometric features of the fold from the isogon pattern. Using the dip isogon lines in conjunction with the traces of the folded layers classify the folds in different parts of the field of the diagram.

Now proceed to the Answer and Comments section; then proceed to Session 18, or to the starred questions below.

STARRED (★) QUESTIONS

t'_α equations describing fold shape

Question 17.3★

Figure 17.8 shows a part of a series of folded multilayers consisting of regular alternations of competent layers (P) with parallel fold Class 1B separated by incompetent layers (Q) with Class 3 fold styles. The competent layers are defined by the relationship $t'_\alpha = 1$ for all values of α. Determine the t'_α/α equation which defines the Class 3 folds.

1. where $t_{P_0} = t_{Q_0}$
2. where the ratio $t_{Q_0}/t_{P_0} = n$

Hint: it is simplest to work by way of the thickness parallel to the fold axial plane, i.e. T_P and T_Q. Show that the fold shape in layer Q varies with n, and that, as n increases, the fold style in layer Q becomes closer to the similar model. Plot t'_α/α graphs for the folds in layer Q with n values of 0·5, 1·0 and 2·0. Why are there values of α in each fold which cannot be exceeded?

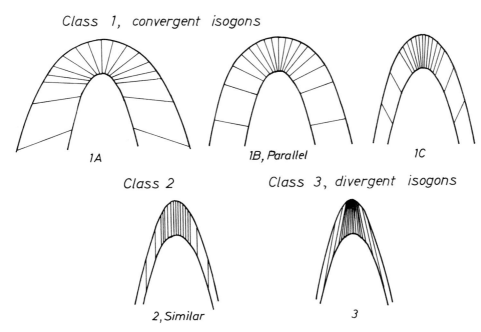

Figure 17.6. *The dip isogon characteristics of the main fold classes.*

Figure 17.7. Ptygmatic vein in granite gneiss. See Question 17.2.

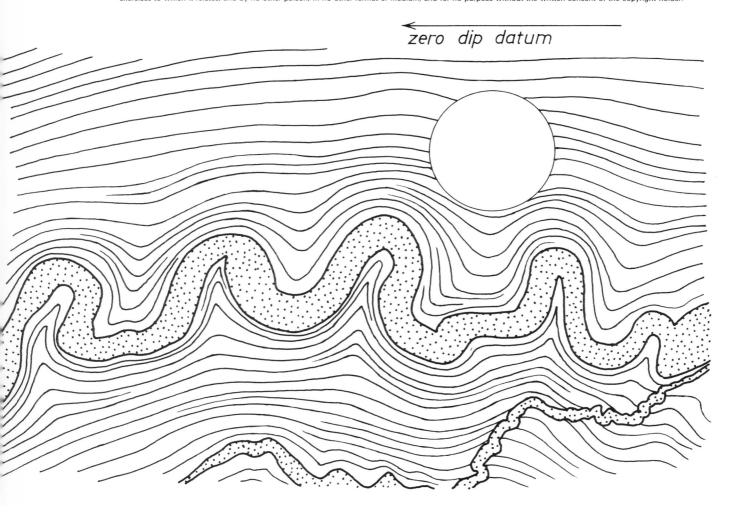

zero dip datum

Flattened folds

Question 17.4★

In Question 17.1 we saw how it might be possible to modify the shape of a previously formed parallel fold by a process of straining over the whole fold and that such a process might lead to a change in fold class. Figure 17.9A shows a simple model of regularly folded multilayers containing competent layers (*P*) with parallel folds alternating with incompetent layers (*Q*) showing Class 3 folds. In Figures 17.9B and C this model has been subjected to a homogeneous strain so that *all* the components in the block undergo an identical superposed finite strain. Determine the fold classes in layers *P* and *Q* graphically from the figure, and determine how the folds progressively change shape with increasing strain.

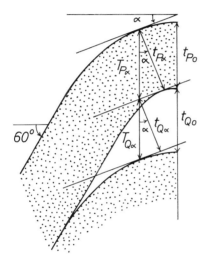

Figure 17.8. *Relationships between layer thickness and dip angles in part of a composite multilayer, P with Class 1B folds and Q with Class 3 folds. See Question 17.3★.*

Determine the equations that define the shapes of the folds in layers *P* and *Q* as a function of the superposed strain ratio $R = (\lambda_1/\lambda_2)^{1/2}$. Express $t'_{\alpha'}$ as a function of R and α' where α' is the angle of dip in the fold *after* straining. Note that the orthogonal thicknesses t_{P_0} and t_{Q_0} are equal in Figure 17.9A.

Laboratory formed folds

Question 17.5★

Figure 17.10 illustrates the results of a laboratory experiment made by Peter Cobbold (1975) in which a single layer of competent wax (with stippled ornament) embedded in a less competent wax matrix was subjected to a layer parallel compression. Determine the isogon pattern and classify the folds, delimiting carefully the positions of change of fold class. Note carefully the changes in fold class *inside* the competent layer as well as in the matrix. Compare the distribution of fold classes in the rather open fold j with those in the more strongly developed fold k and with the strongest amplitude fold l.

Sorby's "fish-hook" folds

Question 17.6★

Figure 17.11 shows a folded and locally extremely thinned layer of crinoidal limestone enclosed in a strongly cleaved calcareous slate from the Devonian of Ilfracombe, N. Devon. These structures were first described by Sorby (1879) and, because of their style, they have been subsequently termed "fish-hook" folds. Describe, using thickness variation graphs, the geometry of these rather strangely shaped folds. What is especially puzzling here is the geometric polarity of the folds; the antiforms are not mirror symmetric with the synforms. What might be the reason for directional polarity of the structures?

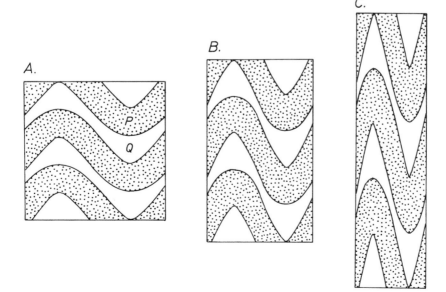

Figure 17.9. A: *Regular multilayer complex of parallel (layer P) and Class 3 folds (layer Q). B and C show the flattened derivatives of these folds after homogeneous strains with ratios R = 1·78 and 4·0 respectively. See Question 7.4★.*

ANSWERS AND COMMENTS

Layer thickness variations

Answer 17.1

The t'_α/α graphs for the six sectors are illustrated in Figure 17.12. The folds in sectors 1 and 2 are of Class 1C with thinned limbs. Note how the inner arcs of these folds are always more strongly curved that the outer arcs. This is a feature of all Class 1 folds. In contrast sectors 3 and 4 show Class 3 folds, a class categorized by having an inner arc curvature greater than that in the outer arcs. To complete this discussion on curvature relationships it should be clear that similar folds (Class 2) form the boundary between the Classes 1 and 3 in that the curvature of both sides of the layer are the same.

In sequences of harmonically folded rocks it is quite usual to find layers with Class 1 fold forms alternating with layers of Class 3. When the fold forms are paired in this way individual folds are able to penetrate long distances through the multilayered sequence along the direction of their axial planes without losing amplitude. One can envisage a Class 1–Class 3 pair averaging out to produce an overall similar model and, as there are no geometric limitations necessitating a similar fold to die away along its axial surface, so there are no necessary geometric restrictions for a Class 1–3 pair to die away.

This overall similar fold geometry of pairs of siltstone–clay-slate layers is seen extremely well in Figure 17.4. For example, the left-hand boundary of sectors 1 and 2 has a form very close to that of the right-hand boundary of sectors 3 and 4. At this stage in our discussion we will only refer to geometric constraints of fold form, but later in Sessions 19 and 20 we will see what mechanical contraints lead to these types of geometry.

The ptygmatic shape of the siltstone layer in the right-hand side of Figure 17.4B is strongly suggestive of mechanical buckling of a competent rock sheet. If this deduction is correct (and it is at this stage based only on inference from experiments—Figure 17.9B) folds in competent layers are characterized by forms of Classes 1B and 1C, whereas those in the incompetent layers are of Class 3. It should, however, be emphasized that this conclusion is a generalization drawn from an analysis of the complete competent (or incompetent) layer taken from its bounding surfaces and that, when we come to analyse the details of the fold forms inside these layers, we will find that the fold styles are much more complex.

Although sectors 5 and 6 are lithologically identical to those of 1 and 2 the fold styles are rather different, being close to the similar fold model (Class 2). A reason for these geometric differences is not obvious. An explanation might be put forward based on the changes of shape of the ptygmatically folded silt layer lying to the right of the sectors under discussion. This layer seems to have suffered less layer shortening where it lies adjacent to sectors 1 and 2 than where it lies close to sectors 5 and 6. This feature suggests that sectors 5 and 6 had fold forms originally like those of 1 and 2, but that these shapes were modified by further compression. They underwent shape modifications as a result of superimposed straining processes which reduced the limb thicknesses and producing geometric characteristics more in accord with those of similar folds. This is the concept originally suggested by **de** Sitter and Ramsay, who termed such structure **flattened parallel folds**. For those with sufficient time we will look into the geometry of these folds in Question 17.4★.

The method of classifying folds by describing thickness variations brings to light very clearly the differences in fold style in different rock layers and it follows that it is not realistic to attribute a single fold style name to a folded multilayered mass. Not only does each layer have its own fold style, but often the style will change from point to point in the same layer. The most important feature of the method is its practicality. It is easy to use, both to give broad subdivisions of the different fold classes and to give very accurate descriptions of fold shapes. This second point is very important indeed when it is required to make detailed comparisons of natural fold shapes with the results of labora-

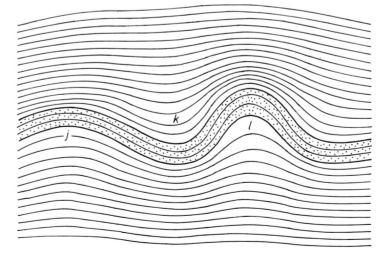

Figure 17.10. *Laboratory formed fold produced by the buckling of a competent layer (stippled) in an incompetent matrix. See Question 17.5★. (From Cobbold, 1975, Figure 9, stage 9.)*

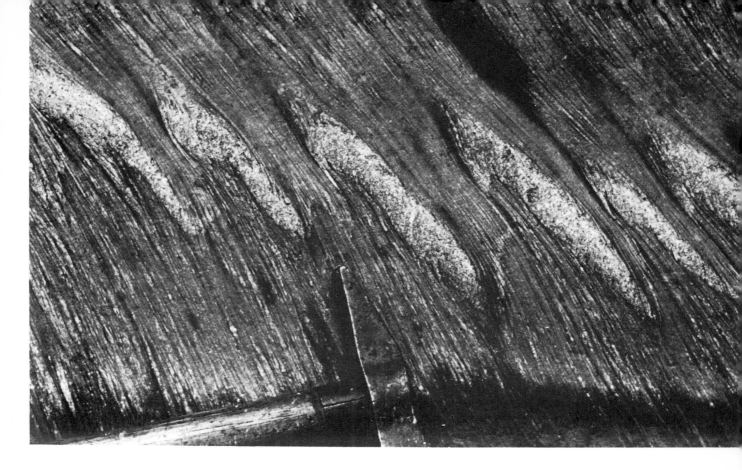

Figure 17.11. "Fish-hook" folds. See Question 17.6★.

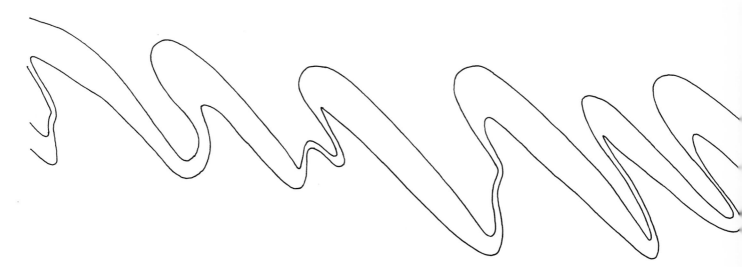

tory experiments and with the predictions arising from mechanical theories of fold formation (see Hudleston, 1973a, b). As a primary classification its main advantage over others is that it is descriptive and non-genetic, having its mathematical basis in a comparison of the curvatures of adjacent surfaces of a folded layer. Clearly the different fold shapes do arise because of different mechanical principles operating during fold formation, but it is most valuable to have a primary classification scheme independent of ideas of fold genesis.

Although folds will be occasionally encountered which have more complex shapes than those described under the main classes developed in this scheme, most will be easily fitted into this system. It is possible, however, for quarter wave fold domains to have characteristics which relate to several of the classes, and it is possible to have t'_α/α plots which cross from one class into another. Methods for describing such rarities can be found in Ramsay (1967, pp. 369–371, 407–410).

Like many schemes of classification, the thickness variation method does sometimes pose problems. You may have already noticed that it is not always possible to make thickness measurements over the whole fold domain. A problem arises when the highest dip α on one surface is exceeded by the highest dip β on the adjacent surface (Figure 17.13A). In this case there is a sector on the fold limb in which it is not possible to construct tangents over the dip range α to β, and in this sector it will not therefore be possible to make thickness measurements. Another problem that occasionally arises is that layer dips at the hinge lines may not be perpendicular to the axial plane of the fold (Figure 17.13B). Hudleston (1973), who was the first to discuss this problem, noted its rarity and suggested another classification procedure closely related to the methods employing dip isogons which are discussed in Question 17.2.

Dip isogons

Answer 17.2

The isogon pattern is illustrated in Figure 17.14. The most striking feature is the succession of closed cell-like regions of high numerical positive and negative dips, all of which show maximum dips coinciding with the limbs of the folded ptygmatic vein. This pattern strongly suggests that the quartz–feldspar vein was the major influence in producing the overall driving mechanism for the folding, and that strong mechanical rotation effects were generated in this, the most competent layer. The high valued dip cells rapidly dissipate away from the median line of the folded vein: the folds in the matrix show a rather rapid amplitude decrease away from the pegmatite layer so that along the top and in part of the lower left of the photograph folding is effectively absent (the limits are marked with a dash–dot line in Figure 17.15). Ptygmatic structures usually imply layer parallel contraction, and the lack of folding at some distance from the vein implies that the contraction taken up by folding passes into a zone where it is taken up by a homogeneous strain. The zone around the folded vein in which folding can be seen and in which heterogeneous strains occur is known as the **zone of contact strain**. We will discuss some of the important mechanical implications of this zone in

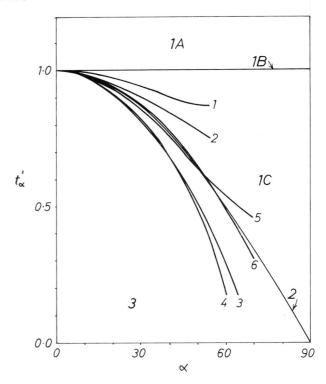

Figure 17.12. Graphs of folds from the six sectors of Figure 17.4.

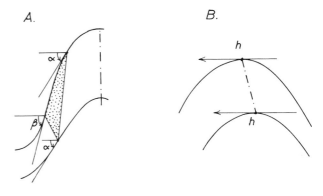

Figure 17.13. Special problems arising with fold classification: A, part of limb not represented in fold classification; and B, dip lines not perpendicular to axial plane joining the fold hinge lines h.

Session 19. There are also cell-like isogon patterns in the thinner vein in the lower part of the diagram, but these have smaller dimensions than those in the main vein. There is also a zone of contact strain around this smaller vein, but it is narrower than that seen in the main vein and the two do not interconnect. This lack of connection relates to the disharmonic nature of the folding in the two competent layers.

The folded gneiss in the inner arcs of the buckled vein shows isogons which converge together so as to lie close or even to coincide with each other, this isogon pattern going together with the development of cusp-like folds. Such cuspate structures are especially common features of folds

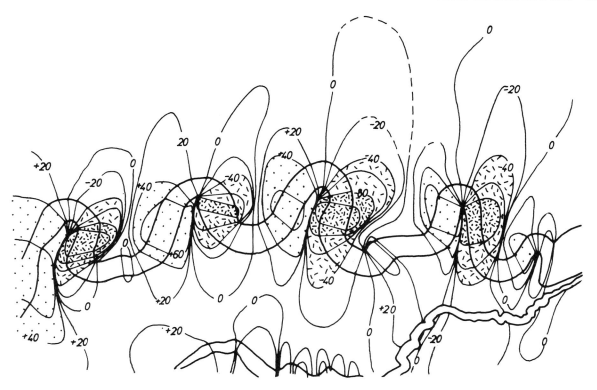

Figure 17.14. Dip isogon plot of the ptygmatic vein of Figure 17.7.

found at or close to the contacts of folded competent and incompetent layers.

Inside the ptygmatic vein the fold shapes all lie close to the parallel model (1B). The inner arcs always show a greater curvature than the outer arcs, the isogons showing convergence and being sub-perpendicular to the layer contacts. In some of the fold hinge zones there is a tendency for the isogons to show more weakly convergent geometry than those of the parallel fold model, and the fold shapes accord more with the Class 1C model. This is quite a common feature of ptygmatic veins, especially where the interlimb angles of the folds are high and where the arc-length to thickness ratio is small (< 10). We have noted this effect previously (Question 17.1, fold sectors 1 and 2) and have attributed it to superposed strain effects.

The great range of fold styles shown by the folded layering in the gneiss will probably have come as a surprise. Examples of practically all the main fold classes can be found within the zone of contact strain, but there are systematic relationships between the fold class and the position within the zone. Close to the boundary of the folded vein Class 1C folds are found in the inner fold arcs and Class 1B (or transitions 1B to 1A) are found along the outer fold arcs. A most striking and regular pattern of alternating Class 1A and Class 3 folds is found in the contact strain zone at some distance from the vein contact, coinciding with outer fold arcs and inner arcs respectively. The alternation of these two classes along individual layers is geometrically connected with the decrease in fold amplitude away from the median line of the folded vein.

It will be clear from this discussion that, in the example of a single layer ptygmatic vein, there is no simple correlation of fold class and rock competence. The overall pattern of fold classes is, however, completely systematic and it is the recognition of the fine details of this pattern that gives special power to this method of shape analysis.

Fold classification methods using thickness variation graphs or dip isogons ought to be made on true profile sections perpendicular to the fold hinge lines. If true profile sections are not used for any reason there will be some misinterpretation of the shape classification. However, in general folds of Classes 1, 2 and 3 will retain their overall class characteristics even in false profiles, providing that the observation plane is sub-perpendicular to the axial surface of the folds.

t'_α equations describing fold shape

Answer 17.3★

1. Let

$$t_{P_0} = t_{Q_0} = k$$

Then from the relationships in Figure 17.8:

$$T_{P_\alpha} = \frac{k}{\cos \alpha} \qquad (17.4)$$

Because the distance measured parallel to the axial plane between the top of one competent layer and the top of the immediately adjacent competent layer remains constant:

$$T_{P_\alpha} + T_{Q_\alpha} = t_{P_0} + t_{Q_0} = 2k \qquad (17.5)$$

Combining (17.4) and (17.5):

$$T_{Q_\alpha} = 2k - \frac{k}{\cos \alpha}$$

$$t_{Q_\alpha} = \left(2k - \frac{k}{\cos \alpha}\right) \cos \alpha$$

$$t'_{Q_\alpha} = t_{Q_\alpha}/k = 2 \cos \alpha - 1 \qquad (17.6)$$

This is the equation of the t'_α/α curve for the fold. It always takes a value of less than $\cos \alpha$ so it lies in the Class 3 field (see Figure 17.16 where $n = 1$).

2. Let $t_{P_0} = k$, then $t_{Q_0} = nk$. Following the method of the previous example:

$$T_{P_\alpha} = \frac{k}{\cos \alpha}$$

$$T_{P_\alpha} + T_{Q_\alpha} = k(n + 1)$$

$$T_{Q_\alpha} = k(n + 1) - \frac{k}{\cos \alpha}$$

$$t_{Q_\alpha} = k\left(n + 1 - \frac{1}{\cos \alpha}\right) \cos \alpha$$

$$t'_{Q_\alpha} = t_{Q_\alpha}/nk = \frac{(n + 1) \cos \alpha - 1}{n} \qquad (17.7)$$

Curves for some values of n are illustrated in Figure 17.15. From the nature of this function, as n becomes increasingly large t'_{Q_α} approaches the value $\cos \alpha$. That implies that the

fold approaches a Class 2 or similar fold. If Equation (17.7) is expressed

$$t'_{Q_\alpha} = \cos \alpha + (\cos \alpha - 1)/n$$

then, because the second term of this function is always negative, it follows that all the fold curves will lie beneath the Class 2 cosine function in the Class 3 field.

The significance of the particular value of α which acts as a curve cut-off point given by:

$$\cos \alpha = 1/(n + 1) \qquad (17.8)$$

can be seen in the diagram of Figure 17.8. At the cut-off angle (in Figure 17.8, $n = 1$ and the angle is 60°) the incompetent layer becomes thinned to zero. If geometric compatibility is to be maintained at angles exceeding this, either the thickness of the competent layer must be decreased on the limbs (with a change of fold Class from 1B to 1C), or layer parallel sliding between the competent layers must occur.

Flattened folds

Answer 17.4★

The t'_α/α curves for the original layers P and Q of Figure 17.9A are indicated in Figure 17.17 as P_A and Q_A. Modifications of the fold styles by homogeneous strains with ratios of $R = 1.78$ (25% extra shortening of the fold) and $R = 4.0$ (50% extra shortening) are shown as curves P_B, Q_B and P_C, Q_C respectively. As a result of this straining the original dips of the fold limbs of Figure 17.9A are increased from α to α'. The original layer thickness may be decreased (on the limbs) or increased (in the hinge zone), but the ratio of thickness at

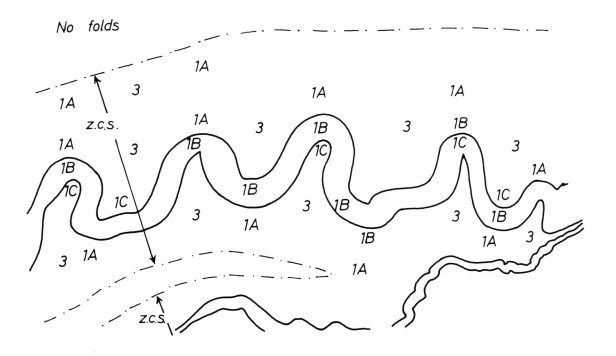

Figure 17.15. *Position of the various fold Classes derived from the isogon plot (Figure 17.14) and layer lines (Figure 17.7). The zone of contact strain is shown between the dash–dot lines (z.c.s.).*

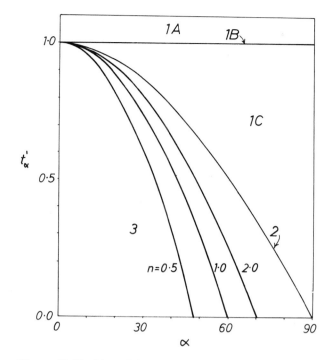

Figure 17.16. *Plots of Class 3 folds situated between parallel folds in a multilayer complex.*

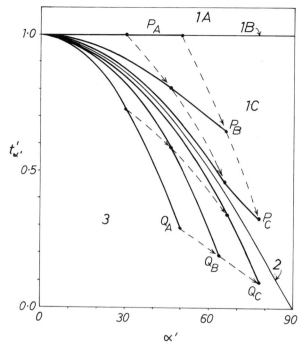

Figure 17.17. *Effects of flattening layers P_A and Q_A of Figure 17.9 to positions $P_B Q_B$ (R = 1·78) and $P_C Q_C$ (R = 4·0). Dashed lines show the movement paths of t'_α / α points with increasing flattening.*

dip α' to hinge thickness is always reduced. The dashed lines show the movements of points representing particular t'_α / α points of the original folds. The graphs show that both the original parallel folds in layer P and the Class 3 folds in layer Q undergo changes that bring them closer to the similar fold model (Class 2).

Fold shapes in flattened parallel folds

In the original layer P, straining by a ratio R leads to a change of dip from α to α' by Equation (D.10).

$$\tan^2 \alpha' = R \tan^2 \alpha$$

or

$$\cos \alpha = (1 + \tan^2 \alpha' / R)^{1/2}$$

The layer thickness across the tangent planes measured parallel to the axial plane is

$$R^{1/2} T_{P_\alpha} = R^{1/2} k / \cos \alpha$$

and the distance t_{P_0} becomes $R^{1/2} k$. Combining these functions we obtain the equation for the new fold

$$t'_\alpha = \cos \alpha' (1 + \tan^2 \alpha' / R)^{1/2} = (\cos^2 \alpha' + \sin^2 \alpha' / R)^{1/2}$$

$$(17.9)$$

This is the equation that has been used to specify the intermediate fold shapes of the Class 1C model lying between the original parallel folds (where $R = 1$, $t'_\alpha = 1$) and similar folds (as $R \to \infty$, $t'_\alpha \to \cos \alpha$) (Ramsay, 1962, 1967; Hudleston, 1973a).

Fold shape in flattened Class 3 folds

Using an exactly similar procedure to that above we can establish how the original fold given by Equation (17.6) is modified by a homogeneous strain of ratio R. The resultant curve is

$$t'_\alpha = 2 \cos \alpha' - \left(\cos^2 \alpha' + \frac{\sin^2 \alpha'}{R} \right)^{1/2} \quad (17.10)$$

In unmodified folds ($R = 1$) the equation reduces to that of Equation (17.6), and with increasing strain ($R \to \infty$) the curves gradually converge towards the similar fold model $t'_\alpha = \cos \alpha'$, but always lie in the Class 3 field because the part of the function under the square root always numerically exceeds the value $\cos \alpha'$.

The shape of the modified folds is not just a function of imposed strain, but also depends upon the ratio n describing the thickness ratio of the layers Q and P at the hinge. From Equation (17.7) we can obtain the most general equation for these folds:

$$t'_\alpha = \frac{1}{n} \left[(n + 1) \cos \alpha' - \left(\cos^2 \alpha' + \frac{\sin^2 \alpha'}{R} \right)^{1/2} \right] \quad (17.11)$$

Equations (17.9), (17.10) and (17.11) all have the potential for providing solutions for a value of superposed straining R on a multilayer complex because all the other variables appearing in the functions are measurable. Most work along these lines has, until now, focused attention on the shapes of the modified parallel fold components, but it would appear that by combining these with the modifications taking place in the original Class 3 fold layer the data from a rock layer pair might provide better results and further checks on the assumption of homogeneity of strain.

Perhaps we ought to ask the question why we wish to evaluate the flattening components superimposed on fold. Understanding the reasons for the geometric configuration of folds is important in its own right and is especially important if we wish to construct fold profiles from a limited amount of data. In the next session we will see that the isogon method of classification may, under certain circumstances, be reversible and be used to predict the unexposed levels of folds. Furthermore, it is often necessary to compute the total shape changes in the Earth's crust and folded layers do offer a method for making such calculations. Measuring along the fold arc length will not produce the correct answer to such a computation where we have folds with flattened geometry, but these equations will enable such structures to be restored to a pre-flattened state.

Laboratory formed folds

Answer 17.5★

The broad features shown by the isogon plot of Figure 17.18 are similar to those of the naturally folded ptygmatic vein of Figure 17.14, especially the closed cell-like isogon lines and the progressive decrease in fold amplitude away from the contact of the buckled layer in the zone of contact strain. So close are the general distributions of types of fold class in the natural and experimental deformation patterns that it seems perfectly justified in correlating the observed buckling mechanism of the experiment with the mechanism that led to folding in the natural rock.

Because we have regularly and closely oriented layer lines in the experiment it is possible to formulate very exactly the locations of the different fold classes in the model and fix the boundaries where one class changes to another. Of special interest in this isogon plot is that we have control on the various fold classes which develop *inside* the buckled layer.

All fold classes of the classification are represented in the model (Figure 17.19). Comparing the distribution of these in the folds *j*, *k* and *l* we see subtle changes as the folds become more strongly amplified. Fold *j* shows mostly parallel folding in the competent layer, and in the matrix folds of Class 1A in the outer arc and Class 3 in the inner arc. There is, however, a narrow zone of folds with changeover class characteristics (2 and 1C) in the incompetent material in contact with the inner arc of the competent layer. In folds *k* and *l* this transitional zone becomes wider and gradually moves through the competent material towards the outer fold arc. These changes of fold class with fold development are related to changes in the incremental strain history and resulting finite strain during the fold evolution, and further discussion will be given to this subject in Session 19.

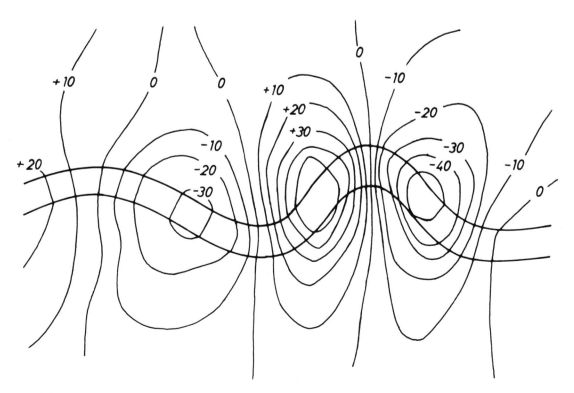

Figure 17.18. Dip isogon plots from Figure 17.10.

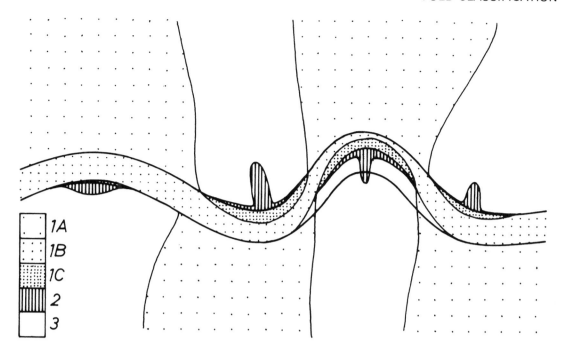

1A
1B
1C
2
3

Figure 17.19. Fold class boundaries in the experimentally formed buckle fold.

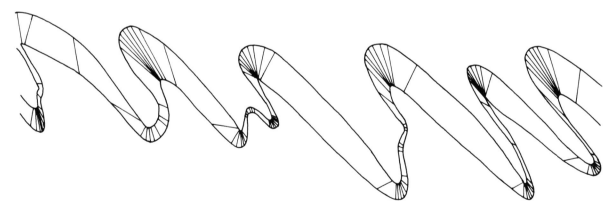

Figure 17.20. Dip isogon plot of the "fish-hook" folds.

Sorby's "fish-hook" folds

Answer 17.6★

The isogon pattern for these series of peculiar structures shows a number of features which repeat periodically through the folds. The isogon pattern in the antiforms is consistently convergent (Class 1C), with isogon spacing wider on the right-hand side of each fold than on the left indicating less rapid overall surface curvatures on the right than on the left. On the right of each antiform the isogon pattern indicates that the crestal 1C pattern becomes closer to the 1B model because the isogon lines come to lie more nearly perpendicular with the layering. In the synformal areas the isogon pattern again indicate Class 1C folds but on the left side of each synform the relationship of the isogon lines and layers changes: the isogons pass through a position where they are oriented 90° to the layer and come to lie in a position more characteristic of Class 1A folds. The marked and consistent differences in limb thicknesses suggest that the thinned zones suffered an abnormally strong localized deformation.

We cannot explain this extraordinary geometric pattern with confidence, but the following deformation history is put forward as a suggestion. The overall pattern of convergent isogons in the limestone layer suggests that it acted as the most competent rock unit. As it is a clastic crinoid limestone it is reasonable to assume that the original bed thickness was more or less constant, and that the thickness variations observed today are the result of tectonic deformation. A most important point to note is the polarity sense of the folds. In situations where a competent layer is enclosed in a uniform host rock the two sides of the buckle statistically have mirror image forms (Figures 1.9B and C; 6.14B; 19.5). The only satisfactory way of accounting for this absence of mirror geometry is to suggest that the initially formed folds also possessed this polarity (Figure 17.21A), and such a geometry would imply a difference between the competencies of the two sides of the limestone layer with contrasts $\mu_1 > \mu_2 > \mu_3$. Careful study of Figure 17.11 reveals that there are slight differences between the matrix on either side of the limestone bed: the lower contact shows more abundant white silt areas (many of which are silt-filled worm burrow trace fossils) than does the upper. Subsequent to this initial folding the principal incremental shortenings must have rotated counter clockwise relative to the layer to produce the fold asymmetry (Figure 17.21B) and then developed further shortening by limb shearing (Figure 17.21C). The very marked slaty cleavage with parallel pressure solution stripes reflects the finite strain XY principal plane which sums these deformation components. A well developed cleavage refraction is seen in the limestone layer because it is more competent that its slaty host.

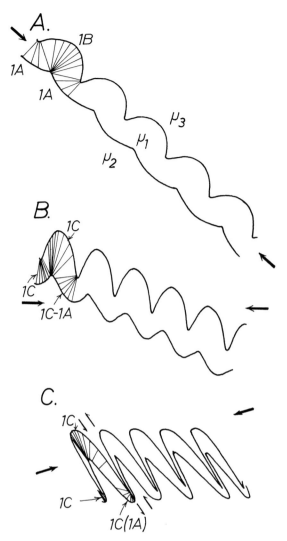

Figure 17.21. *Suggested deformation sequence to account for the "fish-hook" folds. Arrows show suggested directions of maximum incremental shortening; paired arrows with half barbs show shear zones. Dip isogons schematically represented on the left of each part with main fold classification.*

KEYWORDS AND DEFINITIONS

Dip isogons Lines joining locations of equal dip on either side of a folded layer (Figure 17.5).

Convergent isogon folds: Class 1 Folds in which adjacent dip isogons converge when traced from the outer to the inner fold arc. This class can be subdivided into **strongly convergent, Class 1A**, where isogons are rotated in the same rotation sense as the bed perpendiculars; **parallel folds, Class 1B**, where isogons are perpendicular to the layering; **weakly convergent, Class 1C**, where isogons are rotated in the opposite rotation sense to the bed perpendiculars (Figures 17.6, 17.3).

Divergent isogon folds: Class 3	Folds in which adjacent dip isogons diverge when traced from outer to inner fold arcs (Figure 17.6).
Similar folds: Class 2	Folds with parallel dip isogons implying that the bounding surfaces of the fold are geometrically identical (Figure 17.6).
Flattened folds	Folds of any class which have undergone a homogeneous strain over the whole field of the fold (Figure 17.9).
Zone of contact strain	The region on either side of a buckled layer where the matrix shows harmonic but progressively de-amplifying folds (Figure 17.15, z.c.s.).

KEY REFERENCES

Hudleston, P. J. (1973). Fold morphology and some geometrical implications of theories of fold development. *Tectonophysics* **16**, 1–46.

This paper discusses the problems arising from the t'_α/α graphical method of fold classification and suggests a system based on isogon–dip angle relationships (ϕ_α/α graphs).

Ramsay, J. G. (1962). The geometry and mechanics of "similar" type folds. *J. Geol.* **70**, 309–327.

This paper was the first to show how the thickness changes in folded layers could be established and used to compare shape variations of layers in the same fold. It presents a method for calculating the amount of homogeneous strain superposed on a parallel fold developing de Sitter's original concept of "flattened" folds (see below).

Ramsay, J. G. (1967). "Folding and Fracturing of Rocks", 568, pp. McGraw-Hill, New York and London.

In Chapter 7 (pp. 360–372) the ideas previously published in 1962 (see above) were extended so that thickness variations could be used to classify the geometric forms of folds.

de Sitter, L. U. (1958). Boudins and parasitic folds in relation to cleavage and folding. *Geol. en Mijnb.* **20**, 277–286.

This paper discusses many aspects of folding and the relationships of small scale structures to strain in folds, and it was the first to develop, using simple mathematics, the geometry of the folds termed "flattened" folds.

Treagus, S. H. (1982). A new isogon-cleavage classification and its application to natural and model fold studies. *Geol. J.* **17**, 49–64.

This presents a new method of fold classification using the angular relationships between cleavage and bedding planes. It shows how layer thickness variations might be connected with finite strain and therefore with strain history and fold-forming mechanisms. Because of the mechanistic implications it is not a rival to the methods which are based only on layer shape.

SESSION 18

Fold Sections and Profiles

The basic methods for assembling structural data of layer boundaries and dip orientations into cross sections and profiles are summarized. Simple graphical techniques are appropriate to data from sub-horizontal surfaces, and the modifications required for terrain of strong topography are discussed. Extrapolation methods for the prediction of structural forms using the Busk construction methods are analysed and a new technique for fold reconstruction based on dip isogons is put forward.

INTRODUCTION

The construction of geometrically correct cross sections and profiles of folded strata from the limited observations made at the surface or in underground workings is a major problem in structural geology. The accuracy by which we extrapolate observed lithological contacts and orientation data to deeper levels to predict the morphology of the structure at depth plays a key role in tectonics and has most important practical consequences in the oil and mining industries. A field geologist with a broad experience of fold forms in different tectonic environments is often able to produce sketch sections of remarkable accuracy. For example, the cross sections of the often complex tectonic structures of the European Alps made by Argand (1911) and Heim (1921) have not been fundamentally altered by modern work. The only really significant geometric discrepancies appear where extrapolations are made to depths of more than ten kilometres (Hsü, 1979; Müller, 1976). In terrains where the degree of surface exposure and strong topography is not Alpine, however, cross sections must be constructed with much less data. The aim of this session is to examine two main problems. The first is to consider the best methods whereby we can assemble geological information collected from field studies, such as contacts between different lithologies and measured orientations of different structural features at individual outcrops, to produced a *geometrically accurate compilation* and to show how the structural geometry changes with depth and horizontal position. The second problem is to discover the best way we can *extrapolate* from observations we have obtained into regions (usually deeper) where we have no direct information.

Various types of sections can be used to represent the geometry of tectonically disturbed regions. Most published descriptions are accompanied by **vertical or cross sections**. Such sections present the appearance of the structure as it would appear in a vertical plane (Figure 18.1A) with the section plane generally chosen perpendicular to the general trend of the fold hinge lines. In regions of simply folded, non-metamorphic sedimentary rocks these vertical sections give a good representation of the structure. However, in regions with more complex fold orientations where the axes

of the folds are not sub-horizontal, peculiar distortions appear, and where the fold hinges have a high angle of plunge vertical sections may be very difficult to construct, and when constructed give a highly misleading impression of the overall geometry. Vertical sections are not appropriate to evaluate fold styles when the accurate fold classification systems developed in Session 17 are employed. In spite of these drawbacks, vertical sections often convey quite well fold geometry, especially if a series of parallel cross sections are assembled in the form of **serial or coulisse sections** (Figure 18.2). They are often used in this way to demonstrate the geometry of underground mining operations, the vertical data distribution being particularly easy to incorporate in this way.

In a **profile** section the projection plane is oriented perpendicular to the fold axis (Figure 18.1B). With horizontal fold axes the profile corresponds to the cross section, but with inclined fold axes they differ (Figures 18.1D and E). The profile gives the true cross sectional shape of the fold, and conveys the correct interlimb angles and layer curvatures. It shows the correct layer thickness variations used to determine fold classes.

Cross section and profile construction

Beginners often think that it is only possible to construct a section from data directly on the section line. If, however, the fold axes are inclined to the ground surface it is possible to incorporate information from some distance away from the section and project it into an appropriate position on the section plane. The principle used is based on the geometry of cylindrical and sub-cylindrical structures, that geometric features have maximum continuity along the direction of the fold axis (Figures 18.1A and B). How far away from the section we can project information depends upon the degree of cylindricity of the structure. The assessment of the degree of cylindricity comes, in the first instance, from a general impression one has obtained from the field study, and second from a back-up analysis of orientation data using the techniques described in Session 16.

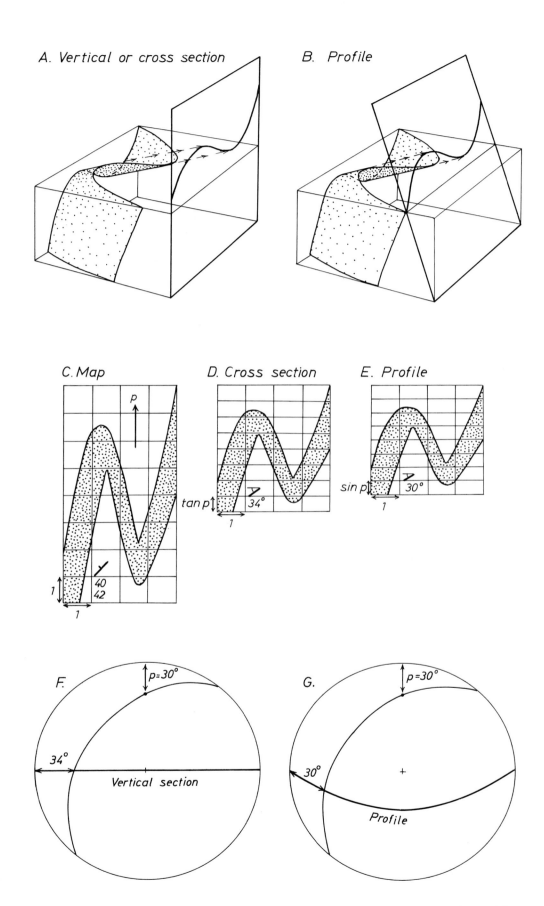

A. Vertical or cross section   B. Profile

C. Map   D. Cross section   E. Profile

F. Vertical section   G. Profile

Figure 18.1. *Relationships between three dimensional geometry of a folded layer, its map representation, and the types of section known as vertical cross sections and profiles.*

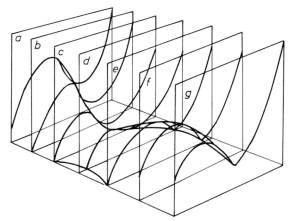

Figure 18.2. A series of vertical cross sections a to g positioned correctly relative to one another to produce a series of coulisse sections.

Once the fold axis orientation has been determined the outcrop distribution and lithological contacts on a map can be transferred to the section plane using an appropriate geometric construction. If the land surface is flat, or nearly so, the simplest method for doing this is by constructing an orthogonal grid on the map and transferring the points in this grid coordinate system to a new system such that the original 1 : 1 grid lines are transformed into a $\tan p$: 1, or $\sin p$: 1 system in a cross section or profile respectively (p is the angle of plunge of the fold axis, Figures 18.1C, D and E). It is also possible to transfer individual dip measurements on to the section or profile: the location of the observation is

carried out using the grid system and the appropriate angle of dip on the projection surface is easily found using an equal area of stereographic projection (Figures 18.1F and G).

QUESTIONS

Question 18.1

Figure 18.3 shows a horizontal outcrop surface of banded gneisses (used for Question 15.4) with the major lithological contacts indicated. Using the grid technique determine the true profile of the folds. The fold axis plunges at an angle of $51°$ in the direction of the arrow.

Question 18.2

Figure 18.4 shows a schematic group of vertically plunging folds in profile. All the main classes of fold shape are represented. Construct the appearance of these folds on planes: (a) dipping north at $70°$; (b) dipping west at $60°$; (c) dipping northwest at $60°$. Determine the apparent dip isogons in each example (lines joining points of equal apparent dip in the section plane) and the "apparent" class of the folds in the sections. In (c) determine the position of the line joining the points of maximum apparent curvature in the section, and compare its position to the axial surface trace. Discuss the significance of this geometry in terms of the practical interpretation of fold shapes in field outcrops.

Figure 18.3. A horizontal outcrop surface of banded hornblende–biotite gneisses (original see Figure 15.25) giving the data for Question 18.1.

Effects of topographic variation on profile construction methods

In hilly and especially in mountainous terrain, it is well known that topographic variation has an effect on the positions of layer boundaries. Figure 18.5A shows the results of topographic variation combined with a constant angle of fold plunge on a point x on the ground surface and its correct location in a profile. In the profile the position of x' contains two components depending upon the height h of point x above a datum plane, and its horizontal distance d from the intersection of datum and profile planes. Simple trigonometry indicates that x' is positioned at a distance $h \cos p + d \sin p$ from the datum–profile plane intersection. If we had decided to construct a vertical section instead of a profile, x' would be located at a vertical height above the datum plane of $h + d \tan p$.

If the fold axes change plunge (Figure 18.5B), a fold profile or cross section can be constructed if we known the manner in which the plunge changes p_1, p_2, p_3, etc., occur. This construction can either be made graphically (as in Figure 18.5B, point x to x'), or it can be carried out with a computer using trigonometric relationships based on successive stepwise changes in the fold plunge.

QUESTION

Question 18.3

The region shown in Figure 18.6 is highly mountainous terrain in strongly folded Mesozoic and Tertiary limestones and shales of the Grand Muveran region of the Helvetic nappes of Valais, Western Switzerland (Morcles nappe). A zero elevation datum line AB has been constructed normal to the fold axes which plunge, on average, 30° towards azimuth 60°.

From a preliminary inspection of the map and the relations of the geological boundary contacts with the topographic contours (interval 100 m) determine the overall structure of the region.

Using successive 100 m elevation locations along the upper and lower contacts of the Upper Jurassic Malm limestone construct a true profile of the area. For example, the point at distance $d = 1.08$ km from AB and at a height of 1800 m will be found in a profile plane at $1.8 \cos 30 + 1.08 \sin 30 = 2.10$ km above the datum. As this type of calculation has to be made several times, a programmable desk computer will speed the work. If time allows complete the profile by constructing the Tertiary–Cretaceous interface.

Discuss the fold geometry (axial plane orientations, lithological thickness variations) and suggest some kinematic model that might account for the geometric variations.

Check the Answer and return to the discussion below.

Extrapolation of structural data

A structural geologist is often asked to solve the problem of deciding the most likely structure in some region where he has no direct data. This often requires him to try and extrapolate information he has obtained at or near the surface to predict the structural forms at depth. To make reasonably

correct predictions it is first necessary to be assured that the structural geometry is likely to have some continuity, with smooth and systematic changes in strain and structural style. If the structure is discontinuous, with unknown fault discontinuities present, it is practically impossible to make predictions with any confidence. With certain types of folds it may be possible to make very good predictions, and with harmonic relationships between successive layers the best results emerge. Folds with polyharmonic relationships are very difficult to extrapolate in any detail, although the geometry of the larger wavelength orders may be determinable. If the folding is disharmonic no geometric solutions are possible.

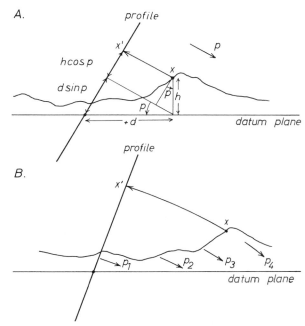

Figure 18.4. Profile through a series of folded layers with the layers showing characteristic thickness variation forms of the main fold Classes discussed in Session 17. See Question 18.2.

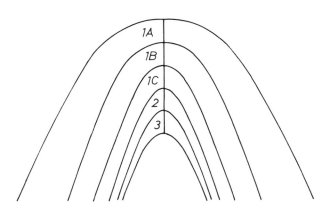

Figure 18.5. Profile construction in regions of topographic relief. A shows the projection of original point x on to the profile plane to position x'. h, height of x above the datum; d, horizontal distance of x from the intersection of datum and profile planes; p, angle of plunge of the fold. B shows how point x' is positioned when the axial plunge of the folds varies from values p_1 to p_4.

The Busk construction

The choice of geometric technique for reconstructing a fold must be governed by the reasonable assumption that the fold behaves in some geometrically ordered and well defined way. In the past such reconstructions have mostly been carried out assuming that all the layers in the structure retain their original thickness at all points in the fold, and that the fold style is parallel throughout. The basic method was suggested in a classic work on folding by Busk (1929). He developed a graphical procedure which allowed the curvature of the beds between any two adjacent dip observations to vary so that the two dips were tangent planes to concentric cylinders. Between two data points (e.g. *b* and *c* in Figure 18.7), the centre of curvature of the sector is established by finding the point of intersection of the two

normals of the dip surfaces. In any sector of the structure the curvature varies from layer to layer, but along any one horizon it is constant. Successive reconstructions of adjacent sectors between each pair of dip observations, each with its own centre of curvature, enables a complete cross section of the structure to be established (Figure 18.7). Slight modifications of this general technique were proposed by Coates (1945) as a result of his practical experience in the oil industry, and his technique is an improvement on the basic Busk method where there is a large amount of dip orientation data and where dips have consistently low values. The average dip of a fold sector is determined, and this value is extrapolated downwards along the bisectors of the mean dip tangents.

Constructions produced by the Busk method show a number of special geometric properties.

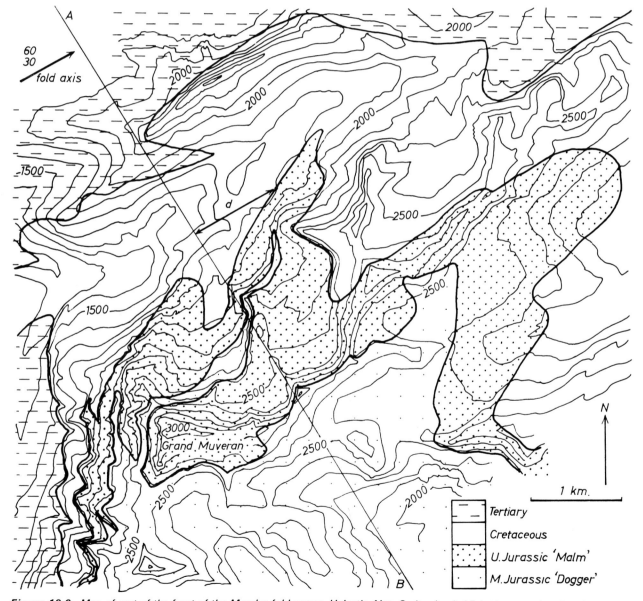

Figure 18.6. Map of part of the front of the Morcles fold nappe, Helvetic Alps, Switzerland. AB is the zero elevation datum line on the profile section. For use with Question 18.3.

1. All layers in the fold retain constant orthogonal thickness irrespective of lithology.
2. Sudden changes of curvature occur in any surface in the fold where it cuts a line perpendicular to an observed dip. In some positions in the structure these curvature changes are very marked and give rise to cusp-like forms (Figure 18.8, *a*). Construction cusps of this type occur at different layer levels in the structure. On the inner arc of the cusps the arc lengths measured along any folded surface are smaller than those on the outside of the cusp.
3. The amplitude on the folds in successive layers always decreases away from the topographic surface where the prime data were collected (Figure 18.8).
4. At some positions in the structure geometrical incompatibility occurs and, over certain sectors, it is not always possible to retain constant layer thickness (Figure 18.8, *b*). At these positions Busk suggested that the folds become non-parallel and that limb thinning takes place; he suggested that freehand constructions should be used as there was no unique geometric solution to the problem by the methods previously used. He was clearly aware of the problems of thickness changes in folds but he thought that there were so many unknown factors that an exact geometric solution of the problem was too complicated or even impossible.

All the features listed above arise because of the geometric assumptions on which the method is based are inapplicable to most fold forms.

Folds with geometric characteristics of the parallel Class 1B are generally uncommon. Folds which have forms approximating to this style are most commonly found in the single layer competent layers of ptygmatic structures and in some multilayered complexes where strong competence contrasts exist between the different components. In these multilayers parallel folds may occur in the most competent layers, especially where they are strongly laminated parallel to their contacts. In this environment, however, the less competent layers take on forms of Class 3 folds. In multilayered complexes which do not show strongly marked competence variations the folds alternate between Class IC and 3, often approaching to the similar (Class 2) form.

The sudden changes in curvature which appear in the reconstructions are clearly an outcome of the methods used, and become less pronounced as more data are incorporated into the construction. A refinement of the Busk technique was suggested by Mertie (1940) in order to smooth the curvature changes along each folded surface. He described a method for establishing positions of continuously changing centres of curvature (an evolute), and how, from this, it is possible to construct the surfaces in the structure showing continuously changing curvature (an involute). This is an improvement on the basic method of Busk, but it is still open

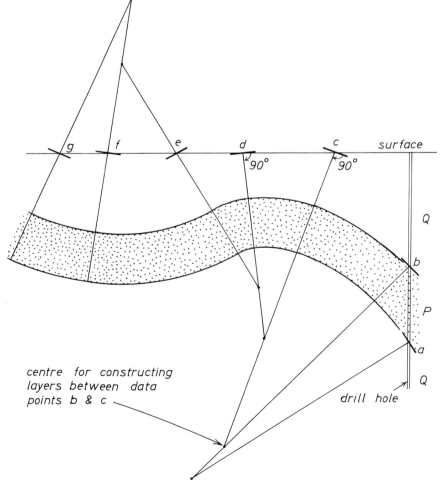

Figure 18.7. The Busk construction method for determining the position of a folded layer P using known dips, at localities a, b, . . . , g.

Figure 18.8. *Busk construction used to reconstruct overall fold form of a multilayer complex from a series of known dips on a horizontal land surface (shown as short lines along the horizontal surface). Construction cusps occur at positions a, and a discontinuity at location b.*

to the criticism that the folding remains parallel throughout all the layers, and it still leads to the appearance of cuspate discontinuities at certain points in the section. Cuspate folds are commonly observed in natural folds (e.g. Figure 17.4) but they are always localized at special positions in the structure close to interfaces between materials of marked ductility contrast. This observation conflicts with the random location of cusps appearing on the graphical constructions made by the Busk–Mertie methods, and which arise entirely as a result of geometric construction procedures.

The changing amplitude of folds away from the data collection surface is an outcome of the assumption of parallelism of the folded layers. It has been suggested that the change of shape downwards implies that the folds pass into unfolded material, perhaps with an intervening decoupling surface or décollement horizon. The same geometric feature also appears in the sections in an upward direction, and so clearly this interpretation is unsound.

We have gone into some detail with the commentary on the geometry of the Busk construction because it is still widely quoted as a method by which fold geometry may be extrapolated from actual observations. The method is, without question, applicable to folded layers which show true parallel form but, in no succession we are aware of, does this style persist throughout all the layers. The special geometric style of the folds constructed in this way (Figure 18.8), in which antiforms always become more rounded upwards and more cuspate downwards, is not corroborated by geological observations. We conclude that, although in special circumstances this construction method may be used, the method is not applicable as a general technique.

If you accept these conclusions, the next problem is to consider if there are other techniques for reconstructing fold geometry on more soundly based ideas. We are of the

opinion that the geometric basis varies with and is dependent upon the geological environment appropriate to the problem. For example, the reconstruction of fold forms which relate to the folding of a single competent layer in an extensive matrix of incompetent material will require a quite different technique from that applicable to the folding of a multilayer complex.

The most appropriate techniques for fold reconstruction require, first, some knowledge of the overall fold geometric potential of a region and, following from this, the setting up of a model fold pattern from which it may be possible realistically to extrapolate fold shapes from some given data input. In Session 17 we found that the fold isogon technique was a particularly simple, yet realistic, one for classifying folds and the techniques described below will show how these basic methods of fold classification can be reversed to reconstruct fold shapes from observed layer dips.

Fold reconstruction using isogons

In a folded multilayered rock, the fold Class in each layer has special characteristics which can be expressed in thickness variation diagrams or in isogon patterns. If the characteristic isogon pattern of the rock under investigation is known it is possible to apply the geometric restrictions imposed by the isogonal lines to constrain the geometry of surrounding surfaces. The technique proceeds as follows:

1. A "standard section" is selected to establish the characteristic isogonal pattern of the particular rocks being investigated. This may be chosen from previously studied sections of the same rocks, perhaps in the way we discussed earlier in this session by projection methods, or it might come from a study of small scale

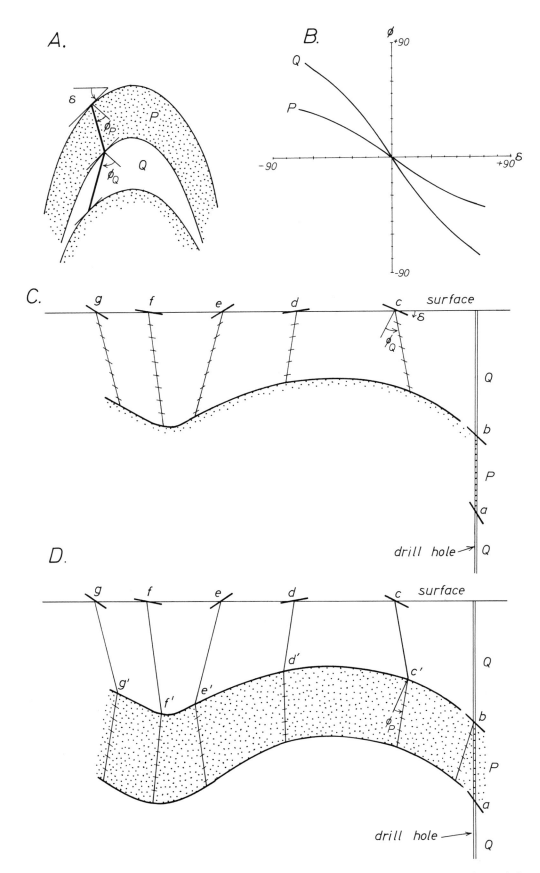

Figure 18.9. Method of determining the underground structure of folded rocks using the isogon construction technique. A shows a "standard section" used to establish the isogon pattern, and B the graphed isogon relationships. C and D show the method used to establish the upper and lower contact of layer P.

folds in rocks of the same lithology using the principle that the fold classes of small scale structures are closely related to those of large folds. The geometry of the isogons is determined from the standard section by measuring the angular deflection ϕ of an isogon line from the layer normal, and tabulating how this angle varies with the layer dip δ (Figures 18.9A and B). With two rock types P and Q, for example, the isogon angles are plotted for layer P surrounded on both sides by layer Q, and for layer Q surrounded by rock type P. The sign convention for ϕ and δ is the same as that we have used elsewhere in these sessions, namely anticlockwise angles are positive, and clockwise are negative.

2. Having established the isogon characteristics we can, for any given dip datum, read off the isogon orientation for that particular rock type and dip, and construct an isogon line downward from the dip locality (Figure 18.9C, locality c in rock type Q, with negative dip δ and positive isogon angle ϕ_Q). Along this isogon line we know that the dips are the same as that of the datum point, and so we mark these directions as a series of parallel lines. This is repeated for the other data points. From the known contact between P and Q at locality b it is possible to extrapolate to the left, making the contact conform with the dip lines where it cuts any isogon (Figure 18.9C).

3. The same technique is applied in layer P. Below locality c' the δ isogon line changes direction according to the graphed relationships of Figure 18.9B, and a new isogon line is constructed at an angle ϕ_P from the bedding normal at c' (Figure 18.9D). The isogon lines

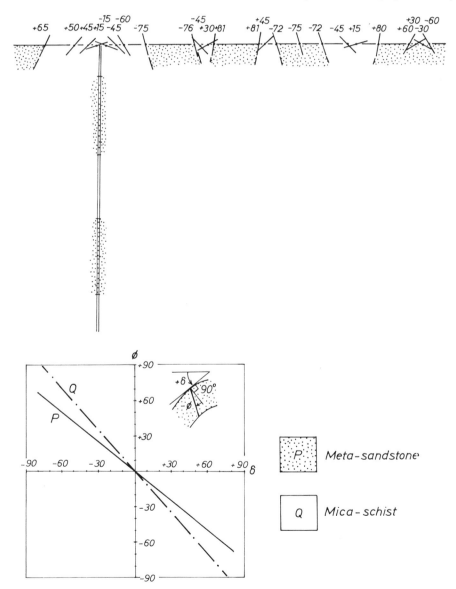

Figure 18.10. *Dip data from a series of folded metasediments, N.W. Scotland. The isogon patterns which characterize the two principal rock types derived from a "standard section" are shown in the graph. Reconstruct the folds beneath surface—see Question 18.4.*

are completed for the other data points d' to g', and finally the lower P_Q contact is extrapolated from datum locality a.

The extrapolated data in Figure 18.9D are the same as those employed to demonstrate the Busk construction (Figure 18.7) and it is instructive to compare the two results. In a way the Busk construction is a type of isogon diagram with all the isogons arranged perpendicular to the bedding surfaces (ϕ is zero for all values of δ, and for both rock types).

QUESTIONS

Question 18.4

Figure 18.10 shows data obtained from a region of metamorphosed sediments from the Highlands of N.W. Scotland. The data consist of surface dips and the rock type distribution together with information from a vertical drill hole. A standard section has been selected from folds in a directly adjacent area, and the isogon line orientation is shown graphically (note sign convention). Extrapolate the surface data downwards to discover the fold structure. If time is limited use only those data in the left half of the figure.

Now proceed to the Answers and Comments section, then continue to Session 19 or to the starred questions below.

STARRED (★) QUESTIONS

Question 18.5★

Show where the main types of fold classes established in Session 17 are located on a δ–ϕ diagram.

Question 18.6★

In Figures 18.1E and G it was shown how individual datum orientations could be incorporated into a fold profile construction. Use the data analysis of the Beinn Caorach fold of N.W. Scotland previously made in Session 16 (map in Figure 16.6) to construct a true fold profile, and incorporate the dip information into the profile. In the analysis of this fold it was shown that the overall geometry was best described as sub-cylindrical, with 80% of the π-poles lying within 20° of the mean π-circle. The specific problem with sub-cylindrical folds is to decide the most appropriate technique for converting the measured dip on to the profile. Probably the most accurate way statistically is to move each π-pole through the smallest angle possible to place it on the mean π-circle. Each π-pole is moved along a great circle containing it and the mean fold axis F (Figure 18.11). The measured dip δ then takes on a value δ' in the profile plane.

ANSWERS AND COMMENTS

Answer 18.1

An orthogonal grid with a 1 cm spacing was drawn on Figure 18.3 with one of the coordinate directions parallel to the fold axis azimuth and the other perpendicular. A new

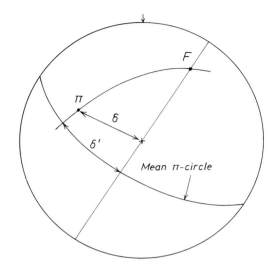

Figure 18.11. Method of determining the value of dip δ' on a profile plane from a π-pole with measured dip δ in a sub-cylindrical fold.

grid was constructed with a reduced spacing in the fold axis direction of $\sin 51° = 0{\cdot}78$, and the layer boundaries transferred to this new grid base (Figure 18.12). Several geometric features of the horizontal section are changed in the profile. The most striking effects are that the extreme apparent thinning of the layers on the lower limb of the uppermost antiform are not real, and that the true interlimb angles of the folds is considerably less in profile than is seen on the horizontal section. In this profile it would be correct to classify the fold shapes using thickness variation diagrams or dip isogon methods.

Answer 18.2

Changing the plane of section through a fold leads to a number of important geometric modifications of the apparent fold shape. It is well known that the apparent amplitude, wavelength and interlimb angles change significantly, and each may increase or decrease depending upon the orientation of the section plane. Perhaps less well known effects relate to changes in "apparent" fold class as a result of apparent layer thickness modifications. In some ways these effects are comparable with the geometric effects of superposing a homogeneous strain on a pre-existing fold (see Question 17.4★). The first feature to notice is that, although the geometry of the true similar fold is modified, the fold retains its Class 2 style. You may recall that we made some mathematical proofs of special cases of this in Volume 1 (Question 4.6★). All the other "apparent" fold classes depart from the "true" fold classes seen in profile. In the planar section of the first example (Figure 18.13A), the limbs of the folds show a thinning relative to the hinge, and the folds of true 1A and 1B classes become modified into apparent shapes of Class 1C. In the oblique section of the fold which dips at 60° to the west, the fold shapes are modified, but now the true 1B and 1C folds move to shapes of apparent 1A and 1B, respectively (Figure 18.13B). The most surprising changes in apparent fold geometry occur in the oblique northwesterly dipping section (Figure 18.13C). The true fold profile is of a symmetric fold, but in the oblique

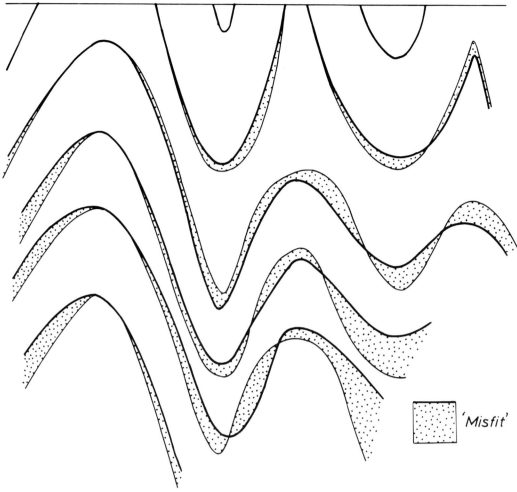

Figure 18.18. *Comparison of the construction lines of Figure 18.17 (fine lines) with the true locations of the geological boundaries (heavy lines). Note how the misfit increases away from the primary data localities.*

'Misfit'

Answer 18.4

Figure 18.17 shows a full reconstruction of the folds using isogon techniques. Because the method uses graphical integration of the dips across adjacent isogon lines, personal differences in detail of result are to be expected, but, if the construction is carried out with maximum accuracy, these differences should not be large. More significant are the variations caused by overall imperfections in the method because the isogon pattern is unlikely to be constant, and because of initial variations in layer thickness across the profile. In this example we do have a complete knowledge of the real geometry, so it is possible to compare the predictions arising from the construction with geological reality. Figure 18.18 compares the layer boundaries arising from the construction (thin lines) with the actual geological boundaries (heavy lines), and the misfit between the two. The misfit is very small close to the horizontal and vertical data lines and it increases, as is expected, as the extrapolation moves into regions away from the primary data sources. However, compared with the results of predictions using other methods, the fit is a remarkably good one. If we had used the Busk construction, for example, practically the whole

area covered by the interpretation would be classified as misfit.

Although the isogon method generally produces good results, a number of problems do arise, some of which might be resolved by refining the basic method.

1. In Session 17 it was shown how the thickness variation (and in consequence the isogon pattern) of the incompetent layers in a multilayered rock sequence might be a function of the relative proportion of competent to incompetent layers (Question 17.3★). It follows from this result that the isogon ϕ/δ graph derived from any standard section should be used with care, for the relationships expressed in these graphs might be more complex than, for example, a graph for rock type P, and another for type Q. If the relative proportions of layers P and Q in the section under construction differs from those of the standard section, the isogonal angles are likely to differ. The greatest discrepancy in ϕ/δ curves is most likely to occur in the incompetent material, and it may be possible to make curve corrections using the geometric compatibility principles developed in Question 17.3★.

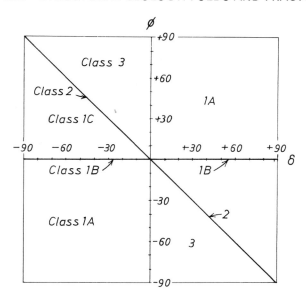

Figure 18.19. A ϕ/δ "standard section" graph showing the locations of the different fold classes of Figure 17.6.

2. In our discussion of limitations of the fold classification procedures we noted that there are likely to be sectors on fold limbs which cannot be classified because no common equal dip can be found on the sides of the layer. Such sectors cannot appear in the ϕ/δ graphs, and will therefore always be absent from constructed sections.

3. If, in the section being reconstructed, variations in shortening occur along the length of the section, it is possible that fold styles and isogon patterns will vary as a result of changes in the amounts of fold "flattening" (Question 17.1). Although difficult to do, it is possible to make corrections for this effect if some information is available as to the variations in "flattening", perhaps with information derived from an analysis of strain makers or of changes in cleavage intensity.

4. If more than two types of rock exist it will be necessary to produce more than two isogon curves from a standard section. For example, if three types of rocks with marked competence contrasts exist (P, Q, R), the isogon pattern for rock type Q in a PQP sandwich will not be the same as that in a PQR sandwich, and the isogon pattern for Q in a synformal PQR sandwich will differ from that in an antiformal PQR sandwich. In regions where several rock types exist the overall accuracy of a reconstruction will be greatly improved if several isogon curves can be derived from the standard section.

Answer 18.5★

The ϕ/δ graph for the main fold classes is shown in Figure 18.19. This scheme was first proposed by Hudleston (1973a).

Answer 18.6★

Figure 18.20 shows the profile with boundaries of the different lithologies, together with the traces of individual dip

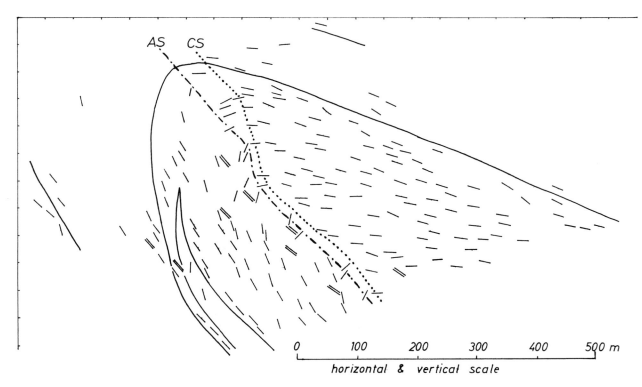

Figure 18.20. True profile of the Beinn nan Caorach fold, N.W. Scotland (Figure 16.6) showing projections of the data of individual lithological planes. Double lines indicate the projections of the axial planes of small scale folds. AS and CS are the traces of the axial surface and crestal surface respectively on the profile.

measurements of lithological layering and axial surfaces of small scale folds. Most of the individual plots accord well with the contacts, although there are discrepancies in the upper part of the western limb of the antiform. These differences could reflect real differences in local layer orientation with the contact, perhaps suggestive of unconformable relationships at the gneiss–sediment interface but, because the sense of misfit is the same on both sides of the interface, this deduction is open to question. You might recall that it was in this region the individual data points used to prepare the π-diagram departed most strongly from the mean π-circle because it was here that the maximum departure from cylindrical geometry occurred. The most likely explanation for the misfit here is that it results from strong non-cylindrical effects in the structure.

The axial surface and crest surface of the fold have been constructed from the dip data. The significance of the crest trace, passing through horizontal dips in the profile plane is clear, and these constructions bring out well the fact that the measured axial surfaces of small folds are sub-parallel to the axial surface of the main fold.

In the analysis we made in Session 16 of the angles between the mean dips of the fold limbs and the axial surface, we concluded that these angles were smaller on the overturned limb than on the normal limb. From the profile it can be seen that the overturned limb is thinner than the normal limb, with a thickness ratio of about 2:3.

How might we proceed further to elaborate other geometric features of this fold? In the profile it is possible to draw smooth lithological plane trend lines and from these it would be possible to develop thickness variation graphs or isogon plots to classify the fold shapes at different levels in the structure.

Concluding comments

In this session we have brought together many of the basic methods whereby geological data may be drawn together and integrated into different types of section. It may be surprising that, although there is general agreement as to the methods to be used for constructing sections and profiles, several of the techniques recommended in current publications and used in industry are based on unsound general principles.

The reconstruction of sections using dip isogons as suggested here is a new one, and is particularly well suited to solving real geological problems. The technique has a high potential because it extrapolates some specific fold model, with the choice of model being made according to the geological nature of the terrain under investigation. It also exploits most effectively the idea that accurate field observations are a key factor in making reconstructions of the structural forms existing beneath the Earth's surface.

KEYWORDS AND DEFINITIONS

Cross section — A reconstruction of geological data on to a vertically oriented plane.

Profile — A section through a geological structure oriented so that the section plane is perpendicular to the mean fold axis.

Apparent axial surface trace — The line joining points of maximum two-dimensional curvature in successive layers exposed in a general plane section through a fold (Figure 18.13C).

Busk construction — A method for reconstructing the shapes of parallel style (Class 1B) folds (see Figure 18.7 and Busk, 1929).

Isogon construction — A general method for reconstructing the shapes of folds of known isogon pattern.

KEY REFERENCES

Busk, H. G. (1929). "Earth Flexures", 186 pp. Cambridge University Press.

This classic book was the first major attempt to devise methods for accurately reconstructing fold shapes from the limited data arising from geological field work. It concentrated particularly on those methods appropriate to concentric and parallel folds. Although some of the geometric features of other fold styles were discussed, no systematic analytical methods for them were developed.

Elliott, D. (1968). Interpretation of fold geometry from lineation isogonic maps. *J. Geol.* **76**, 171–190.

This paper was important in developing new lines of thought on how field data might be more effectively analysed from a geometric viewpoint, and it pointed out the high potential of analytical methods using true and apparent dip isogons.

Gill, W. D. (1953). Construction of geological sections of folds with steep limb attentuation. *Am. Assoc. Pet. Geol. Bull.* **37**, 2389–2406.

Mertie, J. B. (1940). Stratigraphic measurements in parallel folds. *Geol. Soc. Am. Bull.* **51**, 1107–1134.

Ragan, D. M. (1973). "Structural Geology", 2nd edn., 208 pp. Wiley, New York.

Wilson, G. (1967). The geometry of cylindrical and conical folds. *Proc. Geol. Assoc. Lond.* **78**, 179–210.

This presents one of the first serious attempts to modify the Busk parallel fold construction technique to layers with variable thickness.

The construction methods described by Busk were based on discrete fold sectors possessing constant curvature through each sector. This paper showed how it is possible, using the geometrical concepts of involutes and evolutes, to reconstruct parallel fold geometry in situations where progressive changes of curvature occur along the folded layers.

A clear description of fold profile construction from map data together with exercises will be found on pp. 85–90.

The geometrical procedures for constructing profiles are described on pp. 186–194, and problems arising from non cylindrical structures are discussed.

SESSION 19

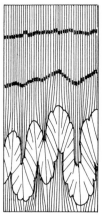

After H. C. Sorby, 1853

Fold Mechanics:
1. Single Layers

The mathematical relationships describing the resistance to buckling in a system of a single competent rock layer in a matrix of less competent rock are examined under the assumptions of linear Newtonian viscosity of the materials. The initial fold wavelength is shown to be controlled by layer thickness and viscosity contrast. The different types of fold shapes and styles arising during the buckling of single competent layers are reviewed. The form of cuspate–lobate fold morphology is shown to be an indicator of competence contrast, and the possibility of competence contrast reversals during geological deformation is discussed. The progressive changes of shape of the contacts of a progressively folding competent layer is analysed using the Fourier series methods of Session 15.

INTRODUCTION

In the investigations of some of the geometrical effects of straining layered rocks that we made in Sessions 1 and 2 of Volume 1 we saw that layer shortening along the stiffer, or more competent layers leads to a periodic, sinusoidal-like sideways deflection of these layers and to the production of folds. One of the features of these folds is that they often show rather constant relationships between the wavelength and the thickness of the competent layers. In this session we will show how such relationships may be accurately established and look into some ideas that have been put forward to account for such geometric regularity. We have not yet built up the necessary mathematical background to make a full analysis of the mechanics of these problems but, because an understanding of the fold forming processes is most important in appreciating many of the features of naturally deformed rocks, we will use certain mathematical expressions without proof, anticipating some of the results which will be fully worked in Volume 3. The theory that we will first investigate was proposed more or less at the same time by Biot (1961) and by Ramberg (1960). We must first set up a geometric model for the folding problem which looks geologically realistic. In this model (Figure 19.1A) a single competent layer of thickness d is embedded in a matrix of less competent material of infinite extent (or very large in comparison with the dimensions of the competent layer). We then allow the two sides of the model to approach each other at a specific strain rate $\dot{e}_x(\mathrm{d}e_x/\mathrm{d}t)$, shortening the layer and matrix in the x-direction by an amount e_x and causing the competent layer to buckle into folds with initial wavelength W_i (Figure 19.1B). To solve the mechanical problems involved here we have to specify the physical properties of the materials. For example we could choose elastic properties such that the strain in the rocks was proportional to the stress (or force per unit area). Such a

choice would not be appropriate to the type of geological deformations we observe in permanently folded rocks because, in a truly elastic material, the strains are reversible on removal of the end loads. Probably the simplest model that is appropriate to geological states consists of rocks with linear viscous or Newtonian properties. Linearly viscous materials undergo flow at rates proportional to the size of the applied stress and, in so doing, can build up large finite and irreversible strains. The constant of proportionality is the viscosity coefficient μ, and we will choose parameters μ_1 and μ_2 for the viscosity coefficients of the more and less competent materials, respectively. Other assumptions that are used in the calculations of the initial fold wavelength are that the fold wave has a very low amplitude, is of sinusoidal form, that the strain is plane (implying no displacements perpendicular to the surface of Figure 19.1) and occurs without volume loss. We also assume that the layer thickness is small compared with the fold wavelength, a feature that is correct for materials which have a relatively high competence contrast.

The resistance to the formation of the folds can be divided into two parts: one depends upon the resistance offered within the competent layer itself, the other relates to the resistance offered by the incompetent matrix as it is pushed aside by the developing buckle fold. The resistant force F_{int} within the competent layer is computed by assuming that the deformations inside this layer are accomplished by stretching of the outer arc of each fold and compression of the inner arc, and that there are no shear strains parallel to the layer boundaries,

$$F_{\text{int}} = \frac{2\pi^2 \mu_1 d^3 \dot{e}_x}{3W_i^2 e_x} \tag{19.1}$$

where e_x is the shortening strain that has taken place along

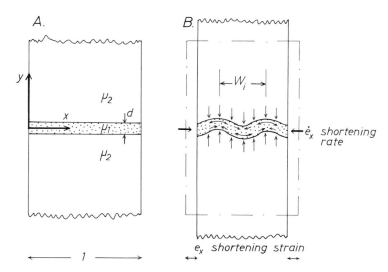

Figure 19.1. *Model of a single competent layer of thickness d and viscosity μ_1 embedded in a matrix of viscosity μ_2. Shortening $\dot{e}_x$ parallel to the layer leads to the formation of buckle folds with a characteristic initial wavelength W_i. Small arrows show resistances to fold formation. Note that the amount of contraction e_x and the fold amplitude have been exaggerated so as to emphasize the geometric features of the model.*

the x-direction. The linear variation of the resistance with competent layer viscosity is to be expected but less obvious are first, the variation of the resistance with the cube of the layer thickness and second, the variation of the resistance with the inverse square of the initial wavelength W_i. This second feature implies that, if the layer were not surrounded and confined by matrix material it would form with *the largest wavelength possible* so as to use the least energy (Figure 19.2A). This latter feature is also correct for an elastic layer and you can verify it by making a very simple experiment. If you push on the two ends of a wooden rule it will buckle elastically into a half wavelength form and will not form into a wave train of shorter amplitude than can be contained by inflexion points governed by the pinned ends.

The force F_{ext} that resists the development of the buckle folds and which arises in the matrix relates to forces acting in the y-direction of the system (Figure 19.1B) and is given by:

$$F_{ext} = \frac{\mu_2 W_i \dot{e}_x}{\pi e_x} \qquad (19.2)$$

The linear dependence of this function on the value of the initial fold wavelength W_i indicates that, from the viewpoint of matrix deflection alone, the folds utilizing least energy for

their formation have *the smallest wavelength that is possible* (Figure 19.2B). We therefore have two effects acting in opposition, and nature seeks a compromise so as to minimize the resistance offered by the sum of these two.

QUESTIONS

Wavelength–thickness relationships: theory

Question 19.1

Using Equations 19.1 and 19.2 find the total resistance to fold formation and, using the standard procedure of differential calculus find an expression for the wavelength of the folds which have the minimum resistance to formation (differentiate the expression for total resistance with respect to W_i, equate to zero and simplify). Now check this result with the Answer section and proceed to Question 19.2.

Wavelength–thickness relationships: measurement

Question 19.2

Figure 19.3 illustrates a layer of quartz-feldspar pegmatite enclosed in a host rock of fairly homogenous biotite-rich granite and the lower diagram shows a part of this wave train enlarged to twice actual scale. The fold train shows a rather regular geometry of the type that we have previously termed ptygmatic structure (Sessions 1 and 2), implying that the folded vein is considerably more competent than the host rock. The problem is to decide how regular is this fold geometry (in the sense of wavelength–thickness relationships), and to see how we can use this geometry to compute the viscosity contrast of the two rocks using the Biot–Ramberg analysis (Equation 19.4) assuming, of course, that the linear viscous model is a correct one!

We have previously noted that, in ptygmatic structures, the arc length W_a measured through the middle layer of the

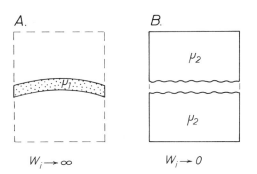

Figure 19.2. *Schematic representation of the initial wavelength inputs,* A *by the competent layer alone and* B *by the matrix alone.*

Figure 19.3. *Ptygmatic vein of pegmatite in a granitic host rock. Chindamora, Zimbabwe. See Question 19.2.*

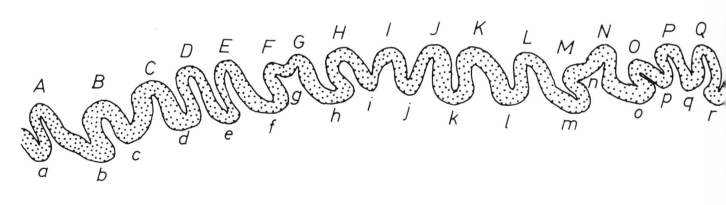

Figure 19.4. Outcrop of granitic gneiss and amphibolite cut by various acidic dykes. Qued Ovadenki, Central Hoggar, Algeria. See Question 19.3

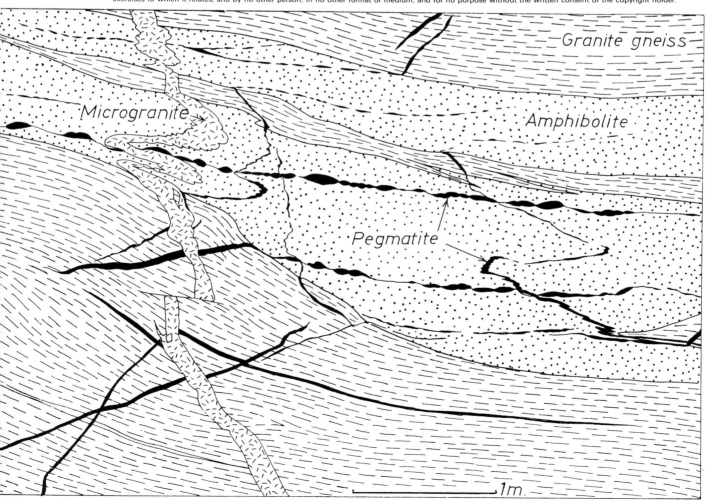

fold gives a close approximation to the arc length of the initially formed folds. If this assumption is correct it follows that arc length measurements can therefore be used to establish the initial wavelength W_i of the folds.

Identify the positions of the antiformal and synformal hinges (Figure 19.3; $A,B,C, a,b,c, \ldots$, etc.). Using a piece of cotton or a wheel measurer determine the arc length between each hinge line pair (a–A, A–b, etc., $W_a/2$ equivalent to $W_i/2$, half the initial wavelength), and determine the average layer thickness d in each sector. Plot $W_a/2$ as ordinate against d as abscissa. Find the value W_a/d from the best fit straight line fit through the origin and the data points. Using Equation 19.4 in the form:

$$\frac{\mu_1}{\mu_2} = 0 \cdot 024 \left(\frac{W_a}{d} \right)^3 \qquad (19.3)$$

compute the viscosity ratio of the two rocks.

Not all the points lie on the best straight line fit to the plotted data, some having a much shorter arc length relative to thickness than the norm. What might be the significance of such short arc length folds?

What is the minimum total shortening along the vein (express this in terms of extension e and as a percentage)? The rock matrix shows very little preferred orientation of the constituent minerals. Can you propose an explanation for the lack of tectonic fabric?

After answering these questions proceed to the Answers and Comments section, then to Question 19.3.

Assessing competence contrasts

Question 19.3

One of the most useful interpretations arising out of a study of small scale structures in deformed rocks is the possibility of determining the competence contrasts between the different components of a mixed assemblage of rock types. We should emphasize that, at this stage of development of the theory of these structures, it is not possible to determine the exact rheological nature of the materials. In the analysis of wavelength–thickness relationships of Question 19.2 you will recall that we *assumed* that the materials were deforming according to a linear viscous flow law. By making this assumption we can use the available theory to compute a value for the ratio of the viscosities of the two rocks. Laboratory experiments on rocks made to determine the stress–strain rate relationships show that the flow laws are generally not those of linear Newtonian viscosity. The extrapolation of the results of these experiments to geological conditions is not firmly established because the tests are generally made at strain rates that are many orders of magnitude faster than those which are thought to be characteristic of natural deformations (laboratory experiments $\dot{e} = 10^{-2}$–$10^{-7}\,\mathrm{s}^{-1}$, geological deformations 10^{-10}–$10^{-15}\,\mathrm{s}^{-1}$). We know from laboratory experiments that different deformation mechanisms operate at different environmental conditions and that the exact nature of the flow law changes with mechanism (see Schmid, 1982). Because of these gaps in our knowledge we often have to fall back on the old concept of general competence contrast—that one rock flows more easily than another in any given set of environmental conditions. In the future, as our analytical understanding of the full development history of the geometry of a particular

fold or boudin form deepens, we shall be in a better position to use geometric analysis of the forms of minor structures to evaluate more exactly rock rheology. We should not forget that, at present, most of the understanding of the mechanical principles of the formation of small and large scale structural forms is mostly confined to the very first instabilities arising in systems with very simple rheology.

Figure 19.4 shows an assemblage of deformed metamorphic rocks (amphibolite facies) in the Precambrian basement of the Hoggar, Southern Algeria. From a study of the contacts between the four rock types discuss the geological relationships and from an analysis of the forms of the small scale structures determine the competence difference arranging the four types in an order from most competent through to least competent. What differential displacements are likely to have taken place between the various rock units?

Now proceed to the Answers and Comments section, then to Question 19.4★ or to Session 20.

STARRED (★) QUESTION

Progressive shape changes during folding

Question 19.4★

The folds we observe in naturally deformed rocks represent the "frozen-in" finite result of the deformation process. It is, therefore, difficult to draw direct conclusions about the actual history of folding and the progressive sequence of shape changes that took place. In the laboratory, however, it is possible to perform mathematical or physical model experiments with materials of known properties and study exactly how individual folds change their geometric characteristics during the folding process (Chapple, 1968; Hudleston, 1973b). The analysis may be carried out using Fourier methods (Session 15) to describe the changing shapes of individual surfaces, or with thickness variation diagrams (Session 17) to evaluate the changes in fold class.

Figure 19.5 shows four stages (indicated 1 to 4) in a laboratory experiment carried out to investigate the development of buckle folds in a single competent layer embedded in a homogeneous less competent matrix. The folds have been systematically lettered (antiforms A to I, synforms a to j). Study the progressive changes of shape which take place along the boundaries of the competent layer between f and h. Locate hinge and inflexion points and determine the Fourier coefficients b_1 and b_3 for the eight quarter wave sectors, separating the shapes characteristic of the outer fold arcs from those of the inner arcs (the techniques are set out on pp. 314–316). If time is restricted, carry out measurements on the two quarter wave sectors on either side of antiform G. Plot the coefficients graphically (Figure 15.13) and determine the nature of the progressive shape changes in terms of variations in the degree of hinge roundness. Compare the shapes of the inner and outer fold arcs with progressive shortening.

ANSWERS AND COMMENTS

Wavelength–thickness relationships: theory

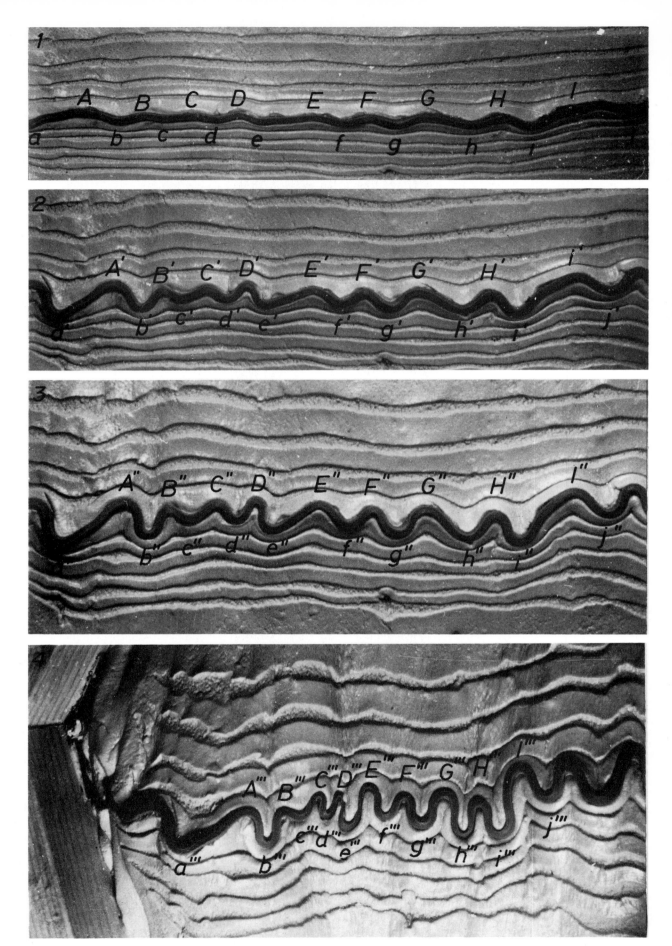

Figure 19.5. *Four views of progressive folding in a laboratory experiment. See Question 19.4★.*

Figure 19.6. Folded pegmatite veins in mica schist, Cristallina, Pennine Alps, Switzerland. The folded vein to the left of the coin shows how fold wavelength is a function of competent layer thickness.

Figure 19.7. Ptygmatic veins from the same locality as Figure 19.3 indicative of an overall constriction type of strain ellipsoid.

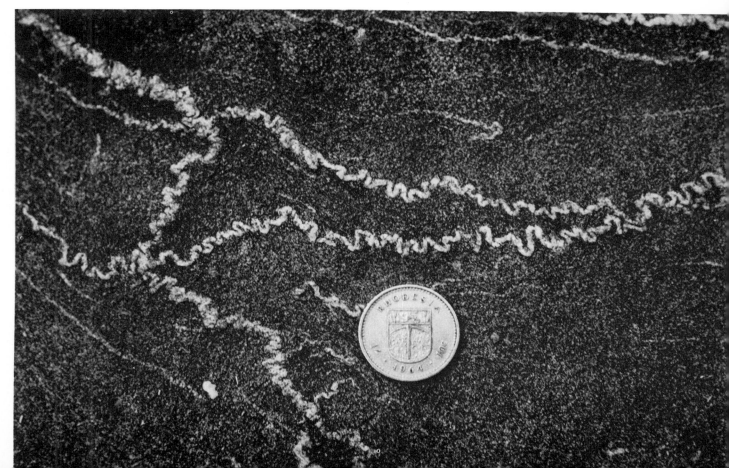

Answer 19.1

From Equations 19.1 and 19.2 the total force F_{tot} resisting the development of the folds is:

$$F_{tot} = F_{int} + F_{ext}$$

$$= \frac{2\pi^2 \mu_1 d^3 \dot{e}_x}{3W_i^2 e_x} + \frac{\mu_2 W_i \dot{e}_x}{\pi e_x}$$

Differentiating with respect to W_i

$$\frac{dF_{tot}}{dW_i} = \frac{-4\pi^2 \mu_1 d^3 \dot{e}_x}{3W_i^3 e_x} + \frac{\mu_2 \dot{e}_x}{\pi e_x}$$

To find the maxima or minima for W_i equate this to zero. On simplifying this gives

$$W_i = 2\pi d(\mu_1/6\mu_2)^{1/3} \qquad (19.4)$$

This value can be shown to be a minimum value because d^2F_{tot}/d^2W_i is always positive. The function shows that the initial wavelength of the fold is a function only of two variables. This wavelength varies linearly with competent layer thickness and a particularly good illustration of this is shown in Figure 19.6. It also varies with the cube root of the viscosity ratio of competent and incompetent materials. Of particular interest geologically is that the wavelength is independent of absolute values of viscosity, implying that rocks at high levels in the crust and of low ductility could behave in a similar way to those flowing much more readily at low crustal levels providing that the viscosity *ratios* are similar.

When Equation 19.4 is expressed in graphical form (Figure 19.8B) the general increase of wavelength to bed thickness ratio with increase of competence contrast is very clear. Consider, however, what happens as μ_1 approaches μ_2 and the competence contrast becomes less marked. The graph shows that even when the materials are identical ($\mu_1/\mu_2 = 1$)

the system should show a characteristic wavelength ($W_i/d = 3.46$). The thickness term d has no meaning where $\mu_1 = \mu_2$ but, even so, it is clear that something is not quite correct here. Equation 19.4 is, in fact, incorrect at low values of μ_1/μ_2 because some of the simplifying assumptions used in its development are unjustified. For example, when the fold wavelength to thickness ratio is low, layer parallel shear plays an important role, and this means that the two expressions used to compute the internal and external resistive forces are in error. The errors in Equation 19.4 are quite small above μ_1/μ_2 values of around 50, but become increasingly large as μ_1/μ_2 decreases. The graphed relationships of wavelength–thickness ratio to viscosity contrast should, in fact, lie somewhere along the dash–dot line of Figure 19.8B. Another problem with this analysis is that the shapes of all ptygmatic structures imply very large viscosity contrasts (20 to 200 are typical), perhaps more than one would anticipate from the actual difference in rock mineralogy. Although the numbers coming from this analysis are correct, suggestions have been put forward that perhaps the assumption we have used of linear Newtonian behaviour is incorrect. At the end of Session 21 we will suggest some publications that will enable the reader to look into some of the predictions arising from theories of flow based on non-linear behaviour. Another problem we will investigate below is that arising from the assumption that the arc length is equivalent to the initial wavelength of the folds ($W_a = W_i$). Now return to Question 19.2.

Wavelength–thickness relationships: measurement

Answer 19.2

The data from the folded vein of Figure 19.3 are shown in

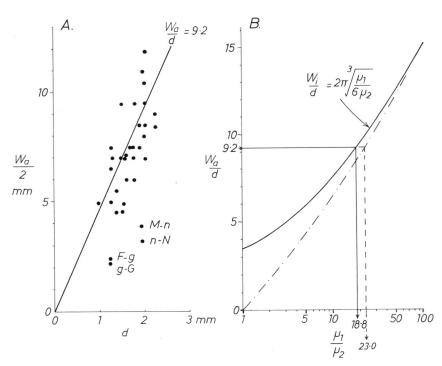

Figure 19.8. A: *Plots of half arc length against thickness for half wavelength sectors of the ptymatic vein of Figure 19.3. B shows the graph of the initial wavelength to thickness ratio and the viscosity ratio μ_1/μ_2.*

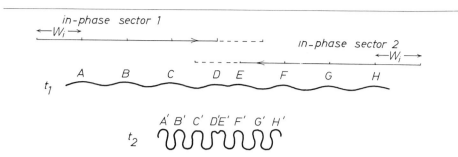

Figure 19.9. *Two in-phase fold sectors ABCD and EFGH overlapping to form an out of phase sector DE. With increasing shortening (time t_2) this out of phase sector D'E' produces a geometric irregularity in the wave train with normal regular wavelength.*

Figure 19.8A. The measurement difficulties arising in this exercise arise from the fact that in any quarter wave sector the layer thickness varies. This is in part due to initial irregularities in the pegmatite vein and in part it results from deformation during folding. In the analysis we have made we have taken the average thickness. Apart from four quarter wave sectors (discussed below), the data points do lie about a straight line through the origin. This can be "eyeballed-in", or fitted more exactly using standard straight line regression methods (query, does the accuracy of the data warrant the use of such sophisticated methods?). The best fit line gives a value of W_a/d of 9·2, and substituting this in Equation 19.3 gives a viscosity ratio of 18·8. Remembering our discussion on the inaccuracies of this function it would be possible to extrapolate the W_a/d value on to the dash–dot curve of Figure 19.8B, giving a viscosity ratio of 23·0.

The general theory of the buckling of a single competent layer in a host rock of lower competence suggests that it is the existence of small irregularities in the layer which initiate the instabilities. These irregularities might be slight undulations along the boundary of the competent layer, perhaps connected with slight thickness variations, or to heterogeneities within the material caused by compositional variations. Both types of imperfections in the competent layer are certain to exist in geological environments. The spacing of these imperfections will be in general rather irregular. When the layer is contracted the imperfections localize sideways deflections of the layer to produce folds. The wavelength input of this process will clearly have a large spectrum, but the theory predicts that only one particular initial wavelength amplifies most efficiently. Folds with this characteristic wavelength (given by Equation 19.4) outpace all others, therefore giving rise to a particular wavelength selection in the system and eventually producing a dominant wavelength throughout the whole fold train. Peter Cobbold (1975) carried out a most interesting series of model experiments to show how folds could propagate progressively along a layer from an initial site of disturbance, the fold trains acquiring more members and becoming increasingly regular. He has also shown how waves arising from geometrical irregularities which were not of the same periodicity as the dominant wavelength of the system can be modified into the dominant wavelength form. The hinges of these non-dominant wavelength folds roll along the competent layer until they come to occupy the correct position for that of the stable dominant wavelength.

Although the general regularity in a ptygmatic wave train

such as shown in Figure 19.3 accords well with this hypothesis, we have seen that there are two pairs of adjacent half wave folds (M–n–N and F–g–G) with markedly shorter wavelengths than those predominant in the system (and about one half of the normal wavelength). Such folds might arise from irregularities in mineral structure or composition in the layer (implying changing values of μ_1) but no obvious indications of such a variation can be seen here. We favour another explanation for this phenomenon. It seems possible that one sector of the compressed layer could develop folds with initially in-phase relationships (Figure 19.9, time t_1, fold trains $ABCD$), and that a second sector of the layer could also develop another train of in-phase folds (Figure 19.9, $EFGH$). It could be that such sectors are out of phase with each other, and that the distance between folds D and E was less than the dominant wavelength W_i. In such circumstances the energy required for the extensive change of fold hinge positions necessary to put sector 1 in phase with sector 2 would be greater than preserving an out of phase fold "hiccup" between the in phase sectors. With increased shortening the ptygmatic fold train would then show sectors in phase (Figure 19.9, time t_2, $A'B'C'D'$ and $E'F'G'H'$) with two adjacent half wave regions ($D'E'$) showing shorter wavelengths than that which predominates in the two in-phase sectors.

The minimum shortening along the ptygmatic layer can be obtained by comparing the arc length of the central line of the competent layer (24·0 cm) with the length of the median surface of the fold wave (9·2 cm). This gives an extension value $e = (9·2 - 24·0)/24·0 = -0·61$, i.e. a shortening of 61%. With such an intense contraction it is somewhat surprising that there is no obvious fabric in the host material around the vein. Two possible explanations for the absence of fabric are:

1. The rock has been strongly recrystallized after deformation with new growth of equidimensional crystals (feldspar and quartz) and random growth of biotite, so that any previous fabric was obliterated.

2. The shortening along the direction of the vein was accompanied by equivalent shortenings in all other directions in the outcrop surface. The three dimensional finite strain ellipsoid is of a constrictional type (Session 10, p. 171, Figure 10.34) with the greatest extension in the X-direction normal to the outcrop surface. In the outcrop of Figure 19.3 there is an intensely developed grain orientation normal to the outcrop surface, and this second explanation seems likely to be correct.

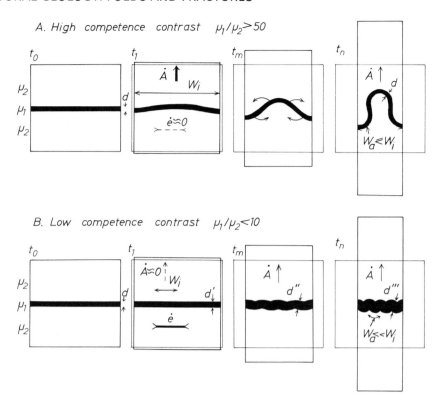

Figure 19.10. *Theory of the progressive evolution of fold shapes in single competent layers. W_i is the initial dominant wavelength of the system and W_a is the arc length between adjacent like points in the fold wave.*

Confirmation of this deduction is seen in nearby outcrops (Figure 19.7) containing pegmatite veins of differing orientations all of which show ptygmatic structure indicative of shortening in a Field 3 type strain ellipse (Figure 4.10).

Dependence of fold shape on viscosity contrast in single layer buckles

The mathematical theory of buckling of single competent layers shows that, for every fixed competent layer-matrix viscosity contrast, there is always a dominant initial wavelength. However, the theory also indicates that, for viscous materials, the growth rate or *amplification rate $\dot{A}$* of this characteristic wavelength varies with viscosity contrast and with the amount of shortening taken up by the competent layer. Although the details of the mathematical theory are too complex to be presented at this stage (they will be investigated in Volume 3), the geometric effects are comparatively simple to appreciate. The different geometric sequences depend upon viscosity contrast, and very high and very low contrasts form end members of a behaviour spectrum.

High competence contrast
($\mu_1/\mu_2 > 50$, Figure 19.10A)

The initial situation is shown at time t_0. At the initiation of deformation (t_1) the system develops folds with a large initial wavelength W_i relative to competent layer thickness d. The initial amplification rate $\dot{A}$ of the fold is very strong, and the sideways deflection of the competent layer is very marked.

Practically no changes of length take place along the middle line of the competent layer W_a. Any shortening effect along the competent layer (known as **layer parallel shortening** $\dot{e}$) is negligible or very small in this system, and this implies that the initial wavelength is directly inherited into, and equivalent to, the competent layer arc length W_a. With increasing shortening of the system (t_m) the amplification rate of the fold begins to decrease and the necessary contraction in the competent layer required for "taking up the slack" is by strong *rotation of the limbs* of the folds. With continued shortening of the system (t_n) this limb rotation can exceed 90° and lead to the development of the typical *ptygmatic fold structures* (Figure 19.11). During the last stages of folding the amplification rate progressively decreases, the finite wavelength W decreases and the limbs become squeezed closer together.

Low competence contrast
($\mu_1/\mu_2 < 10$, Figure 19.10B)

At this initiation of deformation (t_1) the system develops folds with a short characteristic initial wavelength W_i relative to layer thickness. In contrast to the high competence contrast materials, however, these folds are very sluggish in their sideways deflections and their amplification rate is very small. Shortening in the direction of the layer is mostly taken up by strong **layer parallel shortening, $\dot{e}$**, a process which reduces layer length and increases layer thickness (to d'). The closer the viscous properties of the materials are, the greater is the proportion of layer parallel shortening compared to fold amplification. At the limit (where $\mu_1 = \mu_2$) there is no folding, only homogeneous strain of the

Figure 19.11. *Thin section (×50) of ptygmatic structure developed in a strongly shortend siltstone layer embedded in an argillaceous slate matrix of much lower competence (cf. Figure 19.10A, stage t_n). Hope Cove, south Devon, England.*

Figure 19.12. *Cuspate–lobate folds developed in a competent aplite vein embedded in a less competent micaceous gneiss (cf. Figure 19.10B, stage t_n). Cristallina, Lepontine nappes, Switzerland.*

material. With increasing contraction of the block (t_m) layer parallel shortening continues (layer thickness now d''), but gradually the fold amplification becomes greater. At this stage the arc length of the mid line of the competent layer W_a is considerably less than that of the initial dominant fold wavelength ($W_a = W_i d/d''$ if the situation is one of plane strain). Because of the increase of layer thickness with shortening of finite wavelength, folds of ptygmatic form cannot be generated. Instead the folds develop with a marked alternation of rounded and pointed hinge forms and are termed **cuspate-lobate folds** (Figure 19.12). With further total shortening (t_n) the amplification rate of the folds gradually decreases. At very high levels of overall contraction, shortening sub-perpendicular to the axial surface of the folds generally leads to further increases in thickness of the layering at the fold hinge and possibly to reductions of layer thickness on the fold limbs, so that the overall fold form passes from Class 1B towards Class 1C—the type sometimes termed "flattened" folds (see Figure 17.9).

The actual changeover point from layer contraction predominantly by layer parallel shortening to that by side-ways deflection of the layer with the formation of buckle folds is very sharp where competence contrasts are high and moderately sharp even when these contrasts are low. Hudleston (1973b) carried out a series of laboratory experiments in single competent layer viscous models undergoing plane strain and demonstrated these effects very clearly (Figure 19.13). With the experiments performed at viscosity contrasts of $\mu_1/\mu_2 = 100$ the layer parallel shortening only occurred during the initial stages of layer contraction. This is seen where the function of ratio of arc length to initial wavelength (W_a/W_i in Figure 19.13A) flattens after about 13% total model contraction parallel to the layer. The experiment performed with a lower viscosity contrast of $\mu_1/\mu_2 = 11$ shows a marked contrast: layer parallel shortening was initially very active, the competent sheet behaving in a way closely similar to that of a homogeneously strained line. In this experiment even after 50% of model contraction the negative slope of the line shows that layer parallel shortening was still active. Hudleston showed that the change of the active role of layer parallel shortening was connected with the values of the maximum angle of dip of the limbs of the developing buckle fold. Once the fold had developed limb dips of from 10° to 20° only slight changes occurred of the arc length W_a of the layer (shown by the flattening of the curves in Figure 19.13A).

In the course of these experiments Hudleston also investigated the changes in finite amplitude and amplification rates of the folds developing in the more viscous layer (Figure 19.13B). For a particular viscosity contrast μ_1/μ_2 the way that the ratio of finite amplitude A to initial layer thickness d changed with total shortening along the layer was computed. The four curves show that systems with a high viscosity contrast produce folds of greater amplitude than those of low contrast. All four curves show an initial section which is concave upwards and a final section which is concave downwards, implying that the rate of fold amplification first increases, then decreases. The inflexion point of each curve indicates the changeover position where the acceleration of amplification ($\ddot{A} = d^2A/dR^2$) is zero and where the amplification rate is at a maximum value. Systems with high viscosity contrast attain their maximum amplification rates at an earlier stage of total shortening than do those

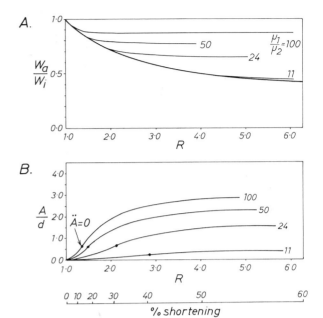

Figure 19.13. *Results of laboratory experiments on the buckling of single competent layers, after Hudleston (1973b). A shows the changes in arc length W_a as a proportion of initial wavelength for different viscosity contrasts. The lower heavy line shows the course of a line unmodified by folding (i.e. $\mu_1 = \mu_2$). B illustrates the ratio of finite amplitude to competent layer thickness. All functions are plotted with abscissa R, the ellipticity of the strain ellipse in the surface of the experiment, and equivalent shortening in the x-direction is given at the base.*

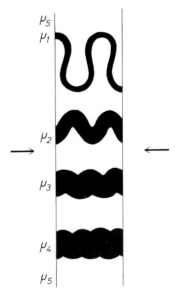

Figure 19.14. *The spectrum of fold shapes produced by buckling a competent layer in a less competent host material. The upper ptygmatic fold shows the maximum viscosity contrast μ_1/μ_5, the lowest cuspate–lobate fold the minimum contrast μ_4/μ_5 and the two middle folds have intermediate viscosity ratios μ_2/μ_5 and μ_3/μ_5. After Ramsay (1982).*

Figure 19.15. *Cuspate–lobate folds developed at the contact of an incompetent pelitic schist layer and surrounding more competent metasandstones. Loch Monar, Inverness-shire, northwest Scotland.*

Figure 19.16. *Detail of a cuspate lobate folds interface between dark incompetent mica schist and competent calcareous sandstone. Nufenenpass, central Switzerland.*

Figure 19.17. Large scale cuspate–lobate folds developed along a deformed unconformity between an underlying metamorphic crystalline basement and an overlying sedimentary cover succession. The height of the hillside is about 500 m. Naukluft Mountains, Namibia.

Figure 19.18. Mullion structure formed by cuspate–lobate folds at an interface between sandstone and slate, North Eifel, Germany.

with low contrast (cf. the positions of the point $\ddot{A} = 0$ on the different μ_1/μ_2 curves).

An important outcome of this theoretical and experimental work is to show that the shapes of folds developed in single layers vary with competence contrast and amount of total shortening. Fold shapes show a spectrum of different forms from the ptygmatic elastica where high competence contrasts exist, to short wavelength (relative to competent layer thickness) cuspate–lobate folds with markedly overthickened layers at low competence contrast (Figure 19.14). The appreciation of the significance of this range of geometric shape spectrum is a most useful tool in aiding rapid deductions about the relative competences of different rock types.

Cuspate–lobate folds

We have seen that buckling instabilities are characteristic of compressed competent layers embedded in an incompetent material. Similar instabilities arise where the reverse situation is encountered, that is where we compress an incompetent layer embedded in a more competent host rock (Figure 19.15). In fact fold formation is the characteristic structural development along single interfaces between competent and incompetent materials where the interface lies along a direction of shortening. The folds developed along such an interface show systematic shape differences between adjacent antiforms and synforms, a feature that we have previously noted in our studies of folded layers (Question 15.1 and analysis Figure 15.28). Along this interface it is always found that the folds which arise by the deflection of the more competent material into the less competent material show a rounded shape and have a relatively large wavelength, whereas those resulting from the deflection of the less competent material into the more competent material are much more pointed and have a relatively small wavelength (Figure 19.16 and Ramsay, 1967, pp. 383–386). The alternating rounded and sharp crested folds are best termed **cuspate–lobate folds**. The boundary layers enclosing a folded competent layer can be envisaged as an interlocking pair of such cuspate–lobate fold interfaces with the inner arc of any individual fold having a cuspate form while the outer arc shows a lobate form (Figures 19.12 and 19.14). Such a geometric interlinking of the two boundaries produces an overall geometric form of the competent layer such that folds will always show a convergent dip isogon pattern and fall into the classification category Class 1. A type of geometry which might be considered as an "inside out" version of this pattern arises when an incompetent layer lies between competent wall rocks (Figure 19.15). The folded incompetent layer boundaries now contain folds which are of Class 3 type. Where the rock materials show some more or less well defined thickness parameter the alternating cuspate–lobate pairs along any one interface generally show a regularity of their combined wavelength (Figure 15.29, $\bar{W}$) related to some characteristic layer thickness in the material. Where the two materials in contact have no characteristic thickness (the interface separates what are mathematically termed "infinite half spaces of the two materials") the cuspate–lobate folds have no characteristic wavelength. Such single interfacial relationships can be found, for example, along "basement-cover" contacts in many orogenic zones over a wide range of scales (Figure

19.17), and in many publications describing large scale fold geometry it is commonplace to find descriptions of so called "pinched synclines" of sedimentary cover between broader adjacent anticlinal regions of basement.

In three dimensions the cuspate–lobate fold forms give rise to linear features parallel to axes of the folds. These structures resemble the exposed wood surfaces in a pile of wooden logs. They have been named after the columnar forms seen in the windows and internal columns of Gothic churches and termed **fold mullion structure** (Figure 19.18).

Now return to Question 19.3.

Assessing competence contrasts

Answer 19.3

The rock types here are typical of the assemblages formed in the middle to lower parts of the continental crust and which are generally described as "gneissic" or "crystalline" basement. Probably all the rocks of Figure 19.4 are of primary igneous origin, later modified by regional metamorphic recrystallization. The granite gneiss is mineralogically and chemically rather homogeneous (main mineral constituents: plagioclase, orthoclase, quartz and biotite), and the strongly developed planar fabric is probably the result of tectonic deformation superposed on an initially more nearly isotropic igneous fabric in the way we have described in Session 3 (Figures 3.3 and 3.15). The granite gneiss comes into contact with an amphibolite (main mineral constituents: hornblende and biotite) of rather homogeneous chemical composition which is schistose but unbanded. The contacts between amphibolite and granite gneiss are everywhere sharp and abrupt. At some localities (Figure 19.19, *a* and *b*) the contact appears discordant to the gneissic banding. These relationships lead to the suggestion that the amphibolite sheets represent metamorphosed basic (basaltic) dykes which were intruded into a previously banded and deformed granite gneiss. The granite gneiss and amphibolite are cut by acid dykes. The geometric relations of these dykes (localities *c* and *d*) indicate that the pegmatite dykes preceded the intrusion of the microgranite and angular xenoliths (locality *e*) suggest that the intrusions were guided by fractures. All four rock types were subsequently deformed together, the small scale structure indicating that all the rock types were ductile but of differing competence.

Striking contrasts in style of small scale structures are seen in the behaviour of the acid dykes depending upon whether the country rock into which they were intruded was granite gneiss or amphibolite. Where the host is granite gneiss the dykes appear only slightly deformed by occasional cuspate–lobate folds (locality *f*) or by slight pinch and swell boudinage (*g*). These features suggest that there were only slight competence differences between acid dyke and acid host rock. In strong contrast, the dykes with amphibolite host rock show either extensive, well formed boudinage (*h*) or ptygmatic folds (*i* and *j*), structures indicating that the acid dykes were much more competent than the surrounding amphibolite. The wavelength–thickness relationships at localities *i* and *j* suggest that the competence contrast between pegmatite and amphibolite was slightly greater than that between microgranite and amphibolite. From these observations we can arrange the rock types in a

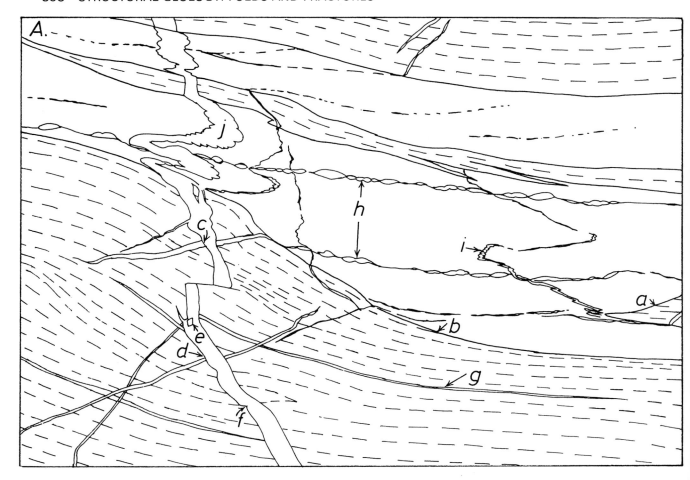

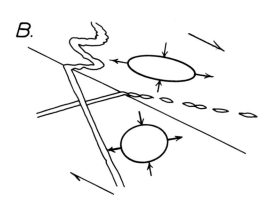

Figure 19.19. A: *Answer to Question 19.3 showing the main structural features a to j described in the text. B shows schematically the most probable interpretation of these structures in relation to the bulk strains suggestive of a general right handed shear displacement.*

competence contrast order: pegmatite ≥ microgranite ≥ granite gneiss > amphibolite. The competence differences in these rocks seem to be controlled by mineralogy (feldspar-rich rocks being more competent than micaceous rocks) and probably by differences in grain size (rocks with coarse grain size being more competent than those with fine grain size).

The development of either folds or boudins in the deformed acid dykes within the amphibolite host appears to correlate with the orientation of the dykes in the adjacent granite gneiss. Where the dykes pass diagonally through the gneiss from bottom left to upper right they take up boudin forms

in the amphibolite, whereas where they pass from bottom right to upper left they take up folded forms in the amphibolite host. The overall deflections in dyke directions when they are followed from granite gneiss into amphibolite are in accord with a strong heterogeneous right handed shear displacement in the amphibolite sheets sub-parallel to the amphibolite–granite gneiss contacts (Figure 19.19B). The shortening by folding or stretching by boudinage of the pegmatite and microgranite dykes corresponds very well with the predictions of the expected changes of length taking place along differently inclined directions (check back to the

Figure 19.20. *Folded basic dyke in a gneiss matrix with fold form indicating that the dyke was more competent than its matrix. Lewisian complex, northwest Scotland.*

Figure 19.21. *Folded basic dyke in a gneiss matrix with cuspate–lobate folds at the dyke interface showing that the gneiss was more competent than the dyke (cf. Figure 19.20). Lewisian complex, northwest Scotland.*

Figure 19.22. Folds and boudins in Triassic limestone (pale) and marl (dark) with structural forms indicating that the limestone was more competent than the marl, Bazena, Adamello, northwest Italy.

Figure 19.23. The same rock units as Figure 19.22 but deformed in the thermal aureole of the Adamello pluton. The metamorphosed marls (dark) are now more competent than the surrounding marble (light). Monte Ferone, Adamello, northwest Italy.

experiments of Session 1, and the behaviour of lines $B'C'$ and $A'B'$, Figures 1.6A and 1.7).

When observations, such as we have made in this example, are carried out from outcrop to outcrop over regions of several km^2 of surface area it is sometimes found that relative competence of the rock types undergo changes. Such changes have a considerable geodynamic importance when it comes to assessing the large scale features of orogenic zones.

Figure 19.20 and 19.21 illustrate the appearance of basic dykes cutting through acid gneissic rocks of the Lewisian Precambrian complex of northwest Scotland. At some localities in this region the basic dykes are planar and unfolded. This relationship is typical of the original geometric features of many basic dyke swarms controlled by regional extension of the crust. In the region of the two outcrops of Figures 19.20 and 19.21, however, the dykes take a complex folded form as a result of a strong regional deformation after dyke emplacement during the so-called Laxfordian orogenic event. At the front of this Laxfordian deformation (Figure 19.20) the form of the folds and the nature of the dyke–gneiss contacts indicate that the dykes were more competent than their acid gneiss host. In contrast, at outcrops situated well within the new orogenic zone (Figure 19.21) the cuspate–lobate fold forms show that the dyke was less competent than the gneissic host. The reason for this reversal in competence difference can be seen when the metamorphic state of the rocks is investigated. Where the dykes are more competent than their host the dyke mineralogy is rather close to that of the original basic intrusions in the terrain outside the influence of the Laxfordian orogenic overprint, the basic rock consisting predominantly of pyroxene and plagioclase feldspar. In contrast, where the dykes are less competent than their host their original mineralogy has been metamorphically transformed by an intense amphibolite facies overprint so that the dyke now consists predominantly of hornblende and biotite. Pyroxene and feldspar are well known to be deformation resistant minerals, and the mineralogical and chemical alterations which led to the formation of biotite and fine grained amphibole are clearly responsible for the change in overall ductility of the dyke rock.

A second example of competence contrast reversal is found around the Tertiary Adamello pluton of northeast Italy. The intrusions making up this pluton were emplaced during a late stage of the Alpine orogeny and the deformations affected the contact rocks around the pluton. These contact rocks consist mostly of well bedded and laminated Triasic limestones and calcareous shales. At localities situated at distances greater than 5 km from the pluton contact the fold forms and boudinage structures seen in these sediments indicate that the limestone layers were more competent than the interbedded marls (Figure 19.22). Lithologically identical Triassic sediments can also be found in a deformed condition within the influence of the thermal aureole of the pluton, the limestone having been recrystallized to a marble and the calcareous shales to a calc-silicate metasediment. The forms of the small scale tectonic structures of these metasediments now indicate a reversal of the competence contrasts seen away from the contact (Figure 19.23). The thermal metamorphism which has led to the recrystallization of the initially mechanically weak clays and fine grained carbonate to much stronger pyroxenes and idocrase is responsible for the reversal of competence of the two main sedimentary components.

Progressive shape changes during folding

Answer 19.4★

Table 19.1 sets out values of the b_1 and b_3 Fourier coefficients calculated from the four experiments shown in Figure 19.5. Figure 19.24 shows plots of these data in two forms: A shows b_1–b_3 plots of all the inner and outer fold arcs together with averages of each of the four experiments joined by tie lines, and B shows a form analysis of the average shapes of inner and outer arcs using the ratio b_3/b_1.

With the very low amplitudes of the folds formed during the first experiment it is somewhat difficult to locate with accuracy the inflexion points on the folded surfaces. As a consequence it is difficult to make an accurate determination of the Fourier coefficients. However, most of the folds show small negative values of b_3 and therefore lie between the chevron and sinusoidal model shapes. As the folds increase in amplitude (general increase in value of b_1) the hinge zones become more rounded, the fold forms moving through parabolic to semi-elliptic shapes. Such a rounding of the fold hinges together with a tendency to equalize the curvatures around the fold is very characteristic of the development of folded single competent layers. In the next session we will see that this tendency is inhibited where several competent multilayers are folded together. In the experiment the inner and outer fold arcs develop slightly differing shapes. The b_1 coefficients of the inner arc are generally greater than those for the outer arc, implying that the standardized amplitude (relative to wavelength) tends to be larger. The ratio b_3/b_1 tends to be largest on the outer arc, implying that the fold forms are more rounded on the outer than on the inner arc.

The results of this analysis should be checked against those of Hudleston (1973b, pp. 203–205). In his experiments the initial fold forms were close to sinusoidal in shape. Our experiments were conducted in plastic materials (plasticene) and it seems possible that the presence of some critical plastic yield stress in the plasticenes might be the reason for the more chevron-like initial folds. Apart from the early fold forms the shape path followed by our experiments takes the same general trend as those of the viscous materials described by Hudleston, although the data points lie a little lower than those described by him.

It seems to us that this type of form analysis offers considerable scope for future research, particularly where the results of the geometrical analysis of laboratory experiments can be interlinked with an analysis of the shapes of folds in naturally deformed rocks.

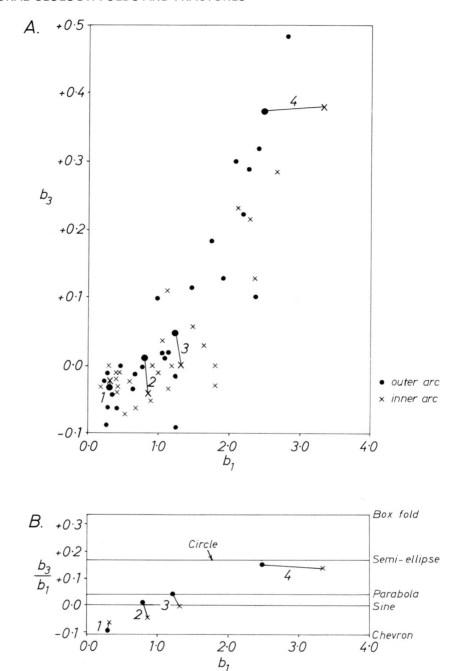

Figure 19.24. *Plots of the Fourier coefficients b_1 and b_3 of the experimentally produced folds of Figure 19.5. Answer 19.4★.*

Table 19.1
Fourier coefficients

Fold sector Coefficient	f-inf b_1	f-inf b_3	inf-F b_1	inf-F b_3	F-inf b_1	F-inf b_3	inf-g b_1	inf-g b_3	g-inf b_1	g-inf b_3	inf-G b_1	inf-G b_3	G-inf b_1	G-inf b_3	inf-h b_1	inf-h b_3	Average b_1	Average b_3	Average b_1/b_3
1 inner arc	0·15	−0·03	0·29	−0·03	0·46	−0·01	0·38	−0·04	0·39	−0·03	0·36	+0·01	0·25	0·00	0·35	−0·02	0·33	−0·02	−0·06
1 outer arc	0·24	−0·09	0·30	−0·06	0·44	−0·06	0·29	−0·01	0·28	−0·03	0·30	−0·02	0·24	+0·04	0·34	−0·04	0·30	−0·03	−0·10
2 inner arc	0·89	−0·05	0·69	−0·06	1·00	−0·10	1·65	+0·03	1·16	−0·03	0·47	−0·01	0·53	−0·07	0·61	−0·02	0·88	−0·04	−0·05
2 outer arc	0·58	0·00	0·82	+0·04	0·80	0·00	1·07	+0·04	1·09	+0·01	0·68	−0·01	0·66	−0·03	0·65	+0·03	0·79	+0·01	+0·01
3 inner arc	1·17	0·00	1·28	−0·14	1·85	−0·14	1·79	+0·06	1·47	+0·06	1·14	+0·11	1·06	+0·04	0·89	0·00	1·33	0·00	0·00
3 outer arc	0·95	+0·06	1·05	+0·02	1·26	−0·09	1·47	+0·12	1·72	+0·18	0·98	+0·10	1·16	+0·02	1·26	−0·01	1·23	+0·05	+0·04
4 inner arc	2·29	+0·22	1·78	0·00	3·79	+0·57	2·34	+0·13	2·10	+0·24	2·89	+0·47	2·71	+0·29	8·62	−1·82	3·33	+0·47	+0·14
4 outer arc	1·92	+0·13	2·05	+0·30	2·33	+0·36	2·66	+0·51	2·36	+0·32	2·21	+0·23	2·25	+0·29	3·87	+0·72	2·46	+0·36	+0·15

KEYWORDS AND DEFINITIONS

Cuspate–lobate folds Folds in which the cross sectional shape of adjacent antiform–synform pairs alternates from a broad rounded hinge zone style (lobate) to a sharp crested hinge zone style (cuspate) (Figures 19.16, 19.17). Such folds are formed as a result of contraction along an interface between more competent and less competent rocks—the lobate folds have cores of competent material, the cuspate folds have cores of less competent material.

Layer parallel shortening The homogeneous or nearly homogeneous strain that can develop when a layered rock is shortened parallel to the lithological layering and where fold forming buckling instabilities are inhibited (Figure 19.10B). This type of strain is characteristic of deformation of layers where the competence contrast between the materials is small.

Mullion structure A linear structure resembling the appearance of the clustered columns which support the arches or divide the separate windows of mullioned windows in Gothic churches. Several types of mullions have been described. **Bedding** or **fold mullions** are developed at the interfaces between rock layers of differing lithology, the surfaces being folded in alternating rounded and cuspate forms (Figure 19.18). **Cleavage mullions** are angular prismatic columns formed at the intersection of lithological layering and tectonically developed cleavage or schistosity (Figure 10.27) and are probably best termed **intersection pencil structure**. Long linear columns with irregular cross sectional form have been termed **irregular mullions** (Wilson, 1953).

KEY REFERENCES

The geological literature on the folding of single layers is very extensive and sometimes involves mathematical techniques which might be found rather tough going. In making our selection, we have chosen those which in our opinion are likely to be most easily assimilated at this stage of the session development and in which the authors make a point of emphasizing the geological significance of the mechanical analysis. In reading articles on the theory of folding be aware that the predictions generally refer to the *initial* instabilities that develop in the system. Sometimes these limitations are not always emphasized in the text or in the diagrams. The theoretical basis for an accurate prediction of the shapes of folds developed after a large bulk strain, and which are generally of prime concern to the geologist, remains practically unknown because the mathematical descriptions of the mechanical processes involved are extremely complex. Investigations of the forms of folds with moderate to high amplitude to wavelength ratios have mostly been made using *laboratory modelling techniques*: with model materials chosen to simulate rock at the reduced scale of the model (viscous gels, waxes or plasticene); or by using *finite element analysis*, a computer based technique which enables fold development to be predicted using an increment iteration program.

Biot, M. A. (1961). Theory of folding of stratified viscoclastic media and its implications in tectonics and orogenesis. *Geol.*

Biot had carried out before this paper was published an extensive analysis of the stability of compressed single and

Soc. Am. Bull. **72**, 1595–1620.

Chapple, W. M. (1968). A mathematical theory of finite amplitude rock folding. *Geol. Soc. Am. Bull.* **79**, 47–68.

Cobbold, P. R. (1975). Fold propagation in single embedded layers. *Tectonophysics* **27**, 333–351.

Fletcher, R. C. (1982). Analysis of the flow in layered fluids at small, but finite, amplitude with application to mullion structures. *Tectonophysics* **81**, 51–66.

Hudleston, P. J. (1973b). An analysis of "single layer" folds developed experimentally in viscous media, *Tectonophysics* **16,** 189–214.

Ramberg, H. (1960). Relationship between length of arc and thickness of ptygmatically folded veins. *Am. J. Sci.* **258**, 36–46.

Ramsay, J. G. (1982). Rock ductility and its influence on the development of tectonic structures in mountain belts. *In* "Mountain Building Processes" (K. Hsü, ed.) 111–127. Academic Press, London and New York.

Schmid, S. M. (1982). Microfabric studies as indicators of deformation mechanisms and flow laws operative in mountain building. *In* "Mountain Building Processes" (K. Hsü, ed.) 95–110. Academic Press, London and New York.

Sherwin, J. and Chapple, W. M. (1968). Wavelengths of single layer folds: a comparison between theory and observation. *Am. J. Sci.* **266**, 167–179.

multilayer systems, but this paper summarizes the results of this basic theory in a way that is particularly relevant to geological situations.

We have not yet developed sufficient mathematical background for a deep understanding of the analytical part of this paper, but the later part, presenting the geological implications of the theory, is important and relatively easily assimilated.

This paper describes the initiation and propagation of folds from a series of very accurately produced wax model experiments. It shows how folds nucleated on initially induced deflections of the more competent layer propagate sideways along the layer. One of the experiments described in this work formed the subject material for Question 17.5★.

The mechanical principles governing the development of cuspate–lobate folds at a competent–incompetent rock interface are presented, and it is concluded that the tendency for the formation of this type of structure is stronger in non linear materials than in linear materials.

This sets out a clear geometrical analysis of a series of model experiments particularly relevant to the formation of ptygmatic structures. It provides examples of the use of simple laboratory techniques to define the relationships between material and mechanical properties.

This sets out the analytical proof of the Biot-Ramberg equation for the wavelength of single layer buckles (Equation 19.4) and discusses the significance of the zone of contact strain around the buckled layer.

This discusses how competence differences may be evaluated from field observations of small scale structures in naturally deformed rocks and establishes lists of competence contrasts in differing geological environments.

A review article which describes different types of deformation regimes and how flow laws and rock rheology are dependent upon the deformation mechanisms of the individual crystal components of the rock. It also describes how stress magnitudes might be determined from observations of the microstructures of rocks. This paper would form a good extension of the problems introduced in Question 19.3.

The wavelength–thickness relationships of over 800 folds in naturally deformed quartz veins embedded in sandstone and phyllite are used to compute the viscosity ratios of the materials. These computations take into account the layer parallel shortening effects in materials of low viscosity contrast.

SESSION 20

Fold Mechanics:
2. Multilayers

A general discussion of the reasons for the great variation in folded multi-layered rocks is initiated by considering the way that folds in single layers can interact with each other to produce disharmonic, harmonic or polyharmonic associations. Spacing distance and viscosity contrast between the competent and incompetent layers in regularly ordered multilayers are shown to be the major factors controlling fold style, and several characteristic model styles are established. Irregular thicknesses of the components of multilayers give rise to special modifications of these styles. Gravitational force acting together with buckling forces is shown to have the potential of developing several orders of fold wavelength. An examination of the properties of anisotropic rocks leads to a discussion of the nucleation and propagation of monoclinal and conjugate kink folds. The geometric variation of these structures is related to the directions of the principal stresses causing folding. The session concludes with an analysis of the geometry and mechanism of development of crenulation cleavage.

INTRODUCTION

This session aims at investigating fold features of considerable complexity and the reader should be aware at the outset that our discussion will only provide an initiation into the problems of folding in multilayered rocks. Field geologists will be aware of the extremely wide variety of geometric form and scale of the folds in naturally deformed rocks. In fact, sometimes these differences seem so wide that one begins to wonder if every observed situation is unique. The wide spectra of fold style and size arises because the mechanical instabilities in a multilayered sequence depend upon a great number of different factors, and nature brings these factors together in many differing permutations and combinations. The main controls on fold geometry are as follows:

1. *The composition of the layers and the primary rheological properties of each rock type.* Each different rock type will have its own characteristic rheological properties dependent upon its mineralogical composition and mineral grain size. In certain types of non-Newtonian materials the rheological properties and flow rates may be non-linear so that strain rates are not simply proportional to the values of the applied stresses. In basically simple mineralogical assemblages (e.g. calcite and clay), different proportions of the end members can produce a range of rock types showing a whole spectrum of varying rheology (e.g. coarse grained limestone, fine grained limestone, marly limestone, marl, calcareous shale, shale).

2. *The change in rheological properties of the layers as a result of changing pressure and temperature conditions*

during the period of fold formation. Most natural folds form over a considerable time period, probably of the order of a million years duration, and the change in environmental conditions over this time span often leads to modifications of grain size or species of the constituent minerals and hence to change in the material properties of the layers.

3. *The development of orientation of mineral grains during deformation.* As a result of strains induced by fold formation certain mineral species may develop preferred alignments as a result of mechanical rotation or recrystallisation processes. The rock layers can develop different types of planar and linear anisotropy (cleavage, schistosity, linear stretching fabrics) in different parts of the same folded layer, and these modifications will lead to irregular changes in the rheological properties of the material in a single layer.

4. *The mechanical properties of the interfaces between layers.* The development of a fold is strongly controlled by the nature of the contacts between adjacent individual layers: whether the layers are effectively welded together or whether the layers become mechanically detached and, in doing so, allow the individual layers to glide past one another.

5. *The thickness of each of the constituent layers in the rock packet, and whether or not the different rock layers are grouped into units.* The range of possible variations here is extremely wide. Probably few natural examples are precisely identical and this factor probably accounts for many of the geometric varieties and wavelength variations of folds.

6. *The nature of the boundary constraints on the rock units undergoing folding.* Some groups of layers in a multi-layered rock complex may undergo free buckling into an incompetent host material without external lateral constraints apart from the resistance offered by the host material, a situation we have investigated in our studies of the development of folds in single layers. In contrast, one or both of the layer packet boundaries may be rigidly constrained (e.g. folding in a sedimentary succession overlying a rigid basement of crystalline rocks, or folding in a thrust sheet moving over a planar sub-stratum). Differing boundary constraints impose different types of mechanical and geometric constraints in the constituent strata of the multilayer.

7. *The overall scale of the multilayer packet being folded.* The scale of the folding is an important factor which decides whether or not gravitational force exerts an influence on the fold geometry. In folds with wavelengths of less than about 100 m gravity plays only a minor role in the control of fold shape, whereas folds with wavelengths greater than 30 km cannot be formed by sideways compression of the layers alone because the buckling forces are too small to uplift the antiforms or depress the synforms against the buoyancy forces induced by gravity (Ramberg, 1970).

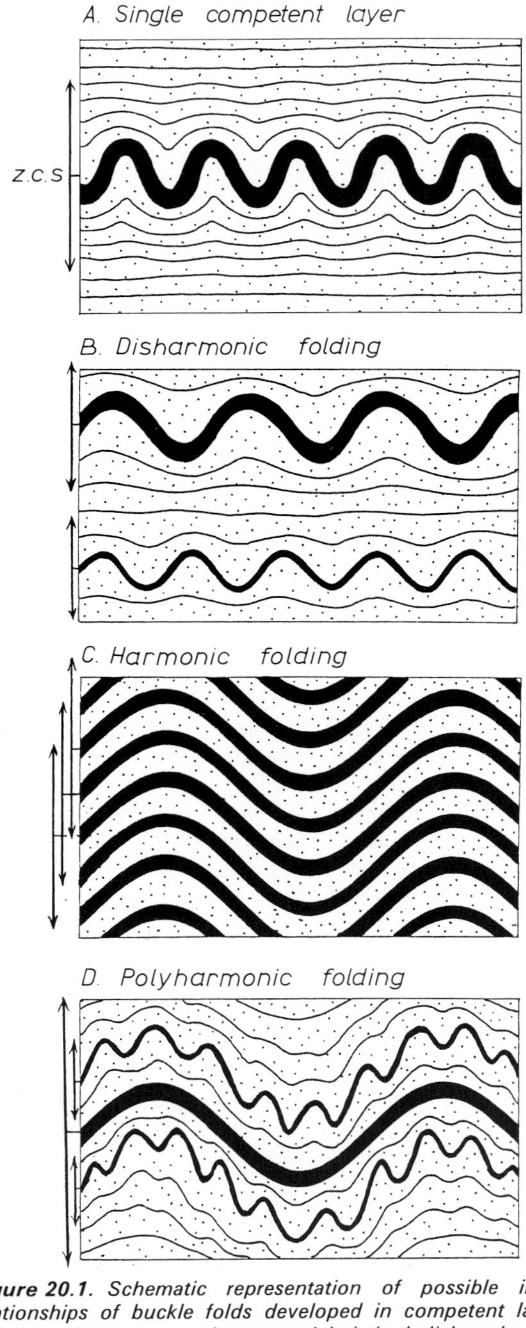

Because of the complex interaction of these factors we cannot always unambiguously interpret the significance of the fold types we encounter in naturally deformed rocks. Investigations of the initial mechanical instabilities using various mathematical techniques have achieved results which have considerable input into the understanding of folded rocks. In this research field work of Hans Ramberg (1961, 1963, 1964, 1968, 1970) has been particularly important in assessing the interaction of several of the fold controlling factors set out above. What needs to be done in the future is to extend this type of mathematically based analysis to folds of large amplitude, and to use more exact comparisons of the predicted fold geometry with the geometry of naturally formed folds and with those folds which can be produced in laboratory experiments using model materials of differing rheological types.

QUESTIONS

Properties of folds in multilayers deduced from the theory of single layer buckling

Question 20.1

In Session 19 we saw how folds arising by compression of a single layer of competent material embedded in an extensive matrix of less competent material form with a characteristic initial wavelength W_i, and that sideways buckling of the competent layer displaced the surrounding material in the *zone of contact strain* of width $W_i/2$ from the median surface of the folded competent layer (Figure 20.1A). Consider what would happen if we had a situation where two competent layers were fairly widely separated from each other such that their separation distance was greater than the sum of their individual half contact strain zone widths derived from the single layer fold theory (Figure 20.1B). In such a system the sideways deflections of the matrix of one competent layer would not penetrate the sideways deflections of the other layer. The folding displacements would not be connected and the two competent layers could fold independently of each other: each layer would have its own characteristic wavelength depending upon its thickness and layer–matrix competence contrast, and the overall form of the folds would accord with the **disharmonic fold** model.

If we allowed the distance between the competent layers to decrease so that the zones of contact strain around each layer overlapped there would be some geometric connection between the folds in the competent layers. Two main types of geometric influence can arise. If the competent layers are

Figure 20.1. Schematic representation of possible inter-relationships of buckle folds developed in competent layers (black) and incompetent host materials (stippled) based on the interactions of the zones of contact strain (Z.C.S.) around each competent layer.

A. Single competent layer

B. Disharmonic folding

C. Harmonic folding

D. Polyharmonic folding

Figure 20.2. *A generally harmonic sequence of folded multilayers. The more competent layers stand up from the outcrop surface and consist of calc-silicate minerals, the less competent material is marble. Sokumfjell, Norway.*

Figure 20.3. *Polyharmonic folding in a multilayer of banded hornblende–biotite–feldspar gneiss. Fusio, Ticino, South Switzerland.*

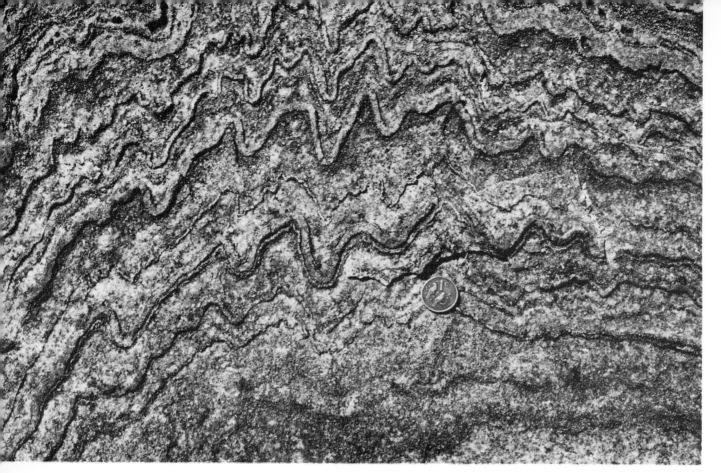

Figure 20.4. Multilayer of alternating calc-silicate layers (A_1 to A_{12}) and marble (B_1 to B_{12}). N. Karibib, Namibia. See Question 20.1.

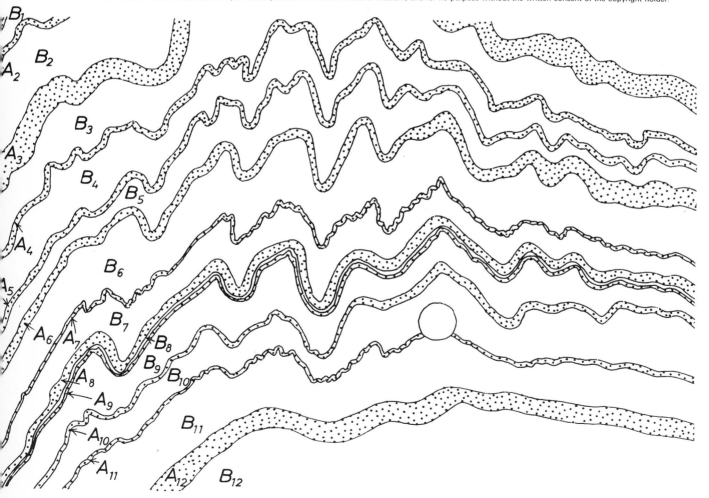

of about the same thickness, the spacing between them is not too variable, and if the competent–incompetent layer ductility contrasts are similar then **harmonic folds** will form (Figures 20.1C, 20.2). If, however, the two competent layers are of markedly differing thickness or show markedly different competent–incompetent layer ductility contrasts each competent layer is likely to induce its own characteristic wavelength into the overall fold pattern. In this second possibility folded multilayers with more than one wavelength, known as **polyharmonic folds**, can be generated (Figures 20.1D, 20.3).

Figure 20.4 shows a folded series of alternating calcsilicate layers (A) and marble (B). Construct axial surface traces of antiforms and synforms. From the shapes of the folds deduce which is the most competent of the two rock types, and from an analysis of the axial surface traces determine the fold style. Discuss the main features of the overall fold style in terms of the widths of the zones of contact strain around each competent layer.

Now check with Answer 20.1, read the general commentary on folding in multilayers, and return to Question 20.2

Geometric development of chevron folds from conjugate kinks

Question 20.2

Figure 20.5A shows two crossing conjugate kink bands developed according to a geometric scheme proposed by Lionel Weiss. Each kink band has an interlimb angle of 120°, and in each the axial surface bisects the angle between the limbs. Kink band 1 formed first and was passive during the formation of kink band 2. Determine the total length of layers a, b, c and d and show that, even though each surface has a very different form as it passes through the kink band pair, the layer length is the same for each surface. If the kink bands have widths (measured perpendicular to their axial planes) of d_1 and d_2 calculate the total shortening in the total conjugate fold across the ends of this model. Assume that no change of length has occurred along the folded surfaces.

On the right hand side of Figure 20.5A the vertical boundaries of the model have been displaced by allowing the left hand axial surface of kink band 2 to migrate into the initially unkinked matrix. Complete the geometry of the upper left hand part of the model by determining the positions of the layers in the blank section of the diagram. Note how the chevron fold style develops from the rekinking of the early kink band 1. Calculate the shear strains of all the differently oriented surfaces in the model assuming that shear displacements are always parallel to the layer surfaces.

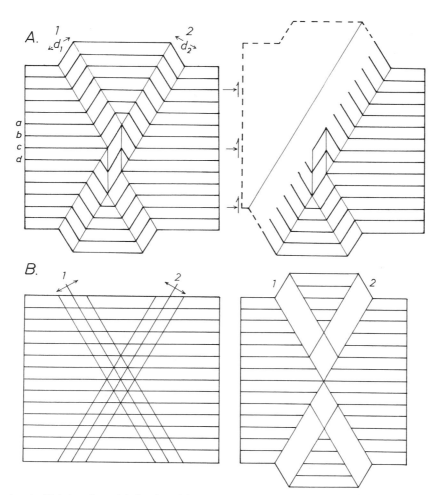

Figure 20.5. A. Conjugate kink band model developed by the sideways migration of an axial surface of one of the kinks. See Question 20.2. B Conjugate kink bands developed by the synchronous development of the two kink zones 1 and 2. See Question 20.3.

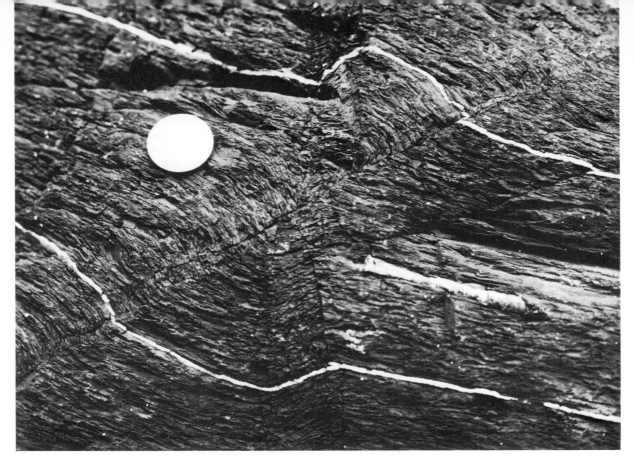

Figure 20.6. *Conjugate kink band developed in strongly cleaved slates. Capo de Penas, North Spain. See Question 20.4★.*

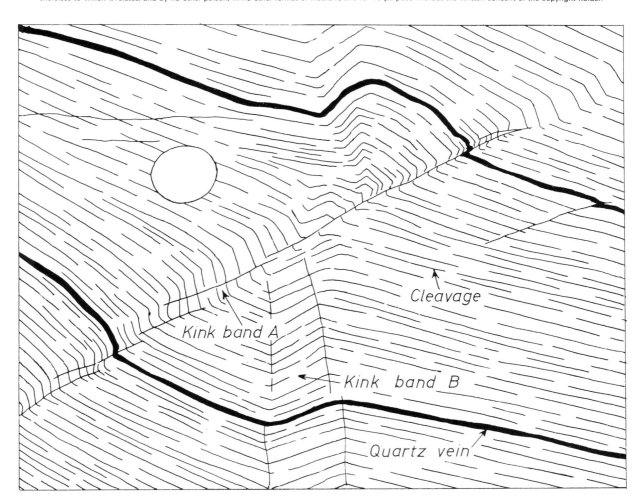

Cleavage

Kink band A

Kink band B

Quartz vein

Figure 20.7. *Answer to Question 20.1. Plots of the axial plane traces (heavy lines) and regions of unfolding (stippled).*

Synchronous development of two kink bands

Question 20.3

The left hand side of Figure 20.5B shows a model with unkinked horizontal anisotropy and the positions (1 and 2) of two potential kink bands. The right hand side of this figure shows the boundaries of this block after *synchronous* development by sideways migration of the axial surfaces of the two potential kink bands 1 and 2. It has been assumed that the regions outside the kinks remain unfolded and undeformed. Draw in the most likely positions of the anisotropy planes through the diamond-shaped intersection zones of the kinks. What special geometric problems arise in these intersection zones which were not found in the model of Figure 20.5A? What do you deduce about the geometric and potential geologic implications of this model?

Now check with Answers 20.2 and 20.3 and the commentary on the geometrical features of conjugate kink folds. Then continue to the section of crenulation cleavage and, if time is available to the starred Question 20.4★.

STARRED (★) QUESTION

Kink band geometry and stress axes

Question 20.4★

Figure 20.6 shows a profile of a conjugate kink fold developed in a strongly deformed argillaceous rock containing an intensely developed slaty cleavage. The following data were measured in the field:

Cleavage surfaces outside kink zone:	strike 17°,	dip 35° NW
Cleavage surfaces inside kink band *A*:	strike 50°,	dip 86° SE
Cleavage surfaces inside kink band *B*:	strike 118°,	dip 32° SW
Axial surface of kink band *A*:	strike 76°,	dip 37° SE
Axial surface of kink band *B*:	strike 66°,	dip 87° SE

Plot these data on a projection. Determine the directions of the fold axes in the two kink bands, the fold interlimb angles and the angle between two kink axial surfaces. What might have been the direction of principal stresses that produced this structure (see Appendix E for a discussion of principal stress axes)? Now continue to the Answers and Comments section of this problem.

ANSWERS AND COMMENTS

Properties of folds in multilayers deduced from the theory of single layer buckling

Answer 20.1

The shapes of the individual folded layers vary with rock type: those in the calc-silicate layers (*A*) are generally close to that of the *parallel fold* model (Class 1B), whereas those in the marble (*B*) are generally folds with *divergent isogons* (Class 3). From this relationship it is deduced that the calc-silicate layers are more competent than the marble layers. Figure 20.7 illustrates the positions of the axial surface traces. The traces of most of the folds with largest amplitude and wavelength can be followed across the whole multilayer.

Folds of small amplitude and wavelength are localized in the rock sandwich between layers B_3 and B_6 and folds with the smallest size are localized in layers A_7 and A_{11}. This geometry suggests that the mutilayer buckled as a whole unit with a characteristic wavelength but that certain of the thinner competent layers imprinted their own characteristic fold wavelength into the system. These shorter wavelengths were only developed where the competent layer had enough space to be relative free to undergo a sideways deflection into the incompetent marble. Note, for example, that the thin competent layer A_9 lay so close to the thicker competent layer A_8 that it had no freedom to develop its individual wavelength, and that the two layers effectively behaved as a single competent unit. The overall style of the folding is *polyharmonic*, but certain groups of competent layers ($A_4 A_5 A_6$, $A_8 A_9$) show more or less *harmonic* features within the overall field.

Certain areas in this outcrop (stippled in Figure 20.7) show either no folds at all, or only very slight local undulations of the layering. These areas all occur on the limbs of the main folds. The absence of folds is probably best interpreted as the result of a late history of unfolding, where an earlier shortened and folded competent layer progressively rotated into a position of stretching (see Session 12, p. 224). A change in the sign of finite extensive strain along the layering from negative to zero and perhaps to a positive value is quite common in regions of tightly folded multilayers. This feature can lead to unfolding of a previously folded layer, to the development of boudinage or to boudinaged or *intrafolial folds* (see Session 12, pp. 226–227). If boudinage is formed the obliquity of the competent layer

to the principal finite strain axes in the fold limb frequently lead to the development of *en-echelon boudinage* (Figure 20.8, and Session 2, Figure 2.14B′). The sense of relative rotation of the boudins is opposed on the two fold limbs of any single fold domain, the rotation of the boudinaged competent layer relative to the fold axial surface being less than that of the overall trend of the layer. The layer lengths of the folded competent layers A_4 to A_{12} were measured over the folded sector and all except that in A_{12} were fairly constant. This feature suggests that each of the layers A_4 to A_{11} had the same amount of initial layer parallel shortening and that each had approximately the same competence contrast with the intervening less competent marble. In layer A_{12} the arc length was only about 75% of that in the other layers, suggesting that the competence contrast between it and the marble was less than that of the other calc-silicate layers.

Change of fold shape with packing distance of competent layers

In our study of folded single competent layers we saw how fold shape in the competent layer varied with ductility contrast between the layer and its matrix. In multilayers an additional factor is also important in controlling fold shape. It is found that, for any two materials of particular competence contrast, the proportion of incompetent to competent material (n-value) strongly affects the curvatures of the folded surfaces. This is because the sideways deflection of any competent layer in a multilayer is constrained by the deflection of adjacent competent layers. Figure 20.9 shows

Figure 20.8. Folded multilayer of calc-silicate (dark) and marble (white). Khan Gorge, Namibia. The competent layers are folded in the hinge zone. In the fold limbs late stretching has led to the formation of boudinaged folds (upper left) or unfolding and en-echelon boudinage (lower centre).

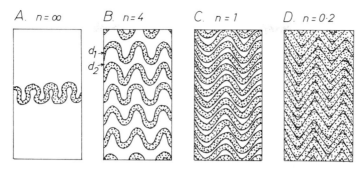

Figure 20.9. Schematic representation of the changes in fold shape in a folded multilayer resulting from a change in the proportion of competent layer thickness d_1 (stippled) to incompetent layer thickness d_2. The proportion of incompetent to competent layer thickness is given by $n = d_2/d_1$.

the differences in fold style that come about in a material with quite marked competence contrast as a result of decreasing the proportion of incompetent material. The ptygmatic form of a single free layer with negative interlimb angles is not possible if other competent layers are close at hand. With a decrease in the proportion of incompetent material high curvatures are concentrated near the hinge zone and the fold limbs become less curved, and as the n-value decreases the overall shape approaches more closely the sharp-hinged, parallel-limbed **chevron fold** model.

Fold styles in multilayers

It should be clear that the number of potential models of multilayers involving different layer thicknesses and rheologies is unlimited. In our discussion we will therefore concentrate on certain specific models which appear to have the widest application or which offer the best possibility of extrapolation to the situations arising most commonly in

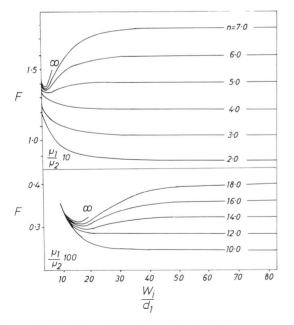

Figure 20.10. Plots of force F resisting buckling in a regular system of alternating multilayers of thickness d_1 (competent) and d_2 (incompetent) for differing $n = d_2/d_1$ values and viscosity constrasts μ_1/μ_2. W_i is the initial fold wavelength.

structural geology. One special and important factor in the analysis concerns the possible effects of gravitational force in controlling fold geometry, a factor which is especially important where we wish to analyse folded multilayers of considerable (kilometric) thickness and folds of very large wavelength.

The mathematical techniques which have been employed have generally aimed at computing the resistance to buckling offered by the component competent and incompetent layers and finding the wavelength of the folds which grow most rapidly along similar lines to those we discussed in the buckling of single competent layers. It should be emphasized that the deductions developed from these methods *describe only the values of the wavelengths of the initially formed folds* (W_i) in the multilayer and rarely is it possible to express these values succintly in a single mathematical formula as we could for the buckle wavelength of a single competent layer. This initial wavelength may, in systems of high competence contrast, be found by measuring the arc length (W_a) between hinge lines of adjacent antiforms or synforms, but the correctness of this deduction depends on the extent of the initial layer parallel shortening in the system. It should also be emphasized that *none of the computations made using these methods can be extended to describe the fold shape as the fold increases in amplitude*, a constraint which is not always obvious from many of the published diagrams. To make predictions as to the finite fold shape we have to resort to either laboratory experiments using model analogue materials or to numerical calculations using a special type of finite element analysis whereby the initial deflections of the layers are grown incrementally using a reiterative computer technique. Although the present state of our analysis gives an incomplete picture of the folding process we would like to stress that the positive results so far attained give us an insight into complex processes which could not have been developed by intuition alone and which do provide us with a strong foundation for understanding the folds developed during tectonic processes.

Three main types of analytical solution to folding in multilayers can be found. The first is where gravitational force is insignificant, and this occurs either when there are no density contrasts between the layers or, if density contrasts do exist, where the thicknesses of the layers are small and the fold wavelengths which arise are less than about 100 m. These are the folds which are developed by pure buckling processes. The second type of situation is where the

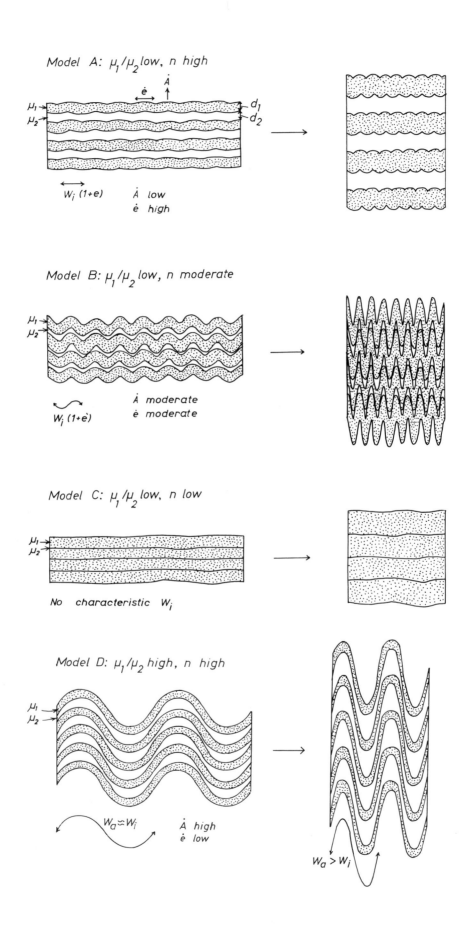

Model A: μ_1/μ_2 low, n high

$\dot{A}$ low
$\dot{e}$ high

$W_i\,(1+e)$

Model B: μ_1/μ_2 low, n moderate

$\dot{A}$ moderate
$\dot{e}$ moderate

$W_i\,(1+\dot{e})$

Model C: μ_1/μ_2 low, n low

No characteristic W_i

Model D: μ_1/μ_2 high, n high

$W_a \approx W_i$

$\dot{A}$ high
$\dot{e}$ low

$W_a > W_i$

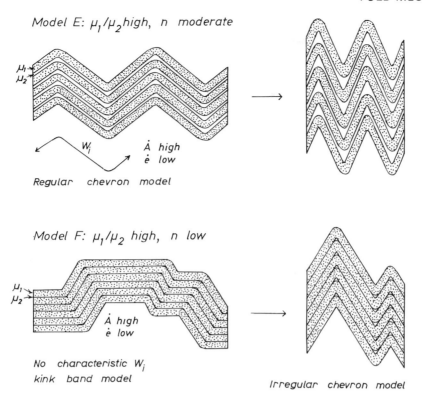

Model E: μ_1/μ_2 high, n moderate

μ_1
μ_2

W_i $\dot{A}$ high
             $\dot{e}$ low

Regular chevron model

Model F: μ_1/μ_2 high, n low

μ_1
μ_2

$\dot{A}$ high
$\dot{e}$ low

No characteristic W_i
kink band model

Irregular chevron model

Figure 20.11. *Models of folds developed in regularly alternating competent layers of thickness and viscosity $d_1\mu_1$ and $d_2\mu_2$ respectively and with $n = d_2/d_1$. W_i is the initial wavelength. $\dot{A}$ is the buckle fold amplification rate, $\dot{e}$ the layer parallel shortening rate, and e the total layer parallel extension.*

forces applied parallel to the layers are insignificant and where gravity can act on layers of different density to produce a buoyancy force across the layers. The third is where both buckling forces and gravity act together. To follow the details of the mechanical analysis requires a strong mathematical background. We will therefore not go into the details of the computations below, but only outline the principal results of the methods with particular emphasis on their practical geological implications.

Folds developed independently of gravitational force:
1. Regularly alternating multilayers

The simplest multilayer model we can propose consists of an infinite stack of regularly alternating layers of thickness d_1 and d_2 with viscosities μ_1 and μ_2 respectively (Figure 20.10). The development of folds in this model is controlled by two parameters: the proportion of incompetent to competent layer thickness ($n = d_2/d_1$) and the viscosity contrast μ_1/μ_2. Ramberg (1961, 1963, 1964) investigated the properties of this system by determining the values of the buckling force F necessary to form folds with a particular ratio of initial wavelength to competent layer thickness (W_i/d_1). This initial wavelength is independent of the magnitude of the externally applied force and independent of the overall strain rate in the system. Two sets of curves for plots of F against W_i/d_1 are illustrated in Figure 20.10 for values of μ_1/μ_2 of 10 and 100. For high n-values these curves show a sharply defined minimum value of F. The W_i/d_1 value at this minimum gives the

wavelength of the folds which can form most easily. Where $n = \infty$ the sharp minimum coincides with the *well defined initial wavelength* developed in a buckled single competent layer deduced from Equation 19.4. As the competent layers come to lie closer together (decreasing n-value) the minimum on the particular n-value curve shifts to the right and becomes less marked. These features imply that, for a given competent layer thickness, the characteristic *initial wavelength becomes larger*, but that the *sharpness of this initial wavelength is not so clearly defined*. With further closing together of the competent layer the n-curve minimum is lost and the curve acquires a zero or even negative slope. These flat curves imply that initial fold wavelengths over a wide range of values are equally favoured and that *the initial wavelength in the multilayer can become very irregular*. Where the curves have a negative slope the folds which are most stable have the *largest wavelength possible* in the system probably governed by the localization of mechanical imperfections in the multilayer. In many geological situations the mutilayer packet does not extend to infinity, as in this model, and in these circumstances the stable wavelength of the multilayer packet is governed by the average properties of the material outside the multilayer packet. Because the resistance to buckling for any given viscosity contrast is less in a multilayer than it is for a single layer, buckle folds tend to grow faster in multilayers than in single layers. There is therefore a lesser proportion of layer parallel shortening in a multilayer compared with that in a single layer system with an equivalent viscosity contrast. However, at low viscosity contrasts layer parallel shortening is still an important factor

Figure 20.12. Folds in metamorphosed sandstone and pelite showing the features of Model B of Figure 20.11. Note the rather irregular wavelength, the almost similar fold shapes in both competent (pale) and less competent (dark) layers and the strong variations in layer thickness as a result of late stage homogeneous strain. The pale cross cutting vertical stripes consist of pegmatite material formed by local rock melting during the fold formation. Loch Monar, Northwest Scotland.

Figure 20.13. Folds in siltstone and argillite showing features of Model C of Figure 20.11. The folds amplitudes are small and their wavelengths are rather irregular. The sedimentary layers are cut by a fairly uniformly oriented slaty cleavage. Tor Cross, South Devon, England.

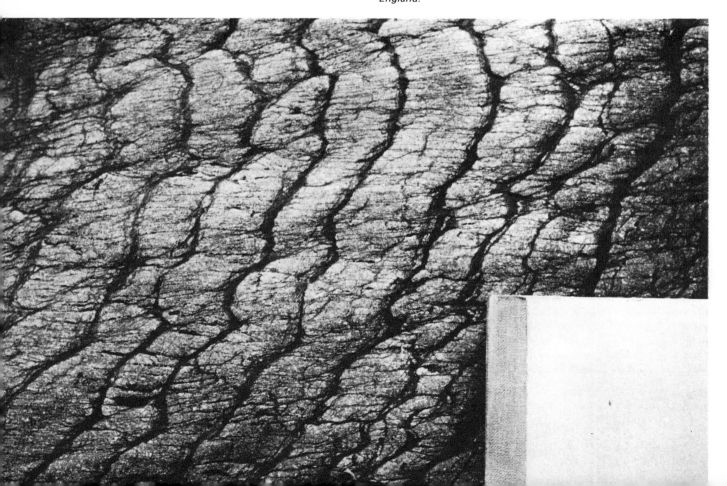

Figure 20.14. Chevron fold in alternating sandstone and shale (Model E) showing strong development of cleavage in the incompetent shale, complex quartz-vein systems in the sandstones and a well developed limb fault. Mullion, Cornwall, England.

Figure 20.15. Chevron fold in alternating sandstone and shale with a well developed quartz filled saddle reef between two thickly bedded sandstone layers. Hartland Quay, North Devon, England.

contributing to the total fold geometry and finite strain. Figure 20.11 illustrates schematically the geometric features of the folds that can arise in a regularly stacked multilayer in situations of low ductility contrast (Models A, B and C) and high ductility contrast (Models D, E and F) depending upon whether the more competent layers are widely (A, D), moderately (B, E) or closely (C, F) spaced.

Model A: Low ductility contrast (μ_1/μ_2), ratio of incompetent to competent layers ($d_2/d_1 = n$) high

The fold wavelength is small relative to the competent layer thickness and the development of the folds leads to characteristic **cuspate–lobate forms** along the interface between the competent and incompetent layers (cf. Figures 19.10B and 19.12 for single layers). The rate of layer parallel shortening ($\dot{e}$) initially strongly predominates over the amplification rate ($\dot{A}$) of the folds, so that the finite arc length W_a is considerably less than that of the dominant initial wavelength W_i by a factor $1 + e$ dependent upon the total layer shortening (e). Because the initial fold wavelength is short relative to the layer thickness, the contact strain effects around each individual buckling competent layer do not penetrate very far, and high n-values may lead to a decoupling of the axial surfaces of the folds in adjacent competent layers and to the formation of **disharmonic folds**. Examples of the folds of Model A type can be seen in Figures 12.9 and 15.15.

Model B: μ_1/μ_2 low, n moderate

As the competent layers come to lie closer than those for Model A they become more closely connected into a **harmonic fold** assemblage. However, the dominant wavelength of this system is less sharply defined than that for Model A. Layer parallel shortening plays an important role in the initial folding stages. At low μ_1/μ_2 values, because of the overall similarity of layer properties more or less homogeneous flattening can produce significant modifications of fold geometry: both competent and incompetent layers are thickened in the fold hinge zone and both are thinned in the fold limbs (see discussion under Answer 17.4★ and Figure 17.9). As a result of these late stage modifications both competent and incompetent layers have modifications in form so that the layers approach that of the **similar fold** model. Examples of this fold style are shown in Figure 20.12.

Model C: μ_1/μ_2 low, n low

The initial fold *wavelengths are likely to be very irregular* and the presence of local imperfections in the layers or unevenness in the initial layer contacts are likely to localise the positions of individual fold hinges. In this situation the closely packed layers of similar viscosity are likely to undergo strong layer parallel shortening, a feature which will tend to predominate geometrically over the development of buckle folds. The multilayer as a whole, although showing some signs of fold development, is more likely to show *rather homogenous strain* throughout the mass, and strain induced fabrics (cleavage and schistosity) are likely to pass with regular orientation through all the layers (Figure 20.13). Where the proportion of competent to incompetent rock is

very high, buckling may be insignificant and a new deformation mode may appear, leading to the formation of *conjugate ductile shear zones* (Sessions 2 and 26). This shear zone mode is especially important where the material has non-linear Newtonian properties and where it becomes **strain softened** (where any deformed region accepts further deformation in preference to the surrounding less deformed region).

Model D: μ_1/μ_2 high, n high

In this type of multilayer, compression leads rapidly to the formation of buckles, the system undergoing very little initial layer parallel shortening. The arc length W_a approximates closely to the initial wavelength W_i. The initial wavelength is fairly large relative to the thickness of the competent layers, and the overlapping contact strain effects from each layer rapidly interact to produce an overall **harmonic fold** pattern. The competent layers usually show a nearly perfect **parallel fold** form (Class 1B), whereas the incompetent layers become preferentially thickened in the hinge zones and thinned in the fold limbs (Class 3). With strong overall shortening the limbs of the folds become rotated into directions of stretching and the competent layers may show the development of **pinch and swell** or **boudinage** structures. Limb stretching may lead to arc lengths exceeding initial wavelengths of the folds. In regions characterized by low grade metamorphism the fold limbs often show the formation of **zones of pressure solution** as a result of transfer of more soluble material (generally carbonates or quartz) from the limbs either to the hinge or into contemporaneously developing vein systems. Individual fold hinges are then separated by generally dark zones known as **pressure solution stripes**. Examples of folding according to this model can be seen in Figure 20.2, in layers $A_4 B_4 A_5 B_5 A_6$ in the upper part of Figure 20.4 and the pressure solution stripe effect is shown in Figure 20.45.

Model E: μ_1/μ_2 high, n moderate

As with Model D, buckles initiate and amplify rapidly with little or no layer parallel shortening and there is a strong interaction between the buckled competent layers producing **harmonic folds**. Because the incompetent material is thinner than that of Model D, physical migration of this material from limb to hinge zone is limited. The folds forming in this situation have narrow hinge zones and straight limbs known as **chevron folds**. The incompetent layers in the fold limbs generally undergo very strong components of *layer parallel simple shear strain*, especially during their later stages of development. If the layer contacts remain welded, high finite strains are localized in the incompetent layers leading to the development of intense local *cleavage* or *extension veins* (Session 21), while in certain environments *shear discontinuities* can develop in the incompetent layer or at the interfaces between competent and incompetent layers. Special types of triangular shaped veins, known as **saddle reefs**, are often produced in the incompetent layer situated at the fold hinge (discussed below). In this model the arc length W_a is generally very close to the initial wavelength W_i. There is generally much less of a tendency for the competent layers to rotate into a stretching orientation as with Model D. Mechanical effects related to the weak incompetent

Figure 20.16. Fibrous silicified asbestos "Tiger's eye" (pale) developed in saddle reefs situated in incompetent layers between folded competent sandstones. Prieska, South Africa, half natural size.

Figure 20.17. Chevron fold in alternating sandstone and shale with development of a bulbous fold hinge in an abnormally thick competent sandstone layer. Hartland Quay, North Devon, England.

Figure 20.18. *Single kink band developed in a laminated siltstone separated by thin incompetent clay rich layers. Bigbury, South Devon, England.*

Figure 20.19. *Conjugate kink folds in laminated competent siltstones separated by thin incompetent clay rich zones. Bigbury, South Devon, England.*

material properties and the marked relative strength of the competent layers generally lead to a lock up of the chevron fold when the interlimb angle reaches about 60°. Examples of this style of fold are shown in Figures 20.14, 20.15 and 20.16.

Model F: μ_1/μ_2 high, n low

Theory predicts that as the competent layers come to lie closer together, so the initial fold wavelength becomes less well defined. The most important feature of the irregular wavelength folds which do form in this situation is that they no longer have their axial surfaces arranged sub-perpendicular to the layering. The multilayer is no longer mechanically active throughout the whole mass. Only in certain zones does the layering rotate relative to its initial orientation. The folds formed by this rotation are straight limbed angular folds often with interlimb angles of about 120° and with their axial surfaces oriented at about plus or minus 60° from the initial layering direction, and are known as **conjugate kink folds** (Figures 20.6, 20.18 and 20.19). The strains set up in the individual kink zones are generally developed by simple shear parallel to the layer surfaces, with the shear displacements concentrated in the thin incompetent layers. In laboratory experiments performed by Lionel Weiss on regular layered card stacks it was found that the kinked limbs increase their lengths by sideways migration of the axial surfaces. The migration of the axial surfaces of one set of kink zones through the axial surfaces of its conjugate set produce **irregular chevron folds** with axial surfaces sub-perpendicular to the initial layering and with interlimb angles of about 60° (see Frontispiece to Session 15). The wavelength W and arc length W_a of adjacent antiforms or synforms of these chevron folds are generally

irregular because their development is produced by a rather special migration process of the sets of conjugate kinks. This style of conjugate folds is not only characteristic of alternating multilayers it is also found in non-layered materials where some well developed planar anisotropic fabric (shale fissility, slaty cleavage and schistosity) has undergone folding. Further discussion of the special features of conjugate folds will be made when the reader has completed Questions 20.2 and 20.3.

Folds developed independently of gravitational force:
2. Irregular alternating multilayers

The first of the more irregularly layered models consists of a unit of alternating mutilayers of thickness d_1 and d_2 and viscosity μ_1 and μ_2 respectively, embedded in a matrix of differing viscosity μ_3 which is of infinite extent on both sides of the multilayer unit. The model, shown in Figure 20.20, is taken from Ramberg (1970) and shows a central unit consisting of five competent layers (d_1) layers and four incompetent layers (d_2), and with viscosity contrasts $\mu_1 : \mu_2 : \mu_3 = 100 : 10 : 1$. Ramberg determined a function K related to the amplification rate of folds with differing initial wavelengths W_i as a function of the spacing of competent and incompetent layers in the central unit ($d_2/d_1 = n$). Where the competent layers are relatively closely spaced ($n = 2 \cdot 0$) the whole competent–incompetent unit shows curves with a single maximum, indicating the preferential amplification of folds with one dominant wavelength W_i''. This wavelength is large relative to the competent layer thickness d_1, and the whole nine-layer sandwich behaves like a single competent unit. As the spacing between the competent layers is

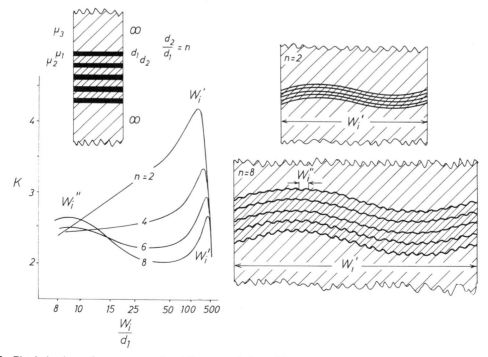

Figure 20.20. *The behaviour of a compressed multilayer consisting of five competent and four incompetent layers (thickness d_1 and d_2, viscosity μ_1 and μ_2 respectively) in a less competent host. K gives the relative amplification rates with differing initial wavelength W_i to thickness ratios. At low n-values the mutilayer produces only one wavelength W_i'', but at high n-values the system develops two characteristic wavelengths W_i' and W_i''. After Ramberg (1970).*

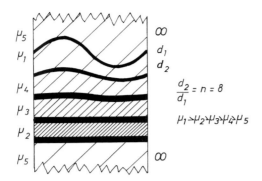

Figure 20.21. *Behaviour of a compressed mutilayer with five competent layers of initial thickness d_1 separated by four incompetent layers of thickness d_2 of variable viscosity. After Ramberg (1970).*

rial between the most competent layer changes its properties. In this nine-layer model (Figure 20.21) the ratio of the incompetent to competent material ($d_2/d_1 = n$) is 8 but the viscosity of the layers of the four incompetent layers varied such that $\mu_1 : \mu_2 : \mu_3 : \mu_4 : \mu_5 = 100 : 8 : 4 : 2 : 1$. With this particular arrangement only one dominant and short wavelength appears, but the amplification rate of the individual competent layers is highest where the local competence contrast is high and decreases very markedly as the incompetent layers take on closer viscosities to that of the competent layer. The pattern of folds developed in a multilayer complex is therefore very sensitive both to the spacing of the competent layers and to changes in competence between the multilayers.

increased ($n = 6$–8), the curves show the appearance of a second maximum, indicating that a second and shorter initial wavelength W_i''' can be developed synchronously with the larger wavelength folds. We therefore have the possibility of contemporaneously developing **polyharmonic folds**.

Another model with considerable relevance to geological studies is similar to that just described, but where the mate-

Folds developed independently of gravitational force:
3. Variations in competent layer thickness

Changes in the thickness of individual competent layers within a more or less regular multilayer sequence can produce very striking modifications in the average fold style, especially in situations where the competence contrast is high (Model E). Figure 20.22 illustrates a symmetric chevron fold with an incompetent layer of thickness d_2

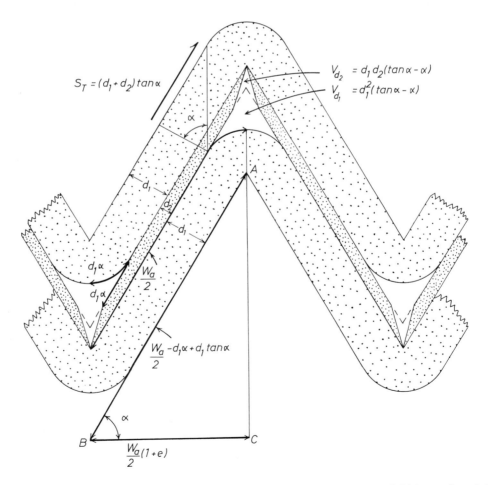

Figure 20.22. *Chevron model consisting of alternating competent and incompetent layers of thickness d_1 and d_2 folded into a symmetric structure with limb dip $\pm \alpha°$. The arc length wavelength is W_a and the amount of shortening (negative extension) is e. S_T is the total slip across any competent–incompetent rock pair and V is the volume increase giving rise to the saddle reef structure at the fold hinge. After Ramsay (1974).*

sandwiched between two competent layers of thickness d_1. This model of alternating competent and incompetent layers is considered to extend indefinitely along the fold axial surfaces. Assuming that the layer contacts on the fold limbs with dip α remain welded, and that the arc length W_a equals the initial wavelength W_i, Ramsay (1974) has shown how it is possible to determine the shear displacement S_T across any competent–incompetent layer pair on the fold limbs.

$$S_T = (d_1 + d_2) \tan \alpha \qquad (20.1)$$

The shear displacement across the competent layer is given by

$$S_{d_1} = d_1 \alpha \qquad (20.2)$$

and that across the incompetent layer is

$$S_{d_2} = S_T - S_{d_1} = (d_1 + d_2) \tan \alpha - d_1 \alpha \qquad (20.3)$$

From the relationships in triangle ABC

$$(W_a/2 - d_1\alpha + d_1 \tan \alpha) \cos \alpha = W_a(1 + e)/2$$

where e is shortening across the structure, or

$$1 + e = (1 - 2d_1\alpha/W_a) \cos \alpha + 2d_1 \sin \alpha/W_a \qquad (20.4)$$

From this function it follows that for any given total shortening, the angle of dip of the fold limb is a function only of the thickness of the competent layer expressed as a proportion of the fold arc length W_a (Figure 20.23A). From these

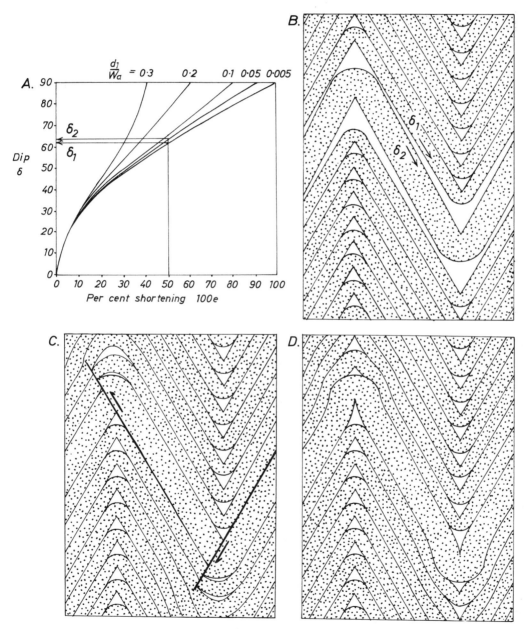

Figure 20.23. A: *Relationship of layer dip δ and shortening in the chevron fold model of Figure 20.22. B shows the geometric problem arising by incorporating an especially thick competent layer into the fold—the thick layer must have a dip δ_2 in contrast to the general dip δ_1. C shows a geometric solution for equalising δ_1 and δ_2 by allowing limb thrusts to develop, and D shows the formation of bulbous hinge zones. After Ramsay (1974).*

curves it can be deduced that if the competent–incompetent pairs are generally regular but contain locally competent layers of abnormally great thickness, then the chevron fold style cannot completely accord with the regularity of Model E (Figure 20.11). For a given total shortening we predict that the thickest competent layer has to attain a higher dip than that of the surrounding, more normal sequence. Such a situation is not geometrically possible if the rock mass as a whole is coherent and does not open up large spaces between the thickest competent layer and its immediate surroundings. Nature generally adopts two solutions to this compatibility problem. If the dips on the fold limb are made to conform there is excess length in the thick competent layer. This excess length can be taken up either locally near the fold hinge to produce **bulbous hinged chevron folds** (Figures 20.23D, 20.17) or by allowing the thick competent layer on one limb to break across and override the layer on the other limb to develop **limb thrusts** (Figures 20.23C, 20.14). Because of these geometric complications the chevron fold model E is not found where there are marked variations in the thicknesses of the competent layers (Figure 20.16). Because Equation 20.4 is independent of the incompetent layer thickness d_2, chevron styles can be geometrically stable in situations where the incompetent layer thickness varies.

We have previously noted the possibilities of forming **saddle reef** openings in fold model E. The features of Figure 20.22 can be used to determine numerically the potential for the development of such structures. The potential saddle reef space owes its geometry to both competent and incompetent layers. In unit depth of material the total volume of the space V_T, volume in the competent layer V_{d_1} and incompetent layer V_{d_2} are given by

$$V_T = d_1(d_1 + d_2)(\tan \alpha - \alpha) \quad (20.5)$$

$$V_{d_1} = d_1^2(\tan \alpha - \alpha) \quad (20.6)$$

$$V_{d_2} = d_1 d_2(\tan \alpha - \alpha) \quad (20.7)$$

These can also be expressed as volume dilatations Δ by dividing the appropriate initial volume of material involved

$$\Delta_T = \Delta_{d_1} = \Delta_{d_2} = 2d_1 (\tan \alpha - \alpha)/W_a \quad (20.8)$$

All *dilatations increase with dip and therefore with shortening*. The dilatation is the same for both competent and incompetent material and is *controlled only by the thickness of the competent layer* and the *inverse of the fold arc length*. Differentiation of function 20.8 with respect to angle of dip shows that the dilatations increase with the angle of dip as a function of $\tan^2 \alpha$, and this shows that the growth of saddle reef is especially strongly marked as the fold develops progressively. The shapes of the different potential spaces at the hinge zone with different $d_2/d_1 = n$ ratio is shown in Figure 20.24. True saddle reefs (Figures 20.15, 20.16) are developed by the crystallization of material from fluid phases existing in the rock during fold development, and if this fluid phase contains metallic components the reefs may be of considerable economic importance (e.g., the gold reefs of Bendigo, Australia). If crystallizing material is not available to fill the potential space, compatibility requires other modifications to take place in the fold geometry. The incompetent material might flow from the limbs towards the hinge zone (situations between Models D and E), or the competent material

bounding the potential space might undergo progressive flow into the developing void. Such **hinge collapse** is a common feature of many chevron folds and leads to somewhat surprising changes of curvature in the competent layers (Figures 20.24C and D, 20.25).

We have seen how variations of thickness of the competent layers in a folded multilayer can, under certain circumstances, give rise to polyharmonic folds. Where the larger wavelength folds have a moderate to large finite amplitude, the finite wavelength and amplitude of the smaller parasitic folds often vary systematically with position in the larger wavelength fold. This variation can often be attributed to the overall strains developed around the larger structure, with stronger shortening taking place on the inner arcs than on the outer arcs of the folds. Examples of such variations are shown in the model experiment forming the frontispiece to this session, and in Figure 20.8.

Folds developed under the influence of gravitational force

Gravitational force acting on a compressed multilayer can act as a strong constraint on the development of folds, particularly those of large wavelength. In general the buoyancy force in synforms and excess weight of the layer pile in antiforms tend to inhibit fold formation. The actual behaviour of any particular system is quite complex and depends upon the values taken by certain dimensionless parameters: the ratios of layer thicknesses ($d_1 : d_2 : d_3 \ldots$, etc.), the ratios of layer viscosities ($\mu_1 : \mu_2 : \mu_3 \ldots$), the ratios of the layer densities ($\varrho_1 : \varrho_2 : \varrho_3 \ldots$) and the proportion of sideways pressure force to the gravitational force (G). Although we do not have space here to examine all the various possibilities, it is instructive to introduce aspects of

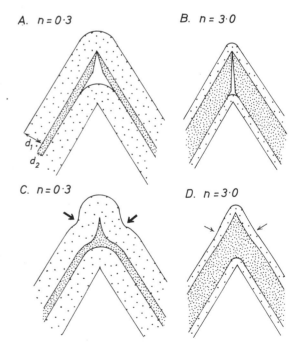

Figure 20.24. A and B *show the relationship of the size of the saddle reef to the bed thicknesses in a chevron fold (n = d_2/d_1).* C and D *show hinge collapse if saddle reef spaces do not develop. After Ramsay (1974).*

Figure 20.25. Chevron folds in sandstones and shales. The upper part shows a hinge collapse zone (cf. Figure 20.24C), the lower part shows the formation of a limb thrust in a particularly thick competent layer (cf. Figure 20.23C). Hartland, southwest England.

this behaviour. Consider a single compressed layer, thickness d, viscosity μ_1 and density ϱ_1 situated in a matrix of infinite extent and with viscosity μ_2 and density ϱ_2. The growth rate of folds with particular W_i/d values is shown in Figure 20.26A. At low values of sideways to gravitational force (ratio G) there is one maximum on the growth rate function K indicating stable folds with a large initial wavelength to thickness ratio (W_i'/d) of about 1500, and which does not greatly differ from the wavelength predicted from that of a system subjected to gravitational force alone (Ramberg, 1968a, b, 1970a, b, 1981). With an increase in the sideways force (and G value), a second maximum appears on the fold growth curves with a much shorter initial wavelength to thickness ratio (W_i''/d) of about 16, close to that of a system undergoing pure buckling (Figure 19.8B). The overall *single layer system shows two characteristic wavelengths*, a feature quite unlike that seen in single layer buckling without

gravitational influence as discussed in Session 19. As the influence of gravity becomes less marked (G large), so the short wavelength buckles become predominant in the system. A further point of great geological significance in this model is that the amplification rates of the larger wavelength folds are uneven and depend upon the density contrasts between the layer and its matrix. At the interface where a denser layer overlies a less dense layer, the synforms amplify faster than those at the interface where less dense material overlies more dense material (Figure 20.26B).

Ramberg (1970) has examined the mechanical properties of many other mutlilayers being buckled under the influence of gravitational force and with differing upper and lower boundary constraints. He has shown that, in general, where the density variation is "normal", that is where the lowest rock layers have a higher average density than the upper-most layers, the characteristic wavelength or wavelengths tend to be smaller than those in situations where buckling takes place without gravitational force. In contrast, where the density variation is inverted, the character-istic wavelengths are generally larger than those of pure buckling.

Anisotropy

So far we have considered folding in terms of material which has layers of differing thickness and of contrasting proper-ties. However, folds can form in materials which contain no specific layers but which do possess a structural and mech-anical *anisotropy*. A good example is a metapelite with a pre-existing tectonic cleavage or schistosity. An anisotropic material is defined as material which has different physical properties in different directions. It may be *homogeneous* if every sub-element is identical, or it can be *heterogeneous*, as is found in a layered rock made up of two or more com-ponents. The geological applications of the concept are very well set out in Cobbold *et al.* (1971) and in Cobbold (1976). Anisotropy implies that the material has differing properties in different directions, for example an anisotropic material has different reactions to compressive or shear stress depending on whether these stresses are applied parallel to or perpendicular to the main plane of anisotropy. For example, if a tensile stress σ_x is applied parallel to the ani-sotropy plane the material will flow at a slower rate $\dot{e}$ than if the stress is applied perpendicular to the plane (Figures 20.27B and C), while a shear stress τ applied parallel to the plane of anisotropy gives a greater shear strain rate $\dot{\gamma}$ than that applied perpendicular to the anisotropy plane (Figures 20.27D and E). Cobbold, Cosgrove and Summers showed how variations in the anisotropic properties could account for a wide variety of folds, types of shear zones and types of boudinage structure depending upon the orientations of the applied stress field to the axes of anisotropy. In general we are enthusiastic about this new look at the development of structures. However, we do not think that this approach supercedes the work carried out by many geologists who have investigated the nature of instabilities in terms of rock layering. Layering is a fundamental feature of so many rocks and, as we have seen above, many of the complex features shown by naturally formed folds seem best explained in terms of thickness and ductility variations of the layers.

Now return to Questions 20.2 and 20.3

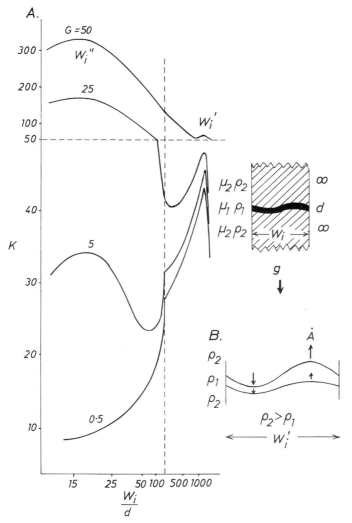

Figure 20.26. Fold development in a buckled single layer under gravitational force. K gives the relative amplification rates of folds with initial wavelength W_i. μ and ϱ are viscosities and densities, G is the proportion of buckling force to gravitational force g, $\dot{A}$ the fold amplification rate. After Ramberg (1970). Note the changes of scale of abscissa and ordinate at the dashed lines.

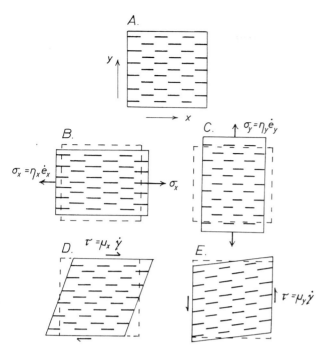

Figure 20.27. The behaviour of an anisotropic material (A) with applied normal stress σ parallel and perpendicular to the anisotropy plane (B and C) and applied shear stress τ parallel and perpendicular to the plane (D and E). η and μ are the normal and shear viscosity parameters.

Geometric development of chevron folds from conjugate kinks

Answer 20.2

Kink folds are always localized between well defined parallel sided or elongate lens shaped axial surfaces sometimes known as **kink planes**. When the primary layering, lamination or anisotropic rock fabric meets a kink plane it makes a sudden deflection or knick like deflection across this surface, generally adopting a regular planar orientation in the **kink band** and then passing across a kink plane defining the second boundary of the kink band to take up its original orientation on the other side of the band. Folds of this type are only found in materials which possess a strong, rather

homogeneous fabric, either a well developed and regular alternation of layers, or in rocks with an initial strongly developed tectonic fabric such as cleavage or schistosity. Although most kink folds show typical half wavelengths (i.e. kink band width) of from 1 mm to 10 cm (1–2 cm being especially common), giant kink bands with half wavelengths measured in hundreds of metres have been described by Collomb and Donzeau (1974). Kink bands are well known in deformed single crystals, such as phyllosilicates, kyanite, clinopyroxene, gypsum and calcite, which possess a well developed mineral cleavage or parting (Orowan, 1942; Turner and Weiss, 1965). In laboratory experiments with rock material kink bands are the dominant structure developed when slates, phyllites or schists are compressed parallel or sub-parallel to their cleavage planes (Donath, 1961; Anderson, 1964, 1974; Paterson and Weiss, 1966, 1968). They have also been produced in laboratory model experiments in compressed card decks (Weiss, 1968), in layered plasticene and artifically produced flaky aggregates (Bailey, 1969; Cobbold et al., 1971; Means and Williams, 1972; Latham, 1979). Individual kink folds usually occur in arrays of parallel bands. In many naturally deformed rocks and in laboratory experiments there are usually two differently oriented sets of kink band arrays producing **conjugate kink folds**, but where the kink planes are oriented sub-perpendicular to the folded surface often only **monoclinal kink folds** are found. In conjugate fold arrays the relative sense of deflection of the folded planar structure is right handed on one set of kink bands and left handed on the other. In conjugate folds the axial planes of each kind set make angles of ±30–70° with the planar structure. Two main types of conjugate arrays can be found (Figure 20.28), and various different names have been suggested for these. We favour the terms **contractional kink band** for Figure 20.28A, and **extensional kink band** for Figure 20.28B, because geometrically the first implies an overall regional shortening of the planar fabric, whereas the second relates to an overall stretching of the fabric. Other terms which have been used for contractional and extensional arrays are: *synthetic* and *antithetic* (Hoeppener, 1955); *reversed* and *normal* (Dewey, 1965) *negative* and *positive* (Dewey, 1969). The term kink band as used by most workers coincides with the contractional kink band array, and some have objected to the use of the term kink band for the extensional array, preferring the more general term *shear band* or *shear zone*

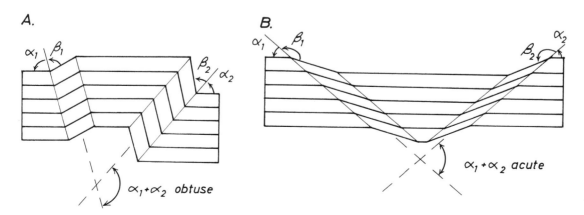

Figure 20.28. Geometric features of contractional (A) and extensional (B) kink bands.

(Weiss, 1969). One particular difference between the contractional and extensional arrays is that in the first the angular conditions of the layering inside and outside the kink zone are that $\alpha \approx \beta$ (Figure 20.28A), whereas in the second type $\beta \gg \alpha$ (Figure 20.28B). These angular differences of the folded surfaces inside and outside the kink band probably do relate to different mechanical constraints of the two types of kink structure. In contractional kink bands the angular relations are usually given by $\alpha \leqslant \beta < 90°$ (Figure 20.29), whereas in extensional kinks $\alpha < 90°$, $\beta > 180° - \alpha$. In practically all natural and experimentally produced conjugate kink folds the angle between the two differently oriented kink band arrays which faces the overall shortening direction is obtuse and that which faces the overall stretching direction is acute (in contractional kink bands $\alpha_1 + \alpha_2 > 90°$; in extensional bands $\alpha_1 + \alpha_2 < 90°$), a feature which is the reverse of the relationships seen in conjugate faults developed during brittle rock fracture (Question 9.5 and Session 25). The axial surfaces of kink planes of conjugate kinks may show mirror symmetric relationships to the folded surfaces. Such **symmetric conjugate kink folds** often show the relationship $\alpha_1 = \alpha_2 \approx 60°$. **Asymmetric conjugate kink folds** are those showing $\alpha_1 \neq \alpha_2$ (Figures 20.19, 20.28A). Laboratory experiments with models and rocks suggest that symmetric conjugate folds arise when the principal compressive stress is aligned parallel to the folded surfaces and that asymmetric relationships develop when this stress direction is obliquely inclined to the surfaces. In asymmetric conjugate folds it has been shown that the axial surfaces of the initially formed folds may be symmetric to the surfaces but that these folds become progressively more asymmetric as deformation proceeds (Cobbold *et al.*, 1971, and the discussion on pro-

gressive deformation and fold development in Session 14, pp. 227–229). In many naturally deformed rocks the degree of development of the individual sets of kink bands is not uniform, one set being more abundant than the other (Anderson, 1968, and Figure 20.29). This feature probably arises because of differing bulk strain constraint imposed during the deformation. For example if only one set of kinks develops the principal directions of the overall bulk strain are markedly oblique to the unkinked sectors of the surfaces, whereas if both sets of kinks develop the principal directions are parallel and perpendicular to the general trend of the surfaces (Figure 20.30).

After this initial general discussion we will now discuss in detail the features posed in Question 20.2. For each of the layers a, b, c and d of Figure 20.5 the horizontal sections are of equal length, while the sectors in the kink bands with dips of $\pm 60°$ are also of equal length $(d_1 + d_2)/\sin 60°$. The total volumes of the rock layers bounded by surfaces ab, bc and cd are also equal. The geometrical features of this model show perfect compatibility. The total shortening depends upon the local shortening that has taken place in the two kink bands $(\delta l = -(d_1 + d_2)/\sqrt{3})$ as related to the total layer length in the kinked sectors $(2(d_1 + d_2)/\sqrt{3})$ and unkinked sectors $(x_1 + x_2 + x_3)$. In our example $e = \delta l/l_0 = -0·12$. If the kink bands had been developed to their maximum potential so as to eliminate all horizontal unkinked sectors *the maximum shortening in a conjugate fold system* would be 50% $(e = -0·5)$. Figure 20.31A shows the geometry of the conjugate kink folds produced by the migration of kink band 2 through kink band 1. The structure becomes a *symmetric conjugate kink fold with unevenly developed kink domains*. The **chevron fold** part of the structure, formed where kink bands 1 and 2 intersect, becomes more elongate, but note that its original width is controlled by the dimensions of kink band 1 and remains constant $(d_1/\sqrt{3})$. Inside each simple kink band the displacements take place by simple shear parallel to the folded surfaces. Relative to these surfaces the shears have values $\gamma = -\sqrt{3}/2$ in kink band 1 and $\gamma = +\sqrt{3}/2$ in kink band 2. These shear displacements induce strains with R values of 2·38 and with orientations of the principal axes of $+33°$ and $-33°$ to the folded surfaces.

The limbs of the chevron fold part of the structure have been evolved in two ways. Those limbs which have a left hand $(+ve)$ shear along the surfaces are essentially parts

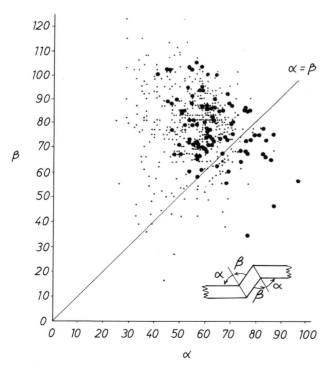

Figure 20.29. *Plot of angles α and β for 513 contractional kink folds; small dots right handed, heavy dots left handed kinks. Ards Peninsula, N. Eire. After Anderson (1969).*

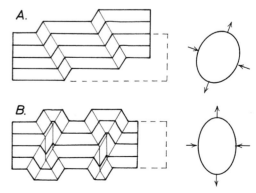

Figure 20.30. *Difference in bulk strain with A single contractional kink folds and B evenly developed conjugate kink folds.*

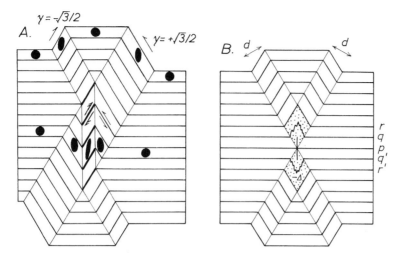

Figure 20.31. *Answers to Question 20.2 (A) and 20.3 (B). A illustrates the finite strain variation in different parts of the conjugate fold model arising from the differently directed and differently valued shear strains γ parallel to the kinked surfaces. B shows the sectors in which compatibility of surface lengths is no longer possible (stippled) and which have a negative dilatation −Δ.*

only of kink band 2. In the region between the two vertical axial surfaces (marked with heavy lines in Figure 20.31A) the chevron fold limb has suffered shear during the formation of kink bands 1 and 2. Along this surface the shear strain is twice that of kink band 1 ($\gamma = -\sqrt{3}$). Although this chevron fold has an axial surface bisecting the fold limbs, one limb has twice as much shear and therefore a differently oriented and valued strain state ($R = 4.8$, principal strains oriented 25° from the folded surface). Not all chevron folds found in naturally deformed rocks are evolved at the intersection of conjugate kink bands, but those that are formed in this special way may be recognized by having certain small scale structural features more developed on one limb than on the other in accord with the strain variation of this model.

If an array of parallel kink bands of differing widths undergoes a reactivation as a result of the later development

of a conjugate array of kinks, the chevron folds so formed will have differing limb lengths proportional to the spacing distances of the first formed kinks. Initially wide first kink bands will produce long limb sectors on the chevron sector, while the smallest first kinks will appear as small **parasitic folds** on the limbs of the larger structure (Figure 20.32). This is a way of producing the geometric features of parasitic folds in a way that is quite independent of variations in layer thickness and ductility.

Synchronous development of two kink bands

Answer 20.3

The geometric features of the meeting or crossover positions of conjugate kink bands are especially important if we wish

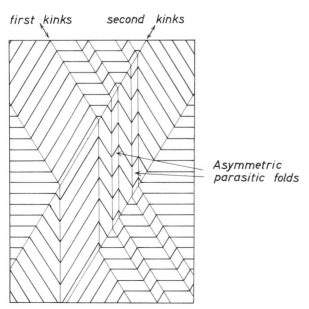

Figure 20.32. *The development of asymmetric parasitic folds as a result of intersecting conjugate kink folds of differing widths.*

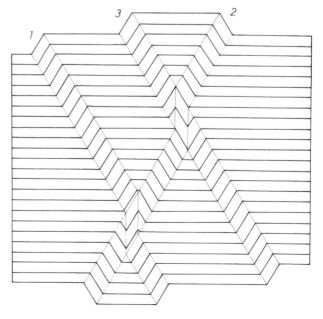

Figure 20.33. *Sequential development of kink domains 1, 2 and 3 in a conjugate kink fold system.*

to understand the nucleation and propagation mechanisms of kink band systems. It is a region where strain compatibility problems may exist and where very complex structures may be found. In the previous question we saw how one kink may be active and offset another which is passive, providing that certain geometric features exist in the system. This is the type of behaviour observed in many laboratory experiments and the geometric features have been thoroughly discussed by Paterson and Weiss (1966, 1968) and Weiss (1968). Where the kink arrays are symmetric with $\alpha = \beta \approx 60°$ the internal surfaces of one kink may be smoothly deflected by another provided that only one of the kink zones is active. Such a geometric interaction is found in many natural systems, the overall total fold pattern suggesting changeovers of the activity of individual kinks of the two intersecting arrays (Figure 20.33, development sequence kink 1, followed by kink 2, followed by kink 3). If the crossing kink bands do not act sequentially, geometric problems arise.

Figure 20.31B shows the completed geometry of the simultaneously developing kink bands. In the regions of the conjugate fold structures at the top and bottom of the block it is possible to accommodate the horizontal shortening easily, without recourse to modifications of layer lengths or changes in area of the fold profile. In the central part of the model (stippled in Figure 20.31B) geometric compatibility problems arise. The centrally situated layer p has a layer length that is $2d/\sqrt{3}$ too short (the two kink bands have widths of d units). The lines (q and q') adjacent to this central line have lengths that are also $d/\sqrt{3}$ too short. It is only when we come to lines r and r' that the line lengths are compatible with the zone outside the crossover sectors. The only way we could adjust the line lengths of p, q and q' to conformity would be to take up the slack, perhaps by forming small scale chevron folds with interlimb angles smaller than 60°, together with a detachment or **décollement** of the two sides of line p at the central crossover point. Another feature of the geometry of this crossover sector is that two diamond shaped stippled regions are derived from an area in the undeformed material that was 1·5 time as large, implying that this sector must suffer a volumetric dilatation Δ of −33%. These necessary changes in layer lengths and volumes show that severe compatibility problems exist if two crossing arrays of kink bands are synchronously active.

Mechanisms of nucleation and propagation of conjugate kink folds

Laboratory experiments are particularly useful in helping us to gain an insight into how kink bands initiate and develop with increasing shortening. The stress–strain curves (Figure 20.34) of rocks compressed parallel or sub-parallel to the anisotropy or planar fabric show an initial section (a) where strain builds up slowly with increasing stress, a typical elastic stress–strain relationship. At some critical stress difference (b) and generally where the overall strain is around 1–2% there is a sudden stress drop (c). With increasing strain the stress–strain curve becomes much flatter (d), but it is also associated with small stress drops (e) leading to further flat sectors (f). When the rock specimens are examined after unloading it is found that the first marked stress drop coincides with the initiation of the first kink and

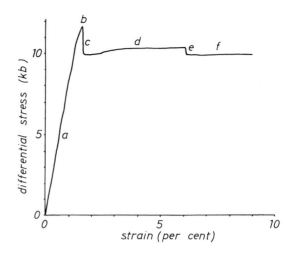

Figure 20.34. Typical stress–strain curve during the experimental development of kink bands in slate (7·24 kb confining pressure). After Anderson (1974).

that the subsequent smaller stress drops mark the initiation of new kink bands either parallel to or conjugate with the first. This correlation of stress drop and kink band initiation is not perfect. It is generally found that the number of kink bands in the rock exceed the number of observed stress drops. Although it is not possible to observe directly what takes place in the rock during the flatter sections of the stress–strain curves, the most likely explanation is that previously formed kinks are taking up the strain by widening, while the sectors between the kinks are being deformed elastically and taking up energy that will be released suddenly during the development of a new kinking event. The initiation of kink bands appears to be a rapid, almost catastrophic, event. In card deck experiments performed by Weiss (1968) the initial stress drop accompanying the rapid nucleation and propagation of the first kinks was accompanied by a loud noise as the result of the release of strain energy.

In some experiments with rocks and model materials the first kink band forms at the corners of the specimen, presumably as a result of the heterogeneous stresses and strains around these positions. Although natural rock situations do not possess "corners", there are often stress risers; irregularities and inhomogeneities in the material resulting from variations in the proportions of different crystal species making up the rock, or geometrical irregularities in the layering or anisotropy in the form of fold-like undulations. Probably the first distortions take place in the form of elastic buckling around an imperfection in the rock. As this buckle develops the shear stress increases on the fold limbs (Figure 20.35A, $\tau = (\sigma_1 - \sigma_3)\sin 2\theta/2$, see Appendix E). At some critical dip value θ this shear stress is sufficient to overcome the mechanical resistance to shear on the buckled surface. All the elastic energy concentrated in the buckle is then transformed into mechanical work in the narrow zone of the first kink. The kink sector rapidly rotates, with strong finite shear strain taking place on the rotating surface (Figure 20.35B). When the surface reaches a position to make an angle of 45° with the principal stress direction the shear stress acting on it attains a maximum value. Providing that the progressive development of the kink does not completely take up the stress difference ($\sigma_1 - \sigma_3$) in the system the

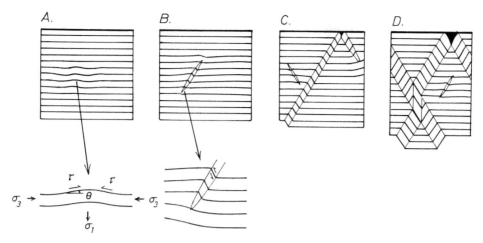

Figure 20.35. Progressive development of conjugate kink folds in a material with one confined boundary (above) and one free boundary (below). σ_1 and σ_3 represent the maximum and minimum normal tensile stresses (minimum and maximum compressive stresses respectively) and τ is the shear stress developed on the limbs of the initial buckle making an angle θ with the σ_3 direction.

surfaces in the kink band continue to rotate, but the shear stress decreases and, at the same time, the incremental shear strains increase. The surfaces in the kink band therefore lock at some angle exceeding $\theta = 45°$. Whether the kink band has a right hand or left hand shear sense depends upon the chance factors of the symmetry of the initiating mechanical imperfection; theoretically both are equally probable where the principal stresses act parallel and perpendicular to the initial surfaces undergoing kink folding. The first kink zone will be an elongate lens shaped zone, the central part of which will, for reasons of volumetric compatibility, show the conditions $\alpha = \beta$ (Figure 20.35B). At the tips of this zone geometric compatibility problems exist because the surfaces ahead of the tips are still more or less planar, and a high stress concentration will exist at these tips. These factors lead to a rapid forward propagation of both kink band tips to stabilize the kink band geometrically and mechanically into a long zone of narrow width and showing geometrical

conformity throughout its length (Figure 20.35C). If the propagating kink band meets an obstacle to its advance, for example a layer of high competence, it may be reflected of this surface to form a conjugate kink zone (Fig. 20.35C and Currie *et al.*, 1962; Weiss, 1969, Cobbold, 1976). Where this occurs there are compatibility problems along the interface between the kinking rock and the reflector, and **décollement** may occur. Further shortening along the kinking surfaces takes place either by sideways migration of the axial surfaces of individual kinks in the way we discussed in Question 20.2 or by the initiation of new parallel or conjugate kinks (Figure 20.35D). Mechanically it is generally easier for the material to take up further strain by the first of these processes, with the surfaces external to the layering being "tipped into" the kink band as the axial surfaces migrate. Conjugate kinks probably result because of the increasing asymmetry of the body as a whole to the applied stresses as discussed earlier (Figure 20.30).

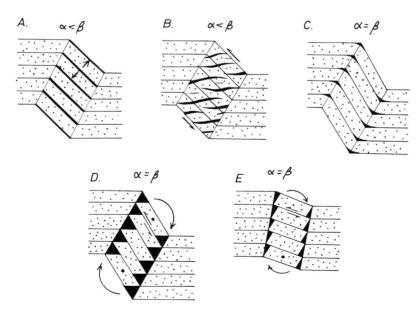

Figure 20.36. Types of dilational openings arising during kink band formation: A *perpendicular to anisotropy:* B *en-echelon extension veins:* C *saddle reef structure:* D *rotation spaces;* E *rotation spaces in high angle joint drags.*

We have seen how during the development kink zones strain is localized inside the kink. The strains set up in the kink zone are generally produced by simple shear parallel to the kinked surfaces. If these surfaces represent a rock anisotropy, then the kink zone geometry can be accomplished with perfect compatibility providing that the angles α and β between the surfaces and the kink band axial planes are equal. It is a commonly observed feature that in many naturally and experimentally formed kink bands $\alpha < \beta$ (Figure 20.29), and this leads to a positive volume change inside the kink (Ramsay, 1967) according to the relationship:

$$\Delta_v = \sin \beta / \sin \alpha - 1 \qquad (20.9)$$

and to the development of vein openings. Volume increase may develop by the formation of spaces parallel to the anisotropy surfaces (Figures 20.36A, 20.37) or by the opening of en-echelon extension veins cross cutting the kinked anisotropy surface (Figures 20.36B, 20.38). If the simple shear displacements are localized on certain of the kinked surfaces there is a rotation of those block-like elements of little or no shear inside the kinked zone, and the development of triangular openings along the kink planes as a result of detachment of the blocks from the unkinked wall (Figures 20.36D, 20.39). This phenomenon can also occur in the monoclinal types of kink bands which have axial surfaces sub-perpendicular to the external surfaces (and often with $\alpha < \beta$) known as **joint drags** (Figure 20.36E). Probably joint drags do not develop in the same way as we have

discussed above for conjugate kink folds but form from a physical rotation of a zone of rock lying between two pre-existing fractures or joint surfaces. An additional type of dilational opening can occur where conjugate kink bands are developed in a uniformly layered rock by the formation of **saddle reefs** between the layers (Figure 20.36C). This dilation is significantly less than that formed in true chevron folds (Ramsay, 1967, p. 454).

In laboratory experiments it appears that the preferential formation of kink bands over other types of structure is a function of confining pressure (Figure 20.40). At very low confining pressures maximum compression parallel to the anisotropy surfaces leads to a splitting and opening of the material along these surfaces. At higher confining pressures rock deformation proceeds with the formation of single or conjugate shear fractures or faults crossing the anisotropy, and at confining pressures above three kb the kink band deformation mode predominates. The controls of kink band width are less clearly established in laboratory experiments. The concept of sideways migration of kink band boundaries predicts that band width should increase with bulk strain. The experimental results are rather conflicting. Paterson and Weiss (1966), Anderson (1974) and Weiss (1969, 1980) have produced confirmatory experiments. In contrast Donath (1961) found no such relationship, his results showing that band width increased with decrease in confining pressure.

Now return to Question 20.4★ or read the comments below on crenulation cleavage and advance to Session 21.

Figure 20.37. *Thin section of kink band in slate with dilation spaces opening along anisotropy surfaces (cf. Figure 20.36A). (×100). Lake district, Cumbria, England.*

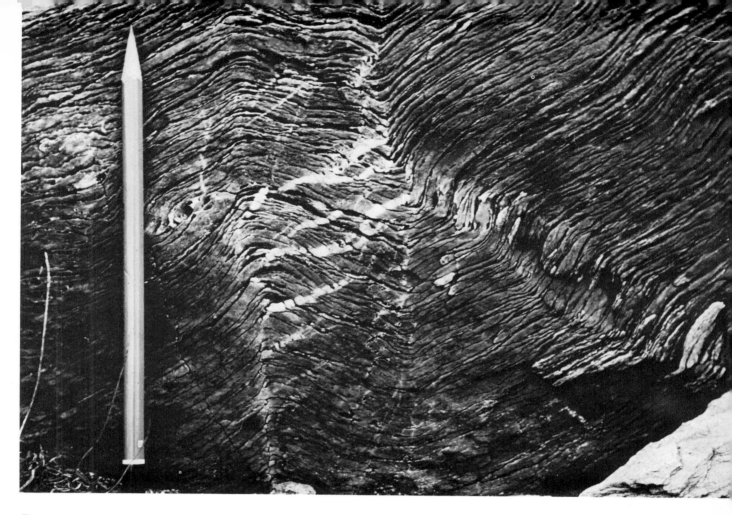

Figure 20.38. *Kink bands with en-echelon quartz filled extension gashes (cf. Figure 20.36B). The anisotropy surface being folded is a pre-existing slaty cleavage accentuated by pressure solution stripes. Rhoscolyn, Anglesey, North Wales.*

Figure 20.39. *Joint drag with triangular quartz filled spaces arising from rotation of kinked sector. Tintagel, Cornwall, England.*

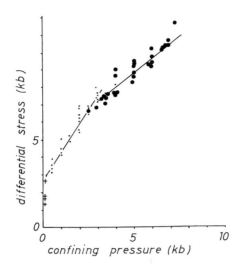

Figure 20.40. *Kink band formation and confining pressure in experimentally deformed slates. Crosses represent experiments where openings occurred along cleavage, small dots where cross cutting fractures formed and heavy dots where conjugate kink bands developed. After Anderson (1974).*

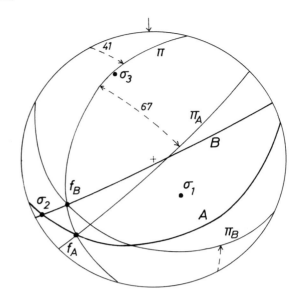

Figure 20.41. *Equal area projection of data of Question 20.4★. π, π_A, π_B give orientations of the cleaved outside and inside kink bands A and B respectively, f_A and f_B are the two fold axis directions, and σ_1, σ_2 and σ_3 are the three principal stress directions.*

Kink band geometry and stress axes

Answer 20.4★

Figure 20.41 shows a projection plot of the data. From this plot we calculate:

Fold axis in kink band A	20–228°
Fold axis in kink band B	26–245°
Angle between kink zone axial surfaces	49°
Interlimb angle in kink band A	67°
Interlimb angle in kink band B	41°

Until now we have considered the geometry of conjugate kink folds from a two-dimensional viewpoint where the fold axes of the two kink arrays are parallel and oriented perpendicular to the profile plane (Figure 20.42A). In this field example we see that the geometry of the conjugate folds, although looking fairly "normal" on the rock surface, is not so simple in three dimensions, the axes of the folds in kink A and B being oblique to each other on the folded cleavage surface. This situation (Figure 20.42D) is quite common in many natural conjugate fold systems (Figure 20.43), and the geometry implies that the intersection of the two conjugate axial surfaces does not lie on the surface undergoing folding.

Crossing kink folds on a single surface might be interpreted as resulting from the interference of two quite different generations of kink forming events. In Figure 20.43 the crossing kinks produce a triangular area of cleavage surface between the two kink bands on the right-hand side of the outcrop which has been displaced away from the observer, and a similar triangular area on the left-hand side which has been displaced towards the observer (cf. Figures 20.42D and E). A narrow dark kink band above the main dark band comes to a stop on the light kink, and in nearby outcrops the reverse relationship can be found. The two arrays therefore appear to be broadly contemporaneous. This geometry shows that the overall maximum extension does not lie perpendicular to the surface undergoing folding and this implies that the direction of minimum stress was oblique to the surface. The next step is to see if we can determine the orientations of the principal stresses at the time of kink formation.

In general geometric terms kink bands show the following characteristics with reference to the surface s outside the kink domains (Figure 20.42):

A. Symmetric conjugate folds with parallel fold axes
B. Asymmetric conjugate folds with parallel fold axes
C. Monoclinal kinks at high angle to s (other shear inhibited—parallel to s?)
D. Symmetric conjugate folds with crossing fold axes
E. Asymmetric conjugate folds with crossing fold axes

The exact relationships of the directions of the axes of principal stress to the fold geometry is not completely defined, except for the special symmetric condition (A), but noting the general shortenings and lengthenings implied by the fold shapes we suggest the following relationships of the stress axes to the surface s (Table 20.1).

It has been suggested that it might be possible, using the orientations of the two sets of axial surfaces and the relative sense of displacement implied by the kink bands, to determine exactly σ_1 (bisector of lengthened sector—generally acute angle), σ_2 (intersector of axial surfaces) and σ_3 (bisector of shortened sector—generally obtuse angle) (Ramsay, 1962). Although this technique probably does give reasonably accurate results ($\pm 15°$) that can be used geologically, there are good theoretical reasons and some experimental evidence that the orthorhombic symmetry of the axial surfaces of the two kink sets does not exactly coincide with the orthorhombic symmetry of the triaxial system producing the folds.

From the field observations plotted in Figure 20.41 we deduce the following directions of principal stress: σ_1 61–145°, σ_2 7–246°, σ_3 27–339°.

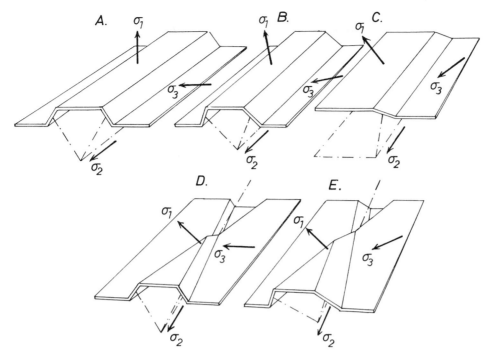

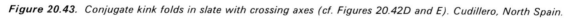

Figure 20.42. The main types of conjugate kink band geometry and the directions of principal stresses σ_1, σ_2 and σ_3 producing them. Dash–dot lines show the main kink band orientations.

Figure 20.43. Conjugate kink folds in slate with crossing axes (cf. Figures 20.42D and E). Cudillero, North Spain.

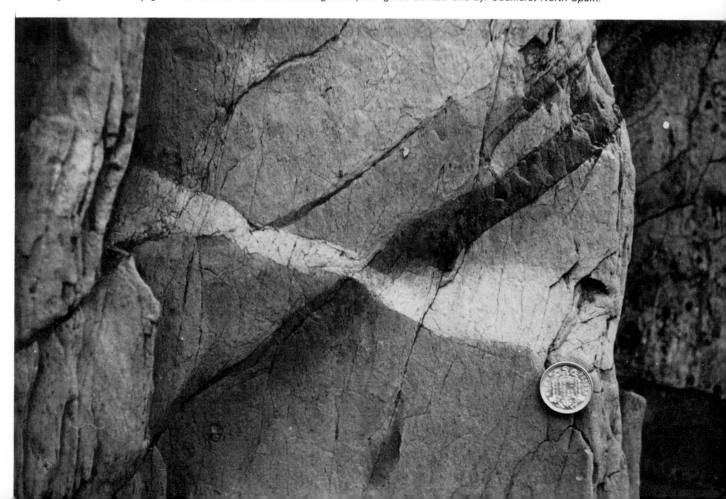

Table 20.1
(see Figure 20.42)

Mimimum compression σ_1	Intermediate stress σ_2	Maximum compression σ_3
A. Perpendicular to s	Parallel to fold axes	Parallel to s
B. High angle oblique to s	Parallel to fold axes	Low angle oblique to s
C. Moderate angle oblique to s	Uncertain	Moderate angle oblique to s
D. High angle oblique to s	Oblique to s, parallel to intersection of kink axial planes	Parallel to s
E. High angle oblique to s	Oblique to s, parallel to intersection of kink axial planes	Low angle oblique to s

Crenulation cleavage

The development of fine scale microfolding can produce rather systematic realignments of layering or pre-existing fabrics into planar and subplanar orientations that often show relationships to larger scale fold forms similar to those we encountered when discussing slaty cleavage and schistosity. One very common type of fabric produced in this way is known as **crenulation cleavage** or, in some older literature, as strain slip cleavage. This structure is formed by the development of regular microfolds in some pre-existing very fine sedimentary or tectonic lamination, or more usually in a pre-existing tectonic fabric such as cleavage or schistosity. The microfolds generally have forms close to the similar fold model and may show symmetric forms (Figures 20.44 and 20.45) or asymmetric forms (Figure 20.48, 20.49).

The overall similar form implies that there are changes of spacing distance between the original laminations or original cleavage planes such that there is a relative widening in the crenulation fold hinges and a reduction in the fold limbs (Figures 20.45, 20.48). The crenulation cleavage structure often gives a striped appearance to the rock because of the regular alternations of limbs and hinges. When hit with a geological hammer the rock usually breaks preferentially along the fold limbs where the platy minerals have been brought into sub-parallel alignment by the folding process. Crenulation cleavage is termed a **non-penetrative structure** (cf. slaty cleavage and schistosity) because of the existence of these discrete cleavage surfaces. The discrete nature of the cleavage is often enhanced by pressure solution processes: it is very common to find that the fold limbs are

Figure 20.44. Fine scale crenulation cleavage in slate. Rheinisches Schiefergebirge, North Germany.

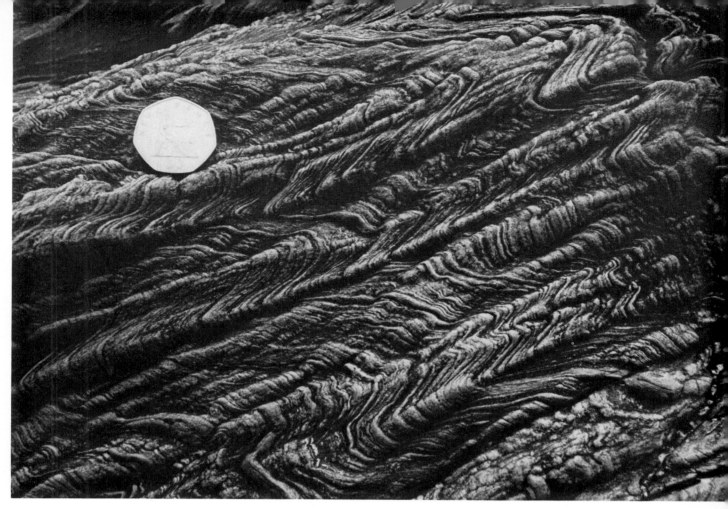

Figure 20.45. *Coarse symmetric crenulation cleavage in a laminated and cleaved meta semi-pelite. Note the strong pressure solution zones (dark) in the limbs of the folds and the light coloured fold hinges where quartz has concentrated. Rhoscolyn, Anglesey, North Wales.*

Figure 20.46. *Typical appearance of the hinge zones of crenulation cleavage microfolds in schist. Col de Galibier, West Alps, France.*

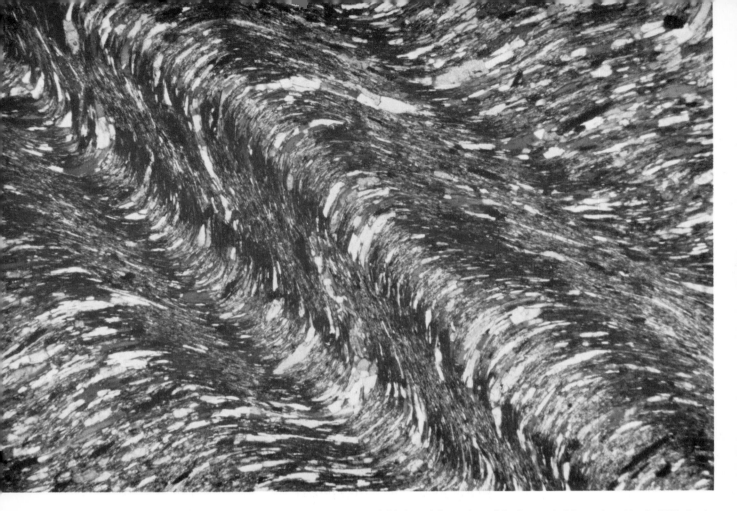

Figure 20.47. *Thin section through crenulation cleavage folds in a deformed amphibolite grade Mesozoic schist (×100). In the limbs of the folds the proportion of quartz has been reduced by the pressure solution process. Lukmanier Pass, central Swiss Alps.*

Figure 20.48. *Coarse asymmetric crenulation cleavage developed in the limb of a major fold (cf. Figure 20.45 in the hinge zone of a fold). Rhoscolyn, Anglesey, North Wales.*

Figure 20.49. Thin section of asymmetric crenulation cleavage in a chlorite magnetite schist. Unst, Shetland Islands, Scotland.

relatively enriched in "insoluble" phyllosilicate minerals and depleted in "soluble" components such as quartz and calcite (Figure 20.47). The structure is therefore often associated with a **tectonically formed striping** or banding as a result of the formation of pressure solution bands and stylolites in the fold limbs (Figures 20.45, 20.48). Although occasionally the axial surfaces of adjacent crenulation cleavage folds may show convergences and divergences that superficially resemble some aspects of kink folds (Figure 20.49) it is only rarely that they are developed as crossing conjugate cleavages.

The spacing of the cleavage planes is clearly a function of finite fold wavelength. In coarse grained rocks, such as schists and mica rich gneisses the spacing distance between fold limbs can be quite wide (c. 1 cm), but in fine grained phyllites, slates and shales the crenulation spacings are generally much smaller (c. 1 mm) and the nature of the fold structure is best appreciated with a hand lens or microscope.

The surface undergoing folding to produce crenulation is usually characterized by a fine crinkling which represents the hinge zones of the microfolds. On a dull cloudy day these may be passed unnoticed in the field, but on a bright sunny day the different degrees of reflection produced by the differently oriented micas can be very striking (Figure 20.46). The axial directions of the folds may be parallel to the axes of large scale folds produced by the same deformation, but this parallelism is not inevitable. Because crenulation cleavage is usually formed on a pre-existing tectonic cleavage, which generally shows cross cutting relationships to the bedding planes, the bedding and early cleavage have the potential of undergoing folding in different axial orientations. The geometrical constraints involved here will be discussed in more detail in Session 22.

A simple model for the development of crenulation cleavage is illustrated in Figure 20.50. This shows the buckling of a single competent layer surrounded by an incompetent material possessing a strong anisotropy. Above the competent layer the geometry of the small scale crenulation is shown, below it the finite strain trajectories are indicated. At the start of the crenulation cleavage process (A) shortening parallel to the anisotropy are likely to produce microfold instabilities in the inner arcs of the competent layer folds. As shortening proceeds (B) these folds will continue to amplify and their wavelengths will be reduced. Evidence from naturally formed folds suggests that the initial fold limbs change their orientations with increasing shortening by line rotation. Although they may have initiated close to the early XY planes of the finite strain ellipsoids, they are unlikely to coincide with the XY planes at high strains. The limbs will, however, be close to the XY planes, so it is often seen that the overall fan-like forms of the crenulation cleavages have geometric similarities with the cleavage fan forms of slaty cleavage and schistosity (Figures 20.51, 20.52). As with slaty cleavage, there are often locations on the outer arcs of competent layers where the intensity of crenulation cleavage is much lower than in the rock as a whole, owing the localisation of regions of low finite strain in these positions (Figures 20.50B and 20.53). In contrast, the inner arcs of the more competent layers often show regions of particularly intense development of crenulation cleavage with the attendant pressure solution sometimes completely overprinting the initial layering with a new tectonic striping (Figure 20.53).

Regions of low grade metamorphism are the characteristic environments for the formation of crenulation cleavage; greenschist facies being typical. In general, crenulation cleavage is found in orogenic regions where the deformation and metamorphic activity is waning. Although the metamorphic peak has generally been passed it should be emphasised that crenulation cleavage can lead to quite severe mineralogical reconstructions and redistributions, and often the most intense tectonic striping can be produced at this deformation stage.

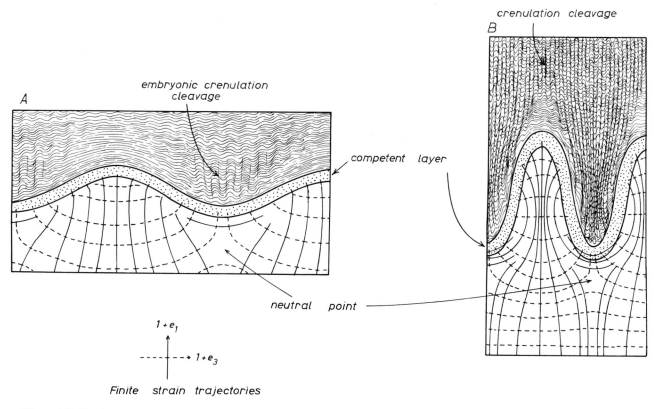

embryonic crenulation
cleavage

A

crenulation cleavage

B

competent layer

neutral point

$1 + e_1$

$1 + e_3$

Finite strain trajectories

Figure 20.50. Scheme of development of symmetric and asymmetric crenulation cleavage in a material with a strong initial fabric as a result of buckling of a competent layer.

Figure 20.51. Crenulation cleavage with variable orientation developed in an amphibolite grade metapelite containing a sandstone band. Compare the cleavage variations with those of Figure 20.50. Columkille, Western Donegal, Eire.

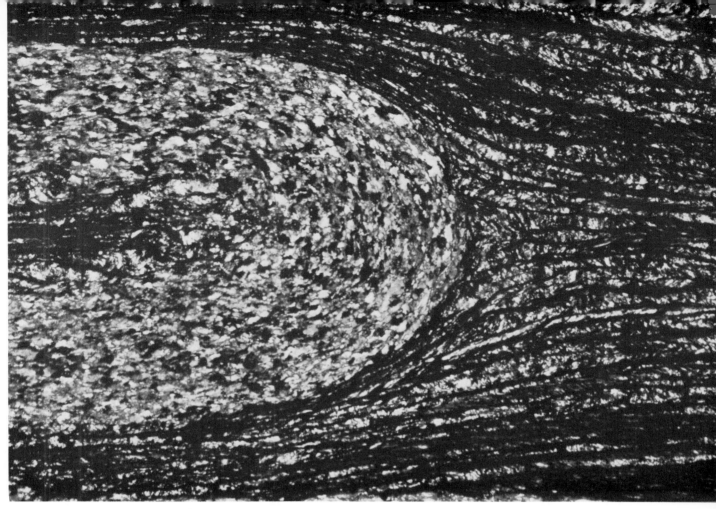

Figure 20.52. *Thin section of crenulation cleavage developed around a buckled siltstone layer in a slate (×100). Note the divergent cleavage fan and the dark pressure solution zones between the crenulation folds. Hope Cove, South Devon, England.*

Figure 20.53. *Folded clastic carbonates and shales showing the development of cleavage fans and pressure solution striping. Blue Ridge Village, Appalachian mountains, U.S.A.*

KEYWORDS AND DEFINITIONS

Chevron fold
Kink fold

A **Chevron fold** (zig zag fold, accordion fold) is a symmetrical fold with straight parallel limbs of nearly equal length and sharp, narrow hinge zones. **Kink folds** have similarly planar limbs and very angular hinges but are markedly asymmetric. The zones defined by their short limbs make up the **kink bands**. The axial surfaces are referred to as **kink planes**. If the kink bands are oriented sub-perpendicular to the folded surface **monoclinal kink folds** are developed, if two differently oriented sets of kink bands intersect each other **conjugate kink folds** are formed. According to their geometric style **contractional kink bands** (synthetic, reversed, negative kink bands) and **extensional kink bands** (antithetic, normal, positive kink bands) are distinguished (Figure 20.28).

Crenulation cleavage

Cleavage developed in a rock with strong preexisting lamination or tectonic fabric (e.g. schistosity), which is transformed by small scale folding into a regular alternation between wide hinges and tight limbs. Often the discrete nature of the crenulation cleavage is enhanced by pressure solution processes which lead to a concentration of "insoluble" components (e.g. phyllosilicates) in the fold limbs.

Harmonic fold

In a folded multilayer where the distance between competent layers is small and these layers have about the same thickness, a regular spacing and similar competent–incompetent layer ductility contrasts, **harmonic folds** (Figure 20.1C) will form. They show a general correspondence of all folded layers in wavelength and symmetry. If a marked difference in thickness or ductility contrast occurs then **polyharmonic folds** will form (Figure 20.1D). In **disharmonic folds** (Figure 20.1B) each layer has its own characteristic wavelength due to the wide separation distance between the adjacent competent layers.

Saddle reef

Special type of triangular shaped vein found in the incompetent layers situated at the hinges of chevron folds (Figure 20.15).

KEY REFERENCES

There is an exceedingly large literature on the mechanisms of folding in multilayers and it has been difficult to make an adequate selection to cover all the points mentioned in this session. In making our choice we have been guided by the following criteria. First the mathematical methods should be understandable by the average undergraduate student. The analytical background necessary to understand many important aspects of folding mechanics is likely to be beyond many readers at this stage, and for those wishing to look over this work we recommend that those references in the source list be investigated together with the named references in this session. Our second consideration in making our choice was that only the publications which fully enter into the geological significance of the results be considered as of key interest. Finally, we have kept this list as short as possible because we do regard it of extreme importance that students should get into the habit of reading the research literature. We could easily have made a list of 50 or 100 key references, but then we should have needed an extensive guide to this list if it was to be of true practical worth.

Multilayers, general

Cobbold, P. R., Cosgrove, J. W. and Summers, J. M. (1971). Development of internal structures in deformed anisotropic rocks. *Tectonophysics* **12**, 23–53.

Cobbold, P. R. (1976). Mechanical effects of anisotropy during large finite deformations. *Bull. Soc. géol. France* **7**, 1497–1510.

These two papers develop the concept of rock anisotropy, showing how a material with a banded structure or a homogeneous internal anisotropy can become unstable during deformation. They are of particular relevance to studies of chevron, similar and kink folds.

Cobbold, P. R. and Watkinson, A. J. (1981). Bending anisotropy: a mechanical constraint on the orientation of fold axes in an anisotropic medium. *Tectonophysics* **72**, T1–T10.

This offers a fascinating new insight to the linear features of a rock fabric and how this might play an important role in guiding fold geometry. This concept might provide a particularly important mechanical constraint in the deformation of rocks which already possess a l-s fabric derived from a previous deformation.

Ramberg, H. (1961). Contact strain and folding instability of a multilayered body under compression. *Geol. Rundschau* **51**, 405–439.

Ramberg, H. (1963). Fluid dynamics of viscous buckling applicable to folding of layered rocks. *Am. Assoc. Petrol. Geol. Bull.* **47**, 484–505.

Ramberg, H. (1964). Selective buckling of composite layers with contrasted rheological properties, a theory for simultaneous formation of several orders of folds. *Tectonophysics* **1**, 307–341.

Hans Ramberg has played the leading role in the development of the mechanical theories of fold formation. The mathematics in these selected papers is not easy at this stage of our session presentations, but here Ramberg points out so many of the implications of the theory to the field worker that the student should at least look at the results even if the mathematics is beyond his capabilities.

Ramberg, H. (1968). Instability of layered systems in the field of gravity I, II. *Phys. Earth Planet. Interiors* **1**, 427–447; 448–474.

Ramberg, H. (1970). Folding of compressed multilayers in the field of gravity I, II. *Phys. Earth Planet. Interiors* **2**, 203–232; **4**, 83–120.

These four papers can be highly recommended to the advanced student, well versed in mathematics. At the present time they probably represent the clearest mathematical statement of the problems, and illustrate the implications with experimental results and natural examples.

Chevron and kink folds

Anderson, T. B. (1974). The relationship between kink bands and shear fractures in the experimental deformation of slate. *J. Geol. Soc. Lond.* **130**, 367–382.

Although this paper mostly describes the results of laboratory experiments the results are presented in a way which clearly reveals their geological significance. This is one of the best discussions of geologically relevant experimental work.

Collomb, P. and Donzeau, M. (1974). Relations entre kink-bands décamétriques et fractures de socle dans l'Hercynien des Monts d'Ougarta (Sahara occidental, Algérie). *Tectonophysics* **24**, 213–242.

We tend to think of kink bands as essentially small scale structural features of rocks. This publication describes giant kink bands of mountain scale dimensions.

Dewey, J. F. (1965). Nature and origin of kink bands. *Tectonophysics* **1**, 459–494.

Dewey, J. F. (1969). The origin and development of kink bands in a foliated body. *Geol. J.* **6**, 193–216.

These two papers can be recommended as providing excellent accounts of the often complex geometrical features of kink bands in naturally deformed rocks, with many suggestions as to their kinematic significance.

Paterson, M. S. and Weiss, L. E. (1968). Folding and boudinage of quartz-rich layers in experimentally deformed phyllite. *Geol. Soc. Am. Bull.* **79**, 795–812.

The structures produced during the laboratory experiments do look remarkably like those of nature. Particularly interesting is the buckling fold mode of quartz layers, producing concentric style folds, and their interaction with kink bands in the phyllitic matrix.

Ramsay, J. G. (1974). Development of chevron folds. *Geol. Soc. Am. Bull.* **85**, 1741–1754.

This paper describes a kinematic development model of chevron folds which can account for many of the special and often unusual geometric complications in these structures

Weiss, L. E. (1980). Nucleation and growth of kink bands. *Tectonophysics* **65**, 1–38.

Lionel Weiss has been one of the most active workers in research into kink band formation. This paper sets out a theoretical model which he thinks best explains the experimental and natural fold geometry.

Crenulation cleavage

Cosgrove, J. W. (1976). The formation of crenulation cleavage. *J. Geol. Soc. Lond.* **132**, 155–178.

This excellent review article describes a wide range of field and laboratory examples, and provides a theoretical background for crenulation cleavage formation.

Gray, D. R. (1979). Geometry of crenulation folds and their relationship to crenulation cleavage. *J. Str. Geol.* **1**, 187–205.

Gray, D. R. and Durney, D. W. (1979). Investigations on the mechanical significance of crenulation cleavage. *Tectonophysics* **58**, 35–79.

The strength and special interest of these two papers is the way naturally formed crenulation cleavage has been approached using many methods of strain measurement and fold geometry analysis.

Means, W. D. and Williams, P. F. (1971). Crenulation cleavage and faulting in an artificial salt-mica schist. *J. Geol.* **80**, 569–591.

These authors describe the laboratory deformation of synthetic foliated material and the geometry of the kink-like folds and crenulation cleavage structures that were produced.

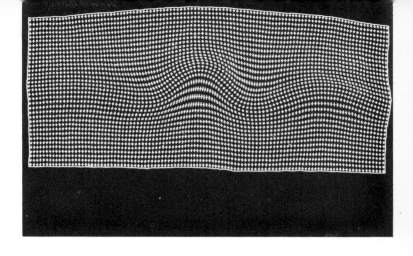

SESSION 21

Strain and Small Scale Structures in Folds

Various different displacement models for fold formation are examined using simple card deck laboratory experiments. Methods of investigating fold geometry using mathematical methods to establish the strain matrix lead to a determination of the overall strain field in a folded layer. The strains within different types of buckle folds are investigated using the models of flexural slip, flexural flow, tangential longitudinal strain and inverse tangential longitudinal strain, and the effects of prefold strains arising from initial compaction, layer parallel shortening and oblique early tectonic strains are predicted. The structural potential of the various incremental and finite strain field is examined and the development of small scale structural features in different parts of a fold is related to the large scale fold geometry. The significance of cleavage fans is examined and the geometric significance of cleavage refraction is fully discussed. It is shown how cleavage–bedding relationships determined in the field can provide valuable information about the large scale tectonic structure.

INTRODUCTION

The topic to be investigated in this session is an extremely wide one. The geometrical features of folds in naturally deformed rocks are very varied, and depend upon many different factors: the thickness, spacing, and rheology of the layers, the nature of the internal rock fabric—whether anisotropic, isotropic or layered, the initial strain state of the rock before folding and the types of boundary displacement that lead to the formation of the folds. In this session we do not have space to examine all aspects of these contributory factors. However, we will develop general methods that are, in our opinion, of special value to the geologist and which enable him to formulate practical and productive ways of thinking about fold geometry. We will develop methods for making models of folds that are geometrically consistent with the most likely constraints on rocks deforming under natural deformation and, in so doing, establish the basic patterns of the overall strain fields in and around a folded layer. The history and final form of this strain field will be used to provide a guide to interpreting the small scale structural features we observe in rock outcrops and to relate these small structures to the overall large scale fold geometry.

Because folds are derived from initially planar layers it is clear that, by definition, the states of strain must be heterogeneous. In this session we aim to acquire some feeling for the possible range of strain states in this heterogeneous field. To do this we will start by making some very simple model experiments developing folds in decks of computer cards. Then we will extend our experience and methodology by showing how it is possible to model fold geometry by mathematical methods; discovering an appropriate displacement scheme and calculating the variations of

the strain matrix that each displacement model implies. The key to this methodology is provided by the constraints of coherent displacement and the principles of strain compatibility introduced in Session 3.

It is well known that a study of the nature and orientation of small scale structural features, folds, boudins, vein systems, cleavage, etc. can provide the basis for positive and practical methods by which the field geologist can not only evaluate the large scale tectonic structures in a region, but also gain an insight into the rheological behaviour of the rocks. In the Answers and Comments section we will relate the geometric predictions of the models we have investigated to the structures that are likely to be developed at different parts of the fold.

QUESTIONS

Strain in flexural folds

Question 21.1

In rocks which are well bedded or which show a regular planar lamination, the strains set up during the folding process are often controlled by simple shear parallel to the planar structure. If the simple shear displacement is distributed continuously through the structure the fold so formed is termed a **flexural flow fold** (Figure 21.1A), whereas if the shear is discontinuous the fold is termed a **flexural slip fold** (Figure 21.1B). Flexural flow folds are generally formed where the rocks have an initially penetrative and fairly

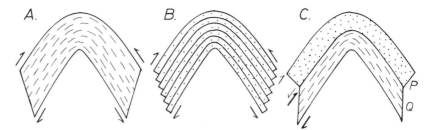

Figure 21.1. A: *Flexural flow fold.* B: *Flexural slip fold.* C: *Type of flexural fold developed in interlayered competent (P) and incompetent (Q) strata. Arrows represent dominant direction and sense of layer parallel slip.*

uniform planar fabric, such as is found in shales, whereas flexural slip folds occur where a number of distinct lithological layers are separated by well marked bedding surfaces. Because the displacements in flexural folds are always parallel to the bedding or lamination surfaces there is no tendency for the orthogonal thickness across the layering to change: the folds therefore always show a **parallel** (Class 1B) form.

We will now investigate the types of strain variation that arise during the development of such folds using a card deck model. Take a stack of computer cards and on one edge draw a series of circles. When we flex the card deck keeping the centre of the cards pinned, some of the circles will be transformed into elliptical shapes as a result of the gliding motion that takes place between the cards (Figure 21.2). We are inducing the fold by a flexural slip mechanism: if we stand back from the card model the discontinuous slip between adjacent cards becomes more difficult to see, and the overall appearance is that of a flexural flow fold. Although you can easily set up such a model yourself, for the convenience of the practical studies and the measurements we wish to make, we have photographed the end of the card deck before and after such an experiment (Figure 21.3A). Although the displacement taking place is simple shear parallel to the card deck surfaces, the resulting geometry is not directly comparable with that of the shear zones we investigated in Session 3. For example, if we construct strain profiles perpendicular to the shear surface the strains vary from profile to profile, and this relationship is different from that of the constant strain profiles of shear zones (cf. Figures 3.2, 3.8 and 3.11 with Figure 21.3A).

In order to investigate the strain variations in flexural folds more closely measure the angle of dip δ of the folded surface (take direction x as the zero datum, with anti-

clockwise angles from this direction as positive, and clockwise angles as negative) and measure the orientation of the long axis of the local finite strain ellipse θ'' at this point. Make a graph of the data with δ as abscissa and θ'' as ordinate. Overlay a transparent sheet of paper on Figure 21.3A and construct the $1 + e_1$ finite strain trajectories of the ellipses. If a cleavage was developed in the rocks as a result of such a strain field discuss the likely orientation (convergent or divergent cleavage fan—see definition in Session 10, p. 193) and the intensity of this cleavage through the fold.

Question 21.2

Figure 21.4 shows an unfolded layer *abcdef* of thickness t and a fold developed from it *ab'c'd'e'f*. This fold has been formed by flexural flow with the sector $e'f$ having constant curvature (radius r_1) and sector $d'e'$ also having a constant but different curvature (radius r_2). Calculate the shear strain γ at e' (dip δ_1) by considering the shear displacement of b' relative to e'. Calculate the γ value at d' by adding together the shear components in the two sectors with radii r_1 and r_2, and show that the orientations and values of the principal finite strains are only a function of dip and independent of the variations in curvature in the fold. Determine the strain matrix and find the equations which give the orientation θ'' and aspect ratio R of the strain ellipse as a function of dip δ.

Question 21.3

In our initial experiment the card deck edge was inscribed with circles and the fold was considered to have developed

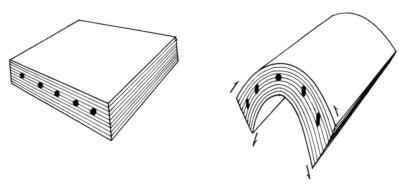

Figure 21.2. *Scheme of the flexed card deck model used for the experiments of Questions 21.1 and 21.3.*

in unstrained material. In many naturally deformed rocks a flexural slip process is often developed in rocks which have been previously strained. For example, before the development of active folding a layer frequently undergoes a homogeneous strain as a result of **layer parallel shortening** and bed thickening (Figure 21.3B). Another possibility is that the rock may have been initially shortened perpendicular to the bedding as a result of **diagenetic compaction** of the sediment (Figure 21.3C). A further possibility occurs where a layer has been involved in some **earlier period of tectonic deformation** and, although unfolded, it has acquired strains in which the ellipses are uniformly shaped but with their principal strain axes oriented obliquely to the surfaces of the layer (Figure 21.3D). We will now investigate what strain fields result when each of these models is deformed by flexural slip. On the left-hand side of Figure 21.3 the edges of the unfolded card decks are illustrated with uniformly oriented strain ellipses having aspect ratios of $3:1$. The right-hand side of this figure shows the appearance of these ellipses after flexing the cards.

Measure the orientation θ'' of the finite strain ellipses at positions with angles of dip of the folded surfaces of value δ. Plot these data on the same graph as you did for the experiment of Question 21.1. On a transparent overlay construct the strain trajectories of the finite strain ellipse long axes. Compare and contrast the strain patterns of these three new experiments (orientations and values of finite strains) with those of the first experiment—you will find some surprising contrasts! What might be the geological significance of these experiments?

Now proceed to the Answers and Comments section and then return to Question 21.4.

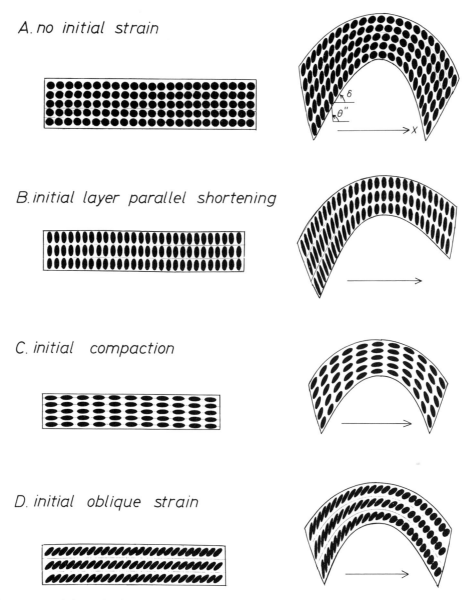

A. no initial strain

B. initial layer parallel shortening

C. initial compaction

D. initial oblique strain

Figure 21.3. *Appearance of the ends of the card deck experiments before (left-hand side) and after (right-hand side) the card flexing experiments. See Questions 21.1 and 21.3.*

Folds with combined tangential longitudinal strain and layer parallel shear

Question 21.4

Figure 21.5A shows a competent (P)–incompetent (Q) rock pair and Figure 21.5B shows an initial stage of folding in which layer P was deformed only by pure tangential longitudinal strain. In contrast the incompetent layer Q was deformed by layer parallel shear together with strains developed by allowing the outer arc to shorten and the inner arc to lengthen. The strains in layer Q, which we will term **inverse tangential longitudinal strain**, lead to a special geometrical effect on the fold limbs whereby the layer suffers a very intense layer parallel shear and a shear displacement or *overshear* which greatly exceeds that which would be expected if the fold was formed by flexural flow alone. If the P–Q pair had been deformed only by uniform flexural flow the initial points X, Y and Z would have been positioned at X', Y' and Z'. Because the deformation has taken place by pure tangential longitudinal strain in layer P and excessive layer shear in layer Q, these initial points are now found at X'', Y'' and Z''. The aspect ratios of the strain ellipses at the hinge zone of this fold have been calculated from Equation 21.10.

In Figure 21.5C the fold has been allowed to develop further, but the displacements in this late development have been taken up *only by* **layer parallel shear**. By comparing the geometry of the strain in the outer fold arc, along the neutral surface and in the inner arc of layer P with the results of experiments C, A and B of Question 21.3, respectively, *sketch* the positions of the $1 + e_1$ finite strain trajectories in the final fold of Figure 21.5C. Sketch also the strain trajectories in the incompetent layer Q, remembering that the outer arc of this layer in the first stage of folding was contracted, whereas the inner arc was stretched. Does the fold of stage C have a neutral surface? Where do we find regions of zero strain and what geometric patterns do the finite strain trajectories show around them? What impli-

cations do these trajectory and strain patterns have in respect of the likely patterns of cleavage intensity and orientation in the fold?

If the initial layer had been subjected to an early period of layer parallel shortening before the development of the first stage of folding by tangential longitudinal strain, what influence would this initial history have on the $1 + e_1$ strain trajectories and what might be the geological implications in respect to cleavage formation?

Now proceed to the Answers and Comments section for this Question, and then continue to Session 22 or to the starred questions below.

STARRED (★) QUESTIONS

The strain matrix in flexural folds

Question 21.5★

Determine the strain matrices for the folds in experiments B and C illustrated in Figure 21.3 in terms of the angle of dip δ. Take the initial aspect ratio $(1 + e_1)/(1 + e_2)$ as R_0 in the case of experiment B with layer parallel shortening, and take the aspect ratio of experiment C as $1/R_0$. Note that the resultant matrix should simplify to that of Equation 21.5 for the example of experiment A ($R_0 = 1 \cdot 0$).

Find the mathematical expressions giving the orientation θ'' and aspect ratio R of the final strain ellipses in the folds.

Using a programmable desk computer plot curves for variations of R and θ'' over initial values of R_0 of 3·0, 1·5, 1·1, 1·0, 0·9, 0·5, and 0·3. Using these data curves discuss the overall strain trajectory patterns and find which folds would produce convergent and which divergent cleavage fans. Which folds have the strongest variations in strain state over ranges of dip δ from $+90°$ through $0°$ to $-90°$?.

Question 21.6★

In Figure 21.3 the initial strain state of the model of experiment D can be expressed in terms of the initial ellipse orientation θ and initial aspect ratio R_0:

$$\begin{bmatrix} R_0^{1/2} \cos^2 \theta + R_0^{-1/2} \sin^2 \theta & (R_0 - R_0^{-1})^{1/2} \sin \theta \cos \theta \\ (R_0 - R_0^{-1})^{1/2} \sin \theta \cos \theta & R_0^{1/2} \sin^2 \theta + R_0^{-1/2} \cos^2 \theta \end{bmatrix}$$

$$(21.1)$$

Determine the strain matrix and equations giving the orientation θ'' and aspect ratio R in the final fold of experiment of Figure 21.3D in terms of the layer dip δ.

Using a programmable desk calculator plot the variations of θ'' and R for folds with an initial strain aspect ratio of $R_0 = 3 \cdot 0$ and for initial orientation $\theta = 30°$, 45° and 60°. Discuss the geometric and geological implications of these data curves.

ANSWERS AND COMMENTS

The geometry of flexural folds
Experiment A: Folds developed in initially unstrained layers

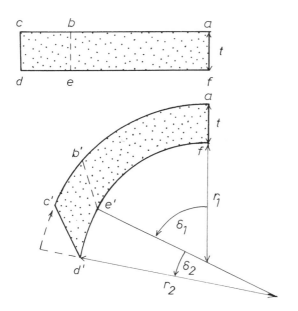

Figure 21.4. Calculation of the shear strain in a flexural flow fold. See Question 21.2.

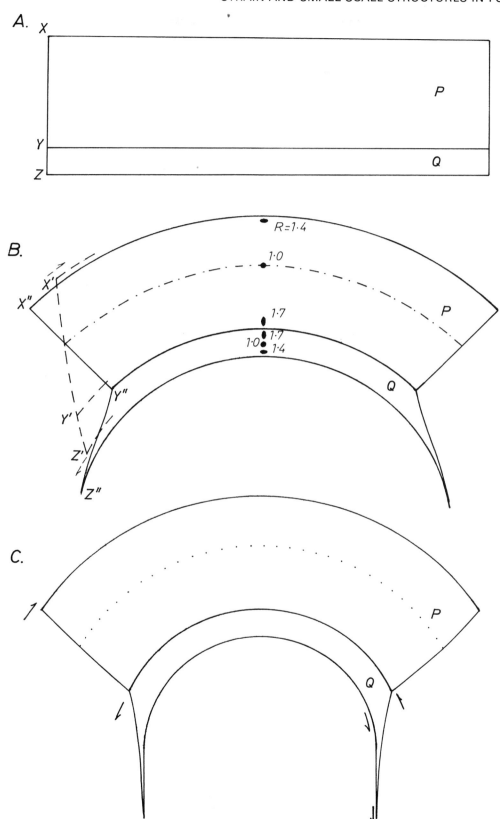

Figure 21.5. *The folding of a competent layer P and incompetent layer Q by combinations of tangential longitudinal strain and layer parallel shear. See Question 21.4.*

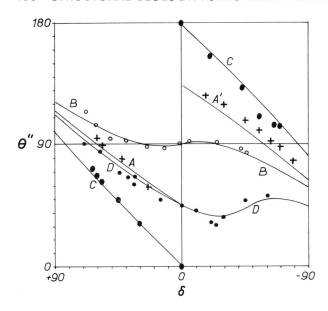

Figure 21.6. *Data plots of variations of orientations of strain ellipse long axes θ'' and layer dip δ for Questions 21.1. (+), and Question 21.3 Experiment B (○), C (●) and D (●).*

Answer 21.1

In a part of a flexural slip or flexural flow fold the displacements which occur during folding consist of a body rotation of the rock element under study together with a relative simple shear motion between the layers. In an antiformal fold the uppermost layers move towards the hinge line, the maximum differential displacement (and γ-values) taking on highest numerical values on the fold limbs and lowest values (zero) at the fold hinge. In a symmetric fold the strain field set up shows a mirror symmetry about the axial surface. On the left-hand fold limb of the antiform of Figure 21.3A (with positive dip values) and regions of low dips showing low strain aspect ratios, and the principal strain axes are oriented at about $+45°$ to the x-coordinate reference direction. The aspect ratios and orientation angles θ'' of the strain ellipses increase as the dip increases (Figure 21.6, data plotted as crosses). On the right-hand fold limb the geometry is almost mirror symmetric, the long axes of the ellipses being oriented at about $+135°$ near the hinge and decreasing with progressive (negative) increase in limb dip. It is possible to calculate and graph the exact functional relationship of the orientation θ'' of the long axes of the strain ellipses with change of dip in a flexural flow fold (Question 21.2, Figure 21.6, the two curves labelled A and A'). The plots of the

orientations of the long axes of the ellipses in our experiment do not lie exactly on these two curves, but systematically deviate so as to lie above the predicted curve. Why is this so? The reason is that, during or before the fold forming process of our experiment, a slight and homogeneous simple shear strain was induced throughout the whole card deck, and the simple shear strains produced as a result of the main flexing of the cards was superposed on this additional strain. A strain history of this type is not uncommon in natural rock deformation processes. For example, Figure 21.7 shows an initial heterogeneous gliding, such as might occur during an overall sub-horizontal displacement of a rock mass in a nappe, followed by flexural slip folding. Each layer would have is own characteristic strain pattern combining different proportions of initial shear zone like simple shear and differing amounts of simple shear produced during folding.

The $1 + e_1$ strain trajectories of this experiment are shown in Figure 21.8A. Note the way that adjacent trajectories converge when they are traced from regions of low strain at the hinge zones to regions of high strain on the fold limbs. In flexural folds found in naturally deformed rocks the deformation may give rise to a planar fabric or **cleavage**. Cleavage is related to the orientation and values of the finite strain state. It forms perpendicular to the maximum finite shortening (and therefore parallel to the $1 + e_i$ strain trajectories) and its intensity increases with the aspect ratio R of the finite strain ellipse (check back to Session 10, pp. 179–185). Around the hinge zone of a flexural fold the $1 + e_1$ finite strain trajectories are divergent from the axial surface (stippled region in Figure 21.8A), but in most rocks the low strains in this region would probably be insufficient to induce a cleavage which could be seen in field outcrops. At dips of around 30–40° R values of about 2:1 occur, and a weak cleavage would probably be observed. In the range of dips 30–55° this weak cleavage would be arranged in the form of a **divergent cleavage fan** (divergent in the sense of adjacent cleavage planes move apart as they are traced from the outer to inner arc of the folded layer, see Figure 10.20 and Session 10, p. 193). Where dips exceed 60° the cleavage would be expected to be quite well developed and arranged in the form of a **convergent cleavage fan**.

A wide variety of small scale structural features, in addition to the cleavage fabric discussed above, can form in flexural folds. In flexural slip folds we have noted that the sense of the displacements concentrated on individual lithological contacts is always systematic and that the structurally uppermost layers always show a displacement relative to the lowermost layers away from a synformal hinge and towards an adjacent antiformal hinge (Figure 21.1). In Figure 21.9 the discontinuous relative displacement between

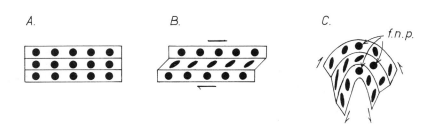

Figure 21.7. *An initial undeformed stage (A) is deformed by heterogeneous layer parallel sliding (B) and followed by flexural slip folding (C). The finite neutral points in the layers (f.n.p.) do not lie adjacent to each other in the final fold.*

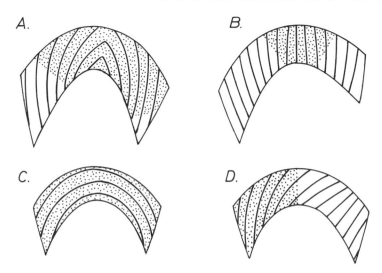

Figure 21.8. *Trajectories of the directions of maximum finite longitudinal strain ($1 + e_1$) in the four folds of Figure 21.3. Stippled areas are those where the trajectories (and possible cleavage) diverges from the position of the axial surface of the fold, unstippled regions where they converge towards the axial surface.*

the sandstone layers in a flexural fold is shown by the shift in position of initially cross cutting quartz veins, and the sense of relative displacement can be used to indicate the positions of the major fold forms. It should be clear than an antiform lies to the right of this outcrop and a synform to the left. The movement of one layer past another is often accompanied by special structural features on the surfaces of strongest displacement. These surfaces may take on a highly polished appearance, known as **slickensides**, and on these slickenside surfaces **movement striae** may be developed

parallel to the direction of relative slip between the layers (Figure 21.11). In many flexural slip folds in rocks deformed under low grade metamorphic conditions (up to greenschist facies) **fibrous shear veins** are developed parallel to lithological contacts containing overlapping fibrous crystals of calcite, quartz or chlorite (Session 13, pp. 257–261). In simple flexural slip folds striae and the **fibre lineation** (Figure 21.10) are oriented perpendicular to the hinge lines of the folds, and the overlapping sense of the fibres can be used to determine the relative movement sense (Figure

Figure 21.9. *Discontinuous slip parallel to the bedding surfaces of sandstone layers in a flexural slip fold. Bude, Cornwall, England.*

Figure 21.10. *Quartz shear fibre veins with fibre lineation parallel to the slip direction of flexural slip movements on a bedding surface in sandstones. Bude, Cornwall, England.*

13.32). Some care should be taken, however, in deducing the direction of the fold axes from striae and fibres. Striae are produced by mechanical grinding of rock particles between the adjacent surfaces and are therefore likely to record only the last displacements along the surfaces, and crystal fibres sometimes initiate before (or after) folding and thus the movement sense they record might not be simply related to fold geometry.

In flexural flow folds, where the shear strains are fairly evenly distributed through a layer of constant dip, **extension vein systems** which cross cut the layering are sometimes developed. The geometry of these vein systems is controlled by the timing of vein initiation during the folding process, and the extent of subsequent straining as a result

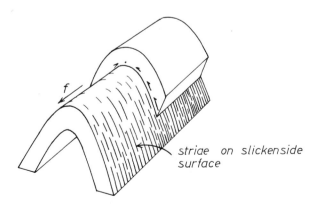

Figure 21.11. *The relationships of bedding slickenside surfaces and movement striae in a simple flexural slip fold. The lengths of the shafts of the slip arrows are proportional to the amount of slip, and this is related to the intensity of development of the striae. The striae are everywhere perpendicular to the fold hinge f.*

of further fold development. The general sequence to be expected accords with the principles established in Session 2 (Figure 2.11), and the types of geometric distribution and appearance of extension veins in flexural folds is shown in Figure 21.12. In simple flexural flow folds the veins will initate at angles of about 45 and 135° to the layering surfaces, depending upon their position in the fold limb. As the development of the fold proceeds (Figures 21.12B and C), the initially formed veins undergo geometric modification while new veins may initiate during the later strain increments. The principal effect is a rotation of early formed veins so that they make higher angles with the bedding surface. This rotation goes together with the need for subsequent geometric shortening of the initial vein. As a result (Figure 21.12D) the vein may widen by internal deformation of the crystalline filling or it may shorten by buckle folding (note how the fold wavelengths will be controlled by the vein thickness). The initial vein may propagate at its tips and acquire a sigmoidal form, according to the model we evolved in Question 2.11. It may open in a direction which is not perpendicular to the rotated wall orientations and therefore develop complex vein fibre geometry, together with cross cutting new veins which offset the inital veins (Figure 21.12D). On fold limbs the vein geometry is likely to show a wide range of form because it is here where the incremental strain sequences show their maximum complexity. Figure 21.13 illustrates the very complex geometric range of quartz veins in compound flexural slip and flexural flow regions in a sandstone–shale sequence. The overall simple shear sense means that the uppermost beds have moved to the left and upwards towards an antiformal hinge relative to the lowermost beds. The pale sandstone layers have suffered less flexural flow than the dark shale layers, while locally fibrous quartz shear veins have been developed between the layers as a result of flexural slip. The quartz

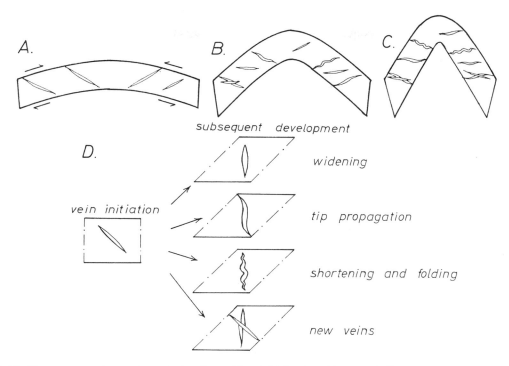

Figure 21.12.   *The geometry of extension veins in a flexural flow fold developed in stages A, B and C. D illustrates the different types of geometric modification of initially formed planar veins as a result of increased shear during folding.*

Figure 21.13. *Rotated and folded quartz filled extension veins in alternating sandstones and shales. Compare with Figure 21.12. The shale layers and some of the sandstone-shale contacts show quartz filled shear fibre veins. Hartland, N. Devon, England.*

veins cross cutting the sandstone layers show a range of orientations indicative of progressive rotation, but the competence contrast between the vein and its sandstone matrix does not seem to have been sufficient to induce folding in the vein. In the shale layers the bedding parallel shear strains appear to have been considerably greater than those in the adjacent sandstone (model Figure 21.1C). The quartz veins appear to have been strongly rotated to produce an overall en-echelon form which is the reverse of that of their probable initial orientation. The strong vein rotation and accompanying shortening has led to the formation of folds, while at some parts of this outcrop the folded layers appear to be rotating into an incremental extension position such that the folds formed earlier are becoming unfolded. These features are exactly those predicted from the model experiments we performed in Session 1 (Figure 1.7). Although the appearance of vein systems in the fold profile looks complicated, it should be clear that some geometric aspects of these structures are quite simple. If the flexural fold has a simple history, like that of our card deck model, the line intersection of any vein, rotated or not, with the bedding surface is parallel to the fold axis, and the folds produced during buckling of early formed veins during the later stages of folding should have axes parallel to the hinge lines of the large scale folds in which they lie. The buckling of quartz veins during later vein shortening processes often produces highly characteristic rippled and corrugated surfaces of the vein, and late stage elongations can lead to boudinaged and rod-like fold forms (Figure 21.14) termed **quartz rods** (Wilson, 1953).

Many geologists who have studied folds have remarked on the relationships of the geometry of small scale folds to that of the large scale fold in which they are found, and we have already discussed some aspects of the form and symmetry of these small scale structures in Session 15. Some workers have suggested that these small scale folds are genetically related to the differential movements taking place between the thicker competent rock units in the flexural slip folding process and have termed these folds **drag folds**. You will have noted that no small folds appeared in the course of our card deck experiments and it should be clear from the geometric constraints of the flexural slip process that there is no tendency for shortening to occur in a direction parallel to the layering. De Sitter (1958) pointed out that such small scale folds often occur most frequently around the hinge zones of large folds where the differential movement between the layers is zero in the flexural model (Figure 21.15). It is clear that these small scale folds in polyharmonic folds (probably best termed **parasitic folds**) cannot have their *origin* in the flexural slip process. They are probably best accounted for by some early layer parallel shortening process which led to the development of buckles of differing wavelengths as a result of thickness and competence variations between the layers as we discussed in Session 20. However, once initiated, any differential shear taking place on the limbs of a large wavelength structure will modify the initial symmetric fold geometry into asymmetric S and Z forms (Figures 15.21), whereas the absence of this shear component at the hinge zone leads to the parasitic folds retaining their M form.

One of the most important geometric characteristics of parasitic folds of particular use in field mapping is that their hinge lines are aligned parallel to the hinge lines of the large scale folds. The hinges of small scale folds often give rise to

Figure 21.14. Strongly folded quartz veins forming quartz rods. Cima di Lago, Lepontine Alps.

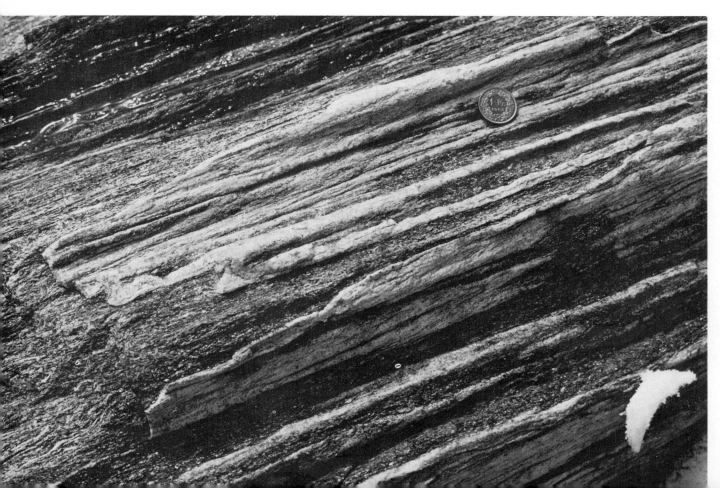

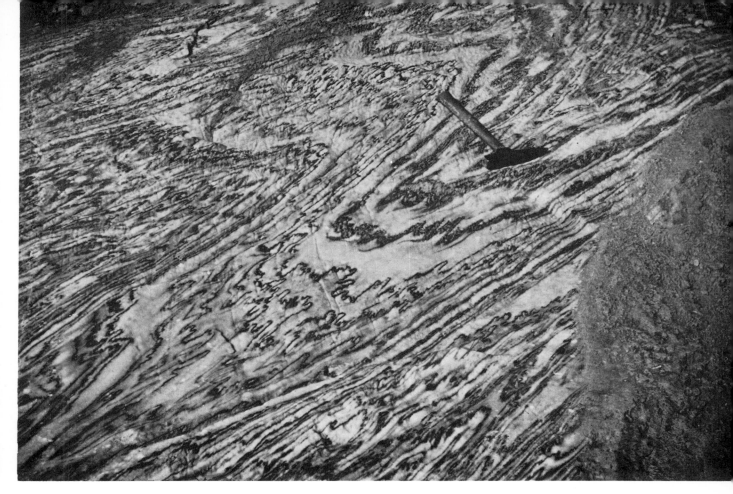

Figure 21.15. Parasitic folds in calc silicate (dark) and marble (light). Note the very intense development of the smallest wavelength folds at the hinge zones of the larger folds indicating that they cannot have originated as a result of a flexural slip process. Khan Gorge, Namibia.

Figure 21.16. Strongly developed fold rodding lineation as a result of the weathering out of the hinges of parasitic folds. Simla, N. India.

an extremely prominent linear fabric or **fold rodding** (Figure 21.16). In very intensely folded rocks the parasitic folds and attendant marked rodding fabric are generally particularly strongly developed near the hinges of major folds and less well developed on the fold limbs. The relative intensity of the fold rodding fabric can therefore be a valuable field guide to help locate the traces of major folds.

Answer 21.2

In the fold sector $ab'e'f$ the amount of slip of point b' relative to e' is the arc length.

$$b'b'' = ab'' - ab' = ab'' - fe'$$

$$= \delta_1(r_1 + t) - \delta_1 r_1 = \delta_1 t \qquad (21.2)$$

The shear strain γ at position e' is

$$\gamma = \lim t \to 0 \quad \delta_1 t/t = \delta_1 \text{ (in radians)} \quad (21.3)$$

Consider the next sector of the fold $b'c'd'e'$ where the radius of curvature has changed to r_2. The total slip at point d' is given by the slip in this new sector plus that of Equation 21.2 ($\delta_1 t + \delta_2 t$) and the shear strain at point d' is given by

$$\gamma = \lim t \to 0 \ (\delta_1 t + \delta_2 t)/t = \delta_1 + \delta_2 \quad (21.4)$$

and is therefore independent of the changes of curvature taking place in the fold.

In a flexural flow fold the strain matrix at a position with dip δ is

$$\begin{bmatrix} 1 & \delta \\ 0 & 1 \end{bmatrix} \qquad (21.5)$$

and the angle θ'' between the bedding surface and the long axis of the finite strain ellipse is given by

$$\tan 2\theta'' = 2/\delta \qquad (21.6)$$

The aspect ratio R of this ellipse is

$$R = (2 + \delta^2 + \delta(\delta^2 + 4)^{1/2})/2 \qquad (21.7)$$

If the fold axial surface is vertical (parallel to the y-coordinate direction) and the flexural flow has been towards this axial surface, then the orientation θ'' of the different strain ellipses with respect to the x-axis, and which defines the finite strain trajectory field, is given by

$$\theta'' = \delta + (\tan^{-1}(2/\delta))/2 \qquad (21.8)$$

This function has been graphed in Figure 21.6, curves A and A'.

Answer 21.3

The superposition of a flexural flow fold on a previously strained layer can lead to a wide variety of finite strain patterns. All the folds produced in the series of experiments shown in Figure 21.3 are classified as *parallel* or *Class 1B* folds. We therefore come to the very important conclusion that *folds which fall into identical classifications* using the layer thickness variation and isogon methods described in Session 17 *can show a wide variety of internal strain patterns and can therefore be formed by many different displacement sequences*. As a further visual check of this important conclusion return to Session 3 and compare the differing strain trajectory plans of the geometrically identical (in terms

of layer shape) similar or Class 2 folds of Figures 3.12C and 3.19.

In each of the models of Figure 21.3B, C and D the initial strains in the unfolded materials were homogeneous. Such a strain homogeneity is not a *necessary* geometric constraint on an initially unfolded layered sequence. In the studies of shear zones made in Session 3 we noted that parallel sided shear zones with identical strain profiles might be produced in a number of ways which were geometrically compatible. In model C, for example, the initial compaction strain could have varied from layer to layer without inducing folds in the layering (check back to Figure 3.17B), and such a complex model is probably very realistic with reference to sedimentary rocks with differing lithologies. In model D with obliquely inclined initial strain ellipses we could have chosen an alternate variable strain state from layer to layer by combining variable compaction with variable layer parallel simple shear, together with any homogeneous strain. Such an initial condition would have been compatible with unfolded layering because it is the general solution for shear zone compatibility described in Session 3 (pp. 46–47). In Experiments B, C and D we have chosen three particular models because they separate most clearly the effects of the various strain components that can exist in unfolded layered material.

Experiment B: Folds developed in layers with initial layer parallel shortening

The variations of the orientations θ'' of the long axes of the finite strain ellipses are shown graphically in Figure 21.6 as open circles. These data points fit fairly well the curve B deduced from theory (see Answer 21.5★ below). Because the initial strain axes were oriented perpendicular and parallel to the layering and because the displacements during flexural flow folding are symmetric on either side of the fold axial surface, it follows that the final strain states also show a geometric symmetry across the fold such that the orientation of θ'' with positive dip δ takes numerically identical but negative values with dip $-\delta$. Around the hinge region of the fold (δ from $+30°$ to $-30°$) the long axes of the finite strain ellipses are almost parallel to the fold axial plane. In detail, however, they show a divergent fan (region shown stippled in Figure 21.8B). Where the fold limbs dip more steeply the influence of the bed rotation and superposed simple shear produces a convergent fan in the $1 + e_1$ trajectories. Although the aspect ratios of the ellipses are high throughout the fold, the variations through the fold are much less strongly marked than those of the simple shear model of Experiment A. In model A the R values from dips of θ to $\pm 60°$ vary from $1:1$ to $2·7:1$, whereas in model B the R values vary from $3:1$ to $6·5:1$. If cleavage developed in a fold with geometry of model B we would expect the cleavage to be strong and of fairly uniform intensity throughout the fold with a slightly *divergent cleavage fan* near the hinge and a marked *convergent fan* on the fold limbs.

Experiment C: Folds developed in layers with initial layer perpendicular compaction

The geometry of folds formed by flexural flow in a material with an initial strain ellipse aspect ratio of $3:1$ and with long axes initially parallel to the bedding shows a number of somewhat unusual characteristics. The orientations θ'' of the long axes of the finite ellipses are graphed in Figure 21.6 as large filled circles (curve C). The $1 + e_1$ strain trajectories are shown in Figure 21.8C. Over the range of our experiment they show a consistent pattern: the trajectory lines diverge from the fold axial surface and adjacent trajectory lines converge together from the hinge region to the limb regions. The long axes of the ellipses change their initial orientations parallel to the layering only slightly, the angles between them and the layering are less than $8°$ on the fold limbs. The variations in aspect ratio R of the ellipses are remarkably small, varying from $3·0 : 1·0$ to $3·4 : 1·0$ over the dip range 0 to $\pm 60°$. These features have important geological implications. They imply that the initial fabrics of rocks showing strong initial compaction fabrics are unlikely to be greatly modified either in intensity or in orientation by the flexural fold forming process. These geometric features probably account for the frequently made observation that shaly sediments in flexural folds developed in the external parts of orogenic zones show fabrics which are either parallel to or only slightly cross cut the primary bedding planes. In these external regions high competence contrasts between layers of differing lithology rapidly set up folding deformations, so there is little layer parallel shortening to produce the effects of Experiment B before rapid fold amplification takes place.

Experiment D: Folds developed in layers with initial obliquely inclined strain states

Because the initial symmetry of the strain ellipse is oblique to the axial plane of the flexural fold the resulting finite strains differ in orientation and values on each fold limb. The orientation θ'' of the long axes of the finite strain ellipses are shown graphically in Figure 21.6 as small filled circles and fit fairly well the theoretical curve D (see Question 21.6★). In the left-hand fold limb the right-handed simple shear displacements increase the intensity of the initial strain whereas on the right-hand fold limb the left-handed shear displacements decrease the initial strain ratios. Although the shear on the right-hand limb reduce the initial strain values they never completely unstrain the layers. Why is this so? The reason is that the orientation and aspect ratio of the initial ellipses were not consistent with a pre-fold simple shear process parallel to the layering, so the simple shear produced by folding can never completely efface the earlier strain. We have discussed previously a similar problem in our discussion of shear zone geometry (Answer 3.6★).

The strain trajectories in the fold are shown in Figure 21.8D, and are markedly asymmetric to the fold form.

Where rocks show a strain sequence such as that of model D they might be expected to show a cleavage which was asymmetrically disposed to the fold axial plane (a convergent fan on one limb, and a divergent fan on the other) and which varied in intensity from limb to limb. A fine example of such a fold in a sandstone–pelite sequence is shown in Figure 21.17. This is a region in which two major fold phases have been recognised using criteria which we will discuss later in Session 22. The fold shown here is a second generation structure and it was superposed in rocks previously strained during a first deformation event. The cleavage clearly cross cuts the axial surfaces of the folds (axial trace

parallel to the pencil) and shows a marked variation in intensity on the different fold limbs. This cleavage geometry would accord with an overall second fold flexural slip process superimposed on layers which initially had an initial strain state such that the initial strain ellipse long axes obliquely cross cut the bedding in a sense opposite to that of the initial card model of Figure 21.3D.

The investigation of the geometric features of the variations of finite strain that can arise in folds brings to light some special features that we have not encountered before. During our study of ductile shear zones in Session 3 we noted that when passing from a region where the ratio of the principal finite strains was low to one where it was high the maximum strain trajectory lines show a convergence. During the studies made in this current Session you may have noted that, although this feature is characteristic of many folds, or parts of folds, it is not an absolute rule. Certain folds show contrary geometric situations (Figure 21.3, B and D). The details of why such variations can take place will be investigated in more detail in Volume 3 when we undertake a general formulation of heterogeneous strain. At this stage we comment that complexities arise because two dimensional finite strain needs four components for its complete formulation. The compatibility constraints of these components show that *there is no unique link between the changes of orientation of the principal finite strains and variation in strain intensity*. With certain types of variation in finite rotation it is possible to have a convergence of adjacent maximum strain directions with decrease in the absolute (or ratio) values of the principal strains even in conditions of plane strain.

Tangential longitudinal strain

In many geological environments interlayered beds of contrasting lithology and ductility are involved in the folding process. Figure 21.1C is an example of a fold developed in a competent layer P which lies next to an incompetent layer Q. Let us first try and apply the simple flexural flow model we have evolved, and study the geometric implications of allowing the total flexural flow to be unevenly distributed through the two materials. The total displacement on the fold limb between the top of the competent layer P and bottom of the incompetent layer Q is identical to that of Figure 21.1A, but we have allowed more shear deformation to take place in Q than in P. This geometry has important differences from that of flexural flow; the outer arc boundary of the competent layer has to be stretched, the inner arc shortened, and vice versa in the incompetent layer. This means that strains additional to those of simple shear have to be incorporated into the layers to allow for strain compatibility. These additional strains have principal axes acting sub-parallel and sub-perpendicular to the layer surfaces in the more competent layer P and are termed **tangential longitudinal strains**. In competent layers within a multilayered rock sequence this type of strain may predominate over the strains induced by layer parallel shear.

We can best analyse the geometric features of folds formed with tangential longitudinal strain by considering the geometry of a folded layer undergoing no layer parallel shear. Figure 21.18 shows an example of such a fold where the curvatures of the upper and lower surfaces are constant and related to a centre of curvature at point c. The fold

Figure 21.17. Folds in sandstones and phyllites showing a cleavage which cross cuts the fold axial surfaces. Holy Island, N. Wales.

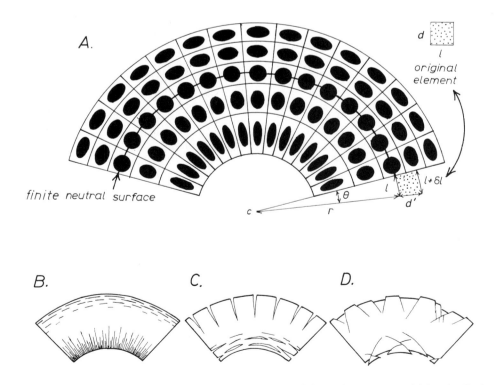

Figure 21.18. A: The geometrical features of a fold formed under conditions of pure tangential longitudinal strain. Structural development of rock strain: B, cleavage; C, extension fissures; D, conjugate shear faults.

would be classified as a **parallel** or **Class 1B fold**, and the example here, with constant curvature, is a **concentric fold**. To produce geometric compatibility in this structure we have to stretch layers in the outer arc and shorten layers in the inner arc so that the principal strains are exactly parallel and perpendicular to the layer surfaces. Between the outer and inner arcs there must be a surface of no finite strain known as the **finite neutral surface**. In this very simple model the original thicknesses of the layers are systematically reduced on the convex side of the finite neutral surface, and increased on the concave side of this surface. If the deformation proceeds so that no area changes takes place in the fold profile it is possible to determine how any surface originally situated at a distance d from the neutral surface becomes modified to a distance d' in the fold. Comparing the areas of an original rectangular element of length l and thickness d with the cylindrical derivative after folding we find

$$lt = 1[\pi(r + d')^2 - \pi r^2]/2\pi r$$

where r is the radius of curvature of the finite neutral surface. Simplifying the expression we obtain

$$0 = d'^2 + 2rd' - 2rd$$
$$d' = -r + (r^2 + 2rd)^{1/2} \qquad (21.9)$$

We can also find the values of the principal finite strains. At distance d' from the neutral surface the layer length is modified by an amount δl. Then

$$l = r\theta$$
$$l + \delta l = (r + d')\theta$$

therefore $\delta l = d'l/r$ and the extension along the layer $e = \delta l/r = d'/r$. Because of the conditions of plane strain $(1 + e_1)/(1 + e_2) = (1 + e_1)^2$ the aspect ratio R of the strain ellipse at any distance t' from the neutral surface is given by

$$R = (1 + d'/r)^2 \qquad (21.10)$$

Structural features of rocks deformed by tangential longitudinal strain

In the outer fold arc layer length is increased and layer thickness decreased, whereas on the inside of the finite neutral surface the opposite changes occur. The types of structures which develop depend on the rheological conditions of the rock, which in turn depend upon the rock composition temperature, pressure and strain rate environment. If the rock is ductile, rock flow is likely to develop cleavage fabrics (Figure 21.18B). Cleavage will probably be most noticeable on the inner arc of the fold and form a strongly **convergent fan** with particularly intense cleavage on the innermost arc of the structure. On the outer fold arc a tectonic fabric will be developed, but this will be parallel to the layer surfaces and is therefore likely to be less obvious in the field. If the rocks are in an environment where extensile cracks can form, then the outer fold arc will show sub-radial fissures widening and becoming more frequent in the outermost arc (Figure 21.19), whereas these veins will form parallel to the layer surfaces in the inner fold arc (Figure 21.18C). If the rock undergoes failure by the development of conjugate shear fracture, then the outer arc will be dominated by layer

Figure 21.19. Quartz carbonate extension fissures in the outer fold arcs in a calcareous sandstone. Champéry, W. Switzerland.

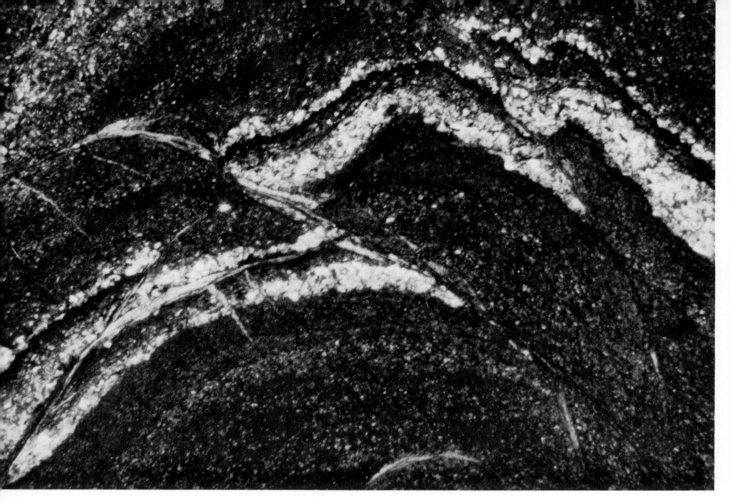

Figure 21.20. *Thin section of an argillaceous sandstone showing contraction of the inner arc of a fold by the development of thrust faults. Torridonian Sandstone, Sleat, Isle of Skye, Scotland.*

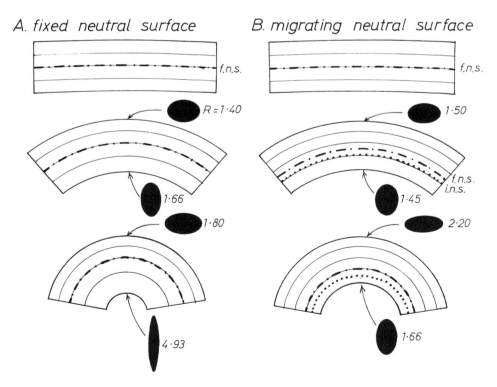

Figure 21.21. *Progressive development of folds by tangential longitudinal strain processes indicating the strains developed A, when the finite neutral surface (f.n.s.) is fixed and B, when the finite neutral surface migrates during fold amplification. The dotted line i.n.s. in B shows the position of the incremental neutral surface at different stages in fold development.*

extension faults and the inner arc by layer contraction faults (Figures 21.18D, 21.20).

The model shown in Figure 21.18A was drawn on the assumption that the finite neutral surface initiated midway in the competent layer, and that this surface continued to act as finite neutral surface throughout the fold history. There are various mechanical and geometrical reasons why this might not occur in naturally formed folds. If the finite neutral surface has a fixed position throughout folding it will be found from Equations 21.9 and 21.10 that there is much greater increase in finite strain on the inner arc than on the outer arc, implying that there is a much higher strain rate in the inner arc than on the outer arc. This high strain leads to a very strong inner arc layer thickening (Figure 21.21A). Mathematically, as the fold curvature decreases to a value of $-1/2d$, so the aspect ratio R approaches infinity. This implies that the overthickening of the inner arc is so strong that the geometry cannot continue to exist according to our model. There are two main ways in which the fold can continue to develop and avoid this geometric impossibility: one is to allow the finite neutral surface to migrate towards the inner fold arc (Figure 21.21B) the other is to constrain the condition of increasing the curvature so that parts of the fold with lowest curvature increase their curvature to conform with that of the highest existing curvature (giving rise to the well known **ptygmatic fold** structure). The solution of migrating the finite neutral surface towards the inner fold arc enables the layer extensions in the outer arc to be more or less equalised to the layer thickenings in the inner arc (Figure 21.21B). This implies that at any stage during the folding there must be an **incremental neutral surface** situated on the inside of the **finite neutral surface** (Figure 21.21B, i.n.s. and f.n.s. respectively). The incremental neutral surface separates zones in the fold which, at that time of fold development, are undergoing active layer stretching and active layer shortening respectively. The migration of the finite neutral surface towards the inner fold arc therefore allows the possibility that layers that were originally contracted could later in the folding be undergoing stretching. Such a sequence might be recorded in the rock by superposed small scale structures which apparently "contradict" each other in the sense that the maximum shortening implied by one structure might be perpendicular to the maximum shortening implied by another (e.g. extension veins running parallel to cleavage surfaces). Such sequences of strain reversals are quite common in many naturally deformed rocks. They clearly do not imply total separation of structural development into separate orogenic events, but they relate to a single deformation process with progressively changing geometric constraints.

So far we have considered a model of tangential longitudinal strain which was highly specialized in that the fold was constructed using the constraints of constant curvature of the neutral surface. If this constraint is relaxed the strains become more complex than those we have computed. You can easily see the problem by inverting the model of Figure 21.18A so that it forms a synform and trying to join it along the fold limb sector to the antiform of Figure 21.18A. Neither the layers nor the neutral surfaces will connect across the join; the two parts do not have compatible links. We have to bend the grid lines away from their circular arc and radial line form, and this implies that we have to allow for shear strain components both parallel and perpendicular

to the layer surfaces. It should also be clear that in this join zone we are likely to develop other neutral strain regions. For example, a layer situated in the inner arc of the antiform, with its maximum strain axis perpendicular to the layering has to take up a position in the outer arc of the adjacent synform with maximum strain axis parallel to the layering (one solution is shown in Ramsay (1967); Figure 7.63).

The "pure" tangential longitudinal strain model we have discussed above is generally linked, in naturally formed folds, with the layer parallel shear model. In fact the way we evolved it was by considering what additional strains we would have to induce in a competent layer which was taking up insufficient layer parallel shear to develop the overall fold. We will now look into some of the effects of combining the two basic strain models together.

Return now to Question 21.4.

Folds with combined tangential longitudinal strain and layer parallel shear

Answer 21.4

The left-hand side of Figure 21.22 shows the shapes and orientations of the finite strain ellipses, and the right-hand side shows the $1 + e_1$ strain trajectory field and the predicted cleavage pattern in the fold. Analogy of the strains in the outer arc of the competent layer P with the results of the experiment (Figure 21.3C) indicates that the layer parallel stretching would become only slightly modified in orientation and in strain ellipse aspect ratio, and that cleavage in this zone would be parallel to layering or show a slightly cross cutting divergent fan form. Along the original finite neutral surface only strains at the fold hinge would remain unmodified, to be retained as a finite neutral point in the final fold. All other strains along the original neutral surface would be modified in according with Experiment A of Figure 21.3. The final fold does not therefore have a finite neutral surface. Along the inner arc of the competent layer

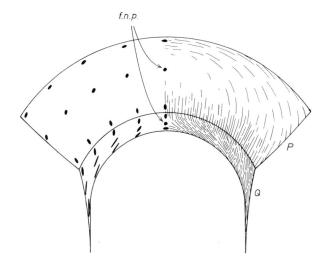

Figure 21.22. *Answer 21.4. The left-hand side shows strain ellipses and the right-hand side illustrates the cleavage predicted from the orientation of the finite strain trajectories and intensity of the aspect ratios of the finite strain ellipses. Two finite neutral points (f.n.p.) are situated in the fold hinge.*

P high strains will be preserved and the cleavage predicted here would be strong and arranged in a **convergent fan** form.

In the incompetent layer *Q* the dominant feature of the strain field is the result of the excess of layer parallel shear result from the strain distribution that we have termed **inverse tangential longitudinal strain** (outer arc contracted, inner arc stretched). The strains and strain gradients will tend to be high with the long axes of the strain ellipses generally lying close to the layer surfaces, especially on the fold limbs. The cleavage that would be predicted in this layer would be much more strongly developed than that seen in the competent layer and the overall pattern would be that of a **divergent fan**. At the fold hinge there will be a finite neutral point situated somewhere between the layer boundaries where the layer parallel shortening of the outer arc passes through a strain-compatible zero strain to link with the layer parallel stretching on the inner arc.

In this model there are two finite neutral points both situated at the fold hinge. Around these neutral points the finite strain trajectories show special geometric patterns. It can be shown that only two types of trajectory patterns can exist around an isotopic strain point, and these enable a separation of a **positive and a negative isotropic point** (see Ramsay (1967); Figures 2.12 and 2.13). Inside the competent layer *P* the $1 + e_1$ and $1 + e_3$ strain trajectories form a series of interlocking curves which define a positive isotopic point (Figure 21.23) whereas the non-interlocking trajectories in layer *Q* define a negative isotopic point (Figures 21.22 and 21.23). Both are regions where cleavage is absent and both have characteristic cleavage patterns in their vicinity which correspond to the two types of possible trajectory patterns. Particularly noteworthy is the triangle-like plan of cleavage around the negative neutral point which is a feature of very many naturally formed folds, either in incompetent layers sandwiched between two competent layers, or in the zone of contact strain around buckled single layers of competent rock (Figure 10.18).

The principal axes of finite strain show a sudden change of direction where they are followed across a lithological contact between a competent and an incompetent layer. This refraction of the strain trajectories is associated in naturally deformed rocks with a refraction of the planar cleavage or schistosity fabric. The only position where refraction does not occur is at the hinge. On the inner arcs of competent layers and on the outer arcs of incompetent layers the cleavage and lithological layer surfaces are perpendicular, and it is only at this position where *cleavage is consistently parallel to the axial surface of the fold* irrespective of rock lithology. We have previously introduced the concept of **cleavage refraction** in terms of variation in strain states (Figure 10.23). Cleavage refraction is a frequently observed but often misunderstood phenomenon. We have seen in Session 10 that penetrative cleavage is parallel to the principal *XY* plane of the finite strain ellipsoid. Cleavage refraction implies that the principal *XY* planes change orientation. If no discontinuous fault-like slip takes place across the contact, the principles of geometric and strain compatibility imply that there must be a connection between the strain ellipsoids on either side of the cleavage refraction surface. Figure 21.24 illustrates the constraints on the adjacent strain ellipsoids in a competent layer *P* and incompetent layer *Q*. The strain directions X_P and Z_P are differently aligned from X_Q and Z_Q, but the half ellipsoids must be connected in such a way that they share an identically shaped and oriented strain ellipse section on the refraction surface (Figure 21.24B). For such a geometric constraint to hold the *strain differences* between the two sides can *only* be accounted for by two effects: either a difference in the amount of layer parallel shear on either side of the contact, or a difference in the amount of bedding normal compaction between the two rock types on either side of the contact (cf. Figure 3.17A and B). Both sides of the strain refraction plane can contain other strain components, but *these components must be identical.*

Proportions of tangential longitudinal strain and layer parallel shear in folds

It has been suggested that components of both tangential longitudinal strain and layer parallel shear are present in folded competent layers and that a special type of "over-

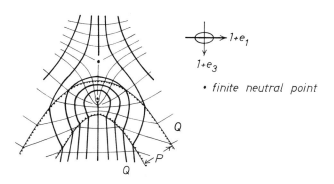

Figure 21.23. *Finite strain trajectories developed around a buckled competent layer P in incompetent matrix Q. The interlocking trajectories around the finite neutral point in layer P define a positive neutral point, the non-interlocking trajectories around the finite neutral point in layer Q define a negative neutral point.*

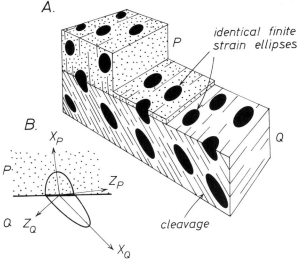

Figure 21.24. *Relationship between the strain states of rock types P and Q along a lithological contact of marked cleavage refraction. The ellipsoids in P and Q differ, but they share a common strain ellipse.*

shear" is developed in incompetent layers as a result of the deformations we have called inverse tangential longitudinal strain. The proportions of the two main straining factors in competent layers depend upon a number of mechanical and geometrical factors. First, materials with a strong layering or anisotropic fabric are more likely to develop deformation by layer parallel shear than those which do not have such initial fabrics. A second factor of importance is the ductility contrast between the differing lithologies making up a layered sequence of rocks. We have seen that, in single layer buckling, if the ductility contrast is high, the size of the initial fold wavelength relative to the competent layer thickness is large. In such folds buckling mostly takes place in the competent layer according to the tangential longitudinal strain plan (Figure 21.25A). In contrast, where the ductility contrast is low the predominant strains are first those of layer parallel shortening without folding. When the folds do start to undergo marked amplification they are of short wavelength relative to competent layer thickness, and the strains induced in the competent layer are predominantly those of layer parallel shear (Figure 21.25B). You can test this yourself by buckling by hand a sheet of plasticene or wax with circular markers drawn on the edge of the sheet which is to become the fold profile: if you produce large wavelength folds you will find that tangential longitudinal strains predominate, and if you force the sheet to form short wavelength folds the same material will deform predominantly by shearing parallel to its contacts. The strain patterns resulting from buckling a single competent layer and its matrix depend greatly on the degree of competence contrast, the large initial wavelength ptygmatic type structures forming in situations of high contrast (Figures 21.25A and 21.26) showing very different cleavage patterns from those where the competence contrast are low (Figures 21.25B, 21.26, 21.27, 21.28).

Effects of layer parallel shortening on the strain fields of folds

In Question 21.4 we modelled a fold developing in layers which had no initial phase of layer parallel shortening such as would be typical of situations of high competence contrast between the layers P and Q. In situations of lower competence contrast we have seen in Session 19 that the initial effect of compressing the layers is to develop layer parallel shortening (and layer thickening) before the amplification of the dominant wavelength folds. In situations where this layer parallel shortening is not too strong, any subsequent tangential longitudinal strain leads to a superposed layer stretching and to an *unstraining* of the outer fold arc of the competent layer and in the surrounding incompetent material. If this unstraining is well developed it can lead to the development of two finite neutral points somewhat similarly located to those predicted in Question 21.4, but probably both lying closer to the incompetent–competent rock interface. Note that the strain history of these finite neutral points is quite complex, the later strains being the reciprocal strain ellipsoid of the earlier layer parallel shortening strain. Such a complex reversal might be identified by sequences of small scale structures which appear to "conflict" with each other, as we discussed earlier in this session. If the extent of the pre-folding layer parallel shortening was greater (lower competence contrast) the

strains induced during folding might not be sufficient to reverse the earlier strain history. In this case no finite neutral points would be developed and the strain trajectories and predicted cleavage pattern would be much simpler than those of the complex loops of Figure 21.23. Figure 21.26 illustrates different types of cleavage patterns that would be expected in folds where no initial layer parallel shortening took place (A), where the initial layer shortening could be partially unstrained (B), and where the fold strains were insufficient to compensate for the layer shortening (C). In general, the cleavage shows overall **convergent fans** in the competent layers, and **divergent fans** in the incompetent layers. (cf. Session 10, p. 181 and Figure 10.21). In Session 17 we discussed the possibility of modifying a fold by a superposition of a homogeneous strain. Figure 21.26D shows how such a modification would effect the cleavage intensity and fan pattern. The divergent and convergent fans can be identified, but are much less marked than those of examples A, B and C. Many situations of so-called axial plane cleavage (Figure 10.17) are of this type. Although at first sight the cleavage appears uniformly oriented through the whole fold, when examined in detail distinct lithologically controlled differences of cleavage orientation can be identified.

It should be emphasized that the predictions that we have made about strain variations and cleavage patterns are almost completely independent of any mechanical analysis of stress–strain history of fold development. The predictions depend upon the concept of geometric constraints implied by the shapes of the folded layers and the requirements of coherence and of strain compatibility between the various parts of a fold. It should be clear that we have chosen to investigate certain displacement fields which appear to have particular geological reality, for example the effects of initial diagenetic or tectonic compaction in a rock, and the extent of the layer shortening before fold formation. There is no doubt that a careful cross reference between the structural features of naturally folded rocks and the predictions of particular displacement and strain models will be an extremely valuable tool to use in interpreting the various strain components that are present in natural folds.

Cleavage and bedding relationships in field interpretations of folds

In geological situations where the cleavage or schistosity is a simple direct result of strain development during the fold forming process, the overall geometric relationships between the orientation and intensity of cleavage are systematically related to fold form. Cleavage refraction on fold limbs and convergent or divergent cleavage fans relate directly to differences in competence between the layers. In these circumstances the relationship between cleavage and lithological layering in a rock outcrop can provide an extremely useful guide in the field to indicate where the particular outcrop being viewed is situated in the overall major structure. Figure 21.29 indicates the variable relationships between cleavage and bedding in a simple antiformal fold with inclined axial surface. Note carefully the following features:

1. On the *normal fold limb* the cleavage has an overall higher angle of dip than the layering.

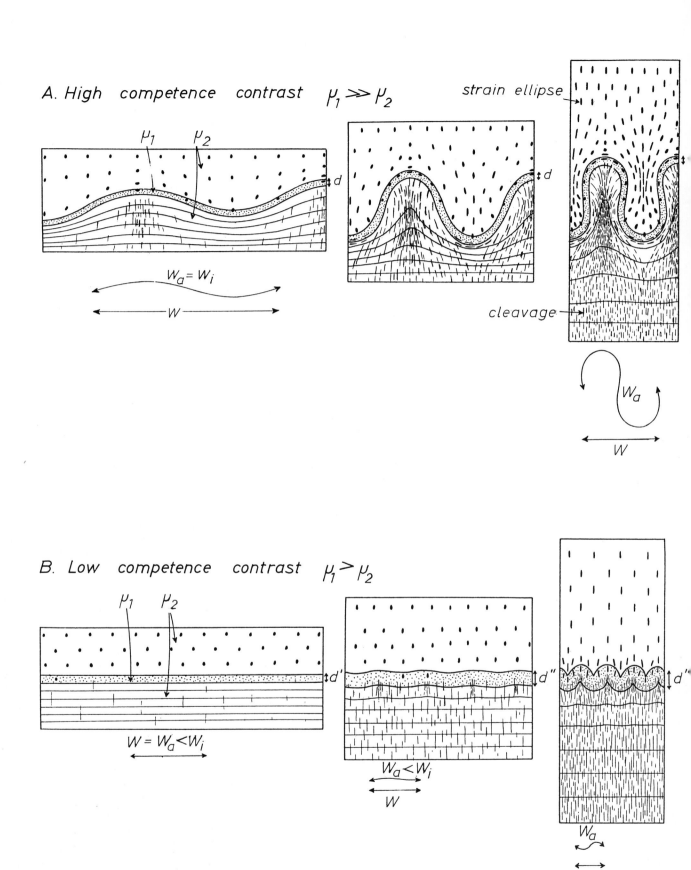

Figure 21.25. *Strain patterns and cleavage variations in buckled single competent layers A where the ductility contrast is high (viscosity $\mu_1/\mu_2 > 50$) and B where it is low ($\mu_1/\mu_2 < 10$). The upper sides of the blocks shows schematically the finite strain states, the lower side show the predicted cleavage patterns.*

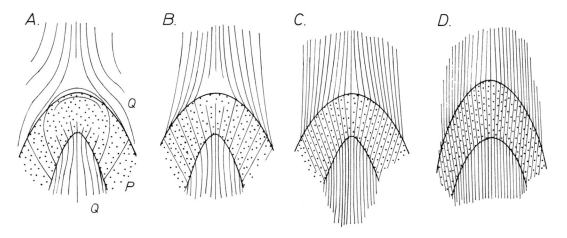

Figure 21.26. *Cleavage patterns in folds with no (A), some (B) and strong (C) initial layer parallel shortening. D shows the modified fold and cleavage geometry arising from a homogeneous strain with shortening normal to the fold axial surface. Rock P is more competent than rock Q.*

Figure 21.27. *Variable cleavage (schistosity) in a folded aplite dyke cutting a xenolithic granite. Maggia nappe, Lepontine Alps, Switzerland. Compare this cleavage pattern with that of Figures 21.25A and 21.26A.*

Figure 21.28. *Cleavage in folded sandstones and shales (slates). Little Rock, Arkansas, USA. Compare the cleavage orientation and intensity with the predictions of Figure 21.26B.*

2. On the *overturned fold limb* the cleavage has an overall lower angle of dip than the layering.

3. At the *fold hinge* the cleavage and layering are perpendicular and there is no cleavage refraction. Here the cleavage is practically always *parallel to the axial surface* of the fold.

4. On both fold limbs *cleavage refraction* occurs, the smaller cleavage-bedding angle α is always characteristic of the less competent layers, the greater angle β is found in the more competent layers. The numerical difference in value of the angle $\beta - \alpha$ is a function of competence contrast and of the location in the fold. If the geometry of the fold is associated with higher strains on the overturned limb than on the normal limb (as is frequent), the cleavage refraction is more marked on the normal limb than on the overturned limb. Generally the *axial surface of the fold* lies with an orientation between the two directions of the cleavage developed in competent and incompetent layers.

5. Note the relationships between refracted cleavage and layering at the fold *crest*, indicating that the crest and hinge lines are not coincident.

6. The geometric relationships described above hold true irrespective whether the fold is an upward or downward facing structure (in Figure 21.29 whether this fold is an antiformal anticline or antiformal syncline—see definitions p. 310). If the polarity of the layers in a fold can be determined (e.g. with sedimentation structures such as cross bedding or lithological grading) then these observations can be combined with cleavage–bedding relationships to *determine the facing direction* of a fold (Figure 21.30). Once this facing direction has

been established then the cleavage (C)–bedding (B) relationships at individual outcrops can be used to *determine the stratigraphic polarity* of the beds at each locality, even when primary indications of bed polarity are absent. For example, Figure 21.31 shows an outcrop of siliceous limestone (pale) and interbedded marl (dark) from the upward facing part of the Morcles fold nappe (Figure 11.10). From the cleavage–bedding relationships it should be clear that the rocks are in a stratigraphically inverted relationship. The slight cleavage refraction shows that the limestone is more competent than the marl, but that the competence contrast is not very great. The elongate shapes of the siliceous concretions on the right-hand side of the outcrop appear to be geometrically related to the orientation of the cleavage. The average aspect ratio of the elliptical forms is $6:1$. This section is an XY plane of the finite strain ellipsoid: are these shapes likely to give directly the strain ratio R_{XY}? Probably the answer is that their forms are not exactly related to the deformation alone because most sedimentary concretions of diagenetic origin have initially non-circular forms with long axes generally lying in the bedding surfaces.

7. The *line of intersection of cleavage and bedding* (trace of cleavage on a bedding surface or trace of bedding on a cleavage surface) is parallel to the hinge line of the folds to which the cleavage is related (Figure 21.32). In any region where cleavage and bedding can be identified the **intersection lineation** should always be measured and recorded on a field map (Appendix F) because these data give a very valuable input into

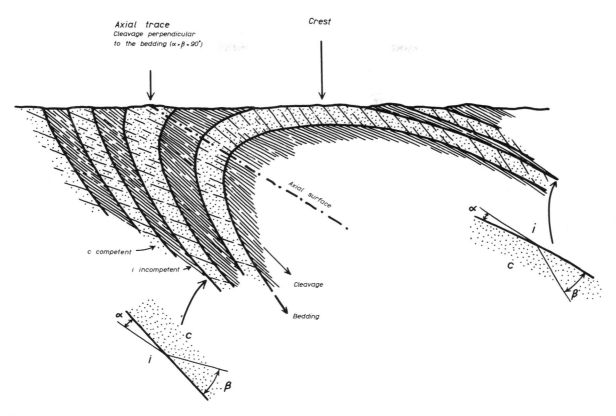

Figure 21.29. *The relationship between cleavage and bedding in an overturned fold.*

establishing hinge line variability in major folds that is not always so quickly recognized using other methods. Practically all folds systems show variations within individual folds and from fold to fold. These variations have many sources:

(i) Variations of the initial dips of layering which are developed into fold hinge line variations. If the folding is accompanied by strong cleavage development, with a pronounced principal strain R_{XY}

aspect ratio, the variations in hinge line orientation become more strongly marked than those of the initial layering (see discussion in Session 10, p. 188 and Figure 10.28). This strong variation in hinge lines sets up the type of folds known as **eyed folds** or **sheath folds**. Such folds are diagnostic of regions which have undergone very high strains with especially marked R_{XY} aspect ratios. Very high strains of this type are

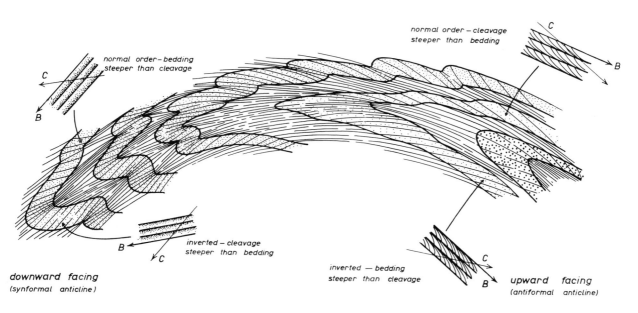

Figure 21.30. *The relationships between cleavage, bedding and bed polarity in upward and downward facing folds.*

Figure 21.31. Cleavage-bedding relationships in the inverted limb of the upward facing Morcles fold nappe. Dent de Morcles, Valais, Switzerland.

Figure 21.32. Cleavage cutting bedding in an antiformal fold. The linear traces of the cleavage on the bedding (left-hand side) and traces of bedding on the cleavage (right-hand side) are parallel to the fold hinge line. South Devon, England.

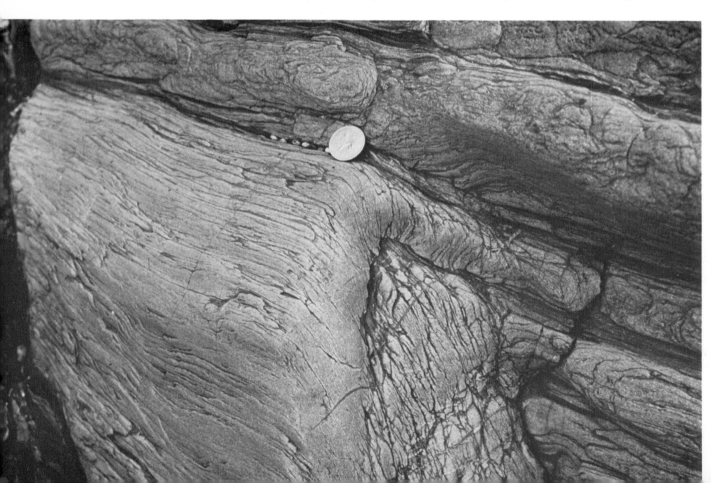

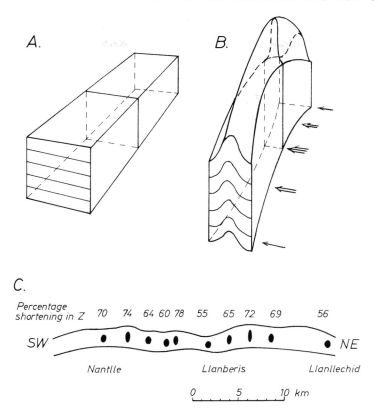

Figure 21.33. A and B show the effect of variation in shortening in the Z-direction on the axis of a fold. An axial culmination coincides with the region of maximum shortening. After Ramsay (1962). C shows a longitudinal section parallel to the cleavage and fold axial planes in the Cambrian slate belt of North Wales After Wood (1974). The two lines represent the changes in fold axial plunge and the variations are related to changes in the degree of shortening perpendicular to this section and the aspect ratios R_{xy} of the strain ellipses.

often found in ductile shear zones (see Session 3) but such strains are not limited to this type of environment.

(ii) Geometric compatibility between adjacent folds developed when one fold "relays" to another (Figure 15.19).

(iii) Variations of shortening sub-perpendicular to the axial surface of a fold (Figure 21.33)

(iv) Late deformations of initially regularly oriented folds under situations of superimposed folding (to be discussed in Session 22).

(v) Movement of folded and transported thrust sheets over irregular thrust surface topography oblique to the initially formed folds (known as side-wall ramps—discussed in Session 23).

Answer 21.5★

The strain matrix for the initial state of Figure 21.3B

$$\begin{bmatrix} R_0^{-1/2} & 0 \\ 0 & R_0^{1/2} \end{bmatrix} \qquad (21.11)$$

Where R_0 is the aspect ratio of the initial strain ellipse with long axis parallel to the y-coordinate direction. Deforming this by simple shear (strain matrix Equation 21.5 using Equation C.22) we obtain the final strain matrix at any point

with dip δ as

$$\begin{bmatrix} R_0^{-1/2} & \delta R_0^{1/2} \\ 0 & R_0^{1/2} \end{bmatrix} \qquad (21.12)$$

The orientation θ' of the finite strain ellipse measured from the rotated layering surface is

$$\tan 2\theta' = 2\delta/(R_0^{-2} + \delta^2 - 1) \qquad (21.13)$$

where δ is measured in radians, and the orientation θ'' measured from the x-coordinate direction can be obtained from

$$\theta'' = \theta' + \delta \qquad (21.4)$$

where δ is in degrees.

The aspect ratio R of the final strain ellipse is given by

$$R = \frac{1}{2}[R_0^{-1} + \delta^2 R_0 + R_0$$
$$+ ((R_0^{-1} + \delta^2 R_0 + R_0)^2 - 4)^{1/2}] \qquad (21.15)$$

with δ measured in radians.

The functions of Equations 21.14 and 21.15 are graphed in Figures 21.34A and B, respectively. For quick reference, the angle α_δ between the X axis of the strain ellipse (and potential cleavage trace on the XZ fold profile) and the bedding (dip δ) can be read off from the graph in question and the dash-dot curve for any particular angle of bedding dip in the fold. Any convergent or divergent relationships of the ellipse long axes with the fold axial surface are indicated in Figure 21.34A in thick or thin lines respectively, the changeover position occurring at $\theta'' = 90°$. Three main types of trajectory relationships can be identified.

Type 1 $R_0 > 1.0$ arising in situations of initial layer parallel shortening

The θ'' curves all pass through the point (0,90), implying that the long axes of the ellipses are always parallel to the axial surface at the hinge. These curves then show turning points at some particular angle of dip generally around 10–20°. These turning points give the angles of dip which coincide with the maximum deviation of potential cleavage from the axial surface of the fold. After passing through the turning point the curves then bow back to cross the $\theta'' = 90°$ line at some numerically higher dip angle and, at dips higher than

this crossing point, the trajectories (and potential cleavage) converge on the fold axial surface. The variation in the orientations of the long axes of the finite strain ellipse increases as the R_0 value approaches the value 1.0. In this type of strain pattern, the strain ellipse aspect ratios R (curves B) vary considerably through the fold and the variation increases with the initial value of R_0.

Type 2 $R_0 = 1.0$ simple flexural folds

The θ'' and R curves are appropriate to the simple flexural fold superposed on unstrained original layers of Figure 21.3A. The θ'' curves are discontinuous across the fold axial surface and each of the two sectors are almost linear. At zero dip they make angles of 45° and 135° (where the strain is zero) and they cross the $\theta'' = 90°$ line at angles of dip $\delta = \pm 58.5°$. At this dip the finite strain ellipses have long axes parallel to the fold axial surface, and at higher dips the trajectories and potential cleavage lines converge with the axial surface.

Type 3 $R_0 < 1.0$ arising in situations of initial layer normal shortening

Although in technical terms R_0, being the ratio of maximum to minimum strain ellipse axis lengths, cannot take on values below 1.0, we have used this nomenclature for R_0 in the sense of the value of the proportion of ellipse y-axis to x-axis lengths. The final graphical representation of R is as normally defined. The plotted curves for $R_0 = 0.3$, 0.5 and 0.9 all pass through the point (0,0), signifying that the long axis ellipse trajectories and layering are parallel at the fold hinge. As the dip values increase numerically, the strain ellipse long axes depart from their initial layer parallel orientation, but the angle of separation is quite small, as shown by the close proximity of the θ'' curves to the dash–dot curve in Figure 21.34A. Over most of the fold surface the long axis trajectories and potential cleavage shows a marked divergent relationship to the fold axial surface. The variation in aspect ratios R of the finite strains is generally small, and decreases as the value of the initial layer normal shortening component becomes more marked (i.e. R_0 decreases in value, compare the decreasing curvature of curves of $R_0 = 0.9$, 0.5 and 0.3).

To conclude, this analysis confirms the results of the limited number of card deck experiments carried out earlier in this session. It demonstrates particularly well how it is possible to investigate the variation in strain patterns in folds using simple mathematical modelling techniques. The following have special geological relevance in situations where flexural slip or flexural flow have played an important role:

1. Initial bedding plane parallel diagenetic fabrics or tectonic fabrics such as would occur in the outer arcs of folds with significant proportions of layer parallel extension exert a very strong influence on the geometric form of the total strain fabric in a rock. In particular, the total fabric shows a strong tendency to be aligned almost parallel to the original fabric component rather than to follow the strain directions of the fold associated deformations, and the intensity of the initial strain is only slightly modified by the folding strains.

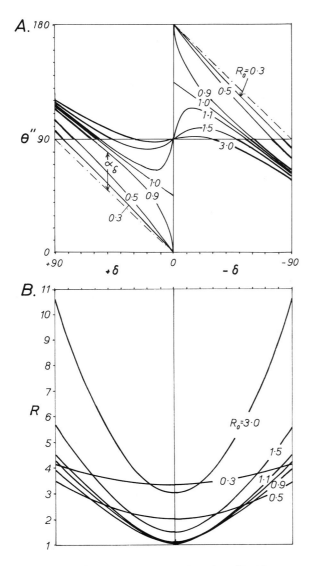

Figure 21.34. A: *Variations in the orientations θ'' of long axes of finite strain ellipses with angle of layer dip δ in flexural folds developed in layers with initial layer parallel strains. For any particular R_0 curve the angle between layering and potential cleavage is given by values like that of α_δ. B: Variations in aspect ratio R of finite strain ellipses.*

2. Fabrics produced by early tectonic layer parallel shortening or intense layer shortening during tangential longitudinal strain processes tend to lead to situations where the overall fabric shows a general alignment with the axial surface of the fold. In detail, however, there is a zone of axial surface divergent fabric near the fold hinge zone and convergent fabrics on the fold limbs. Although the fabric orientations are less marked than in point 1 above, the aspect ratios of the strain ellipses do show more extreme variations.

Answer 21.6★

The terms a, b, c and d of the final strain matrix are derived from the matrix product of Equations 21.1 and 21.5.

$$a = R_0^{1/2} \cos^2\theta + R_0^{-1/2} \sin^2\theta$$
$$+ \delta(R_0^{1/2} - R_0^{-1/2}) \sin\theta \cos\theta$$

$$b = (R_0^{1/2} - R_0^{-1/2}) \sin\theta \cos\theta$$
$$+ \delta(R_0^{1/2} \sin^2\theta + R_0^{-1/2} \cos^2\theta) \quad (21.16)$$

$$c = (R_0^{1/2} - R_0^{-1/2}) \sin\theta \cos\theta$$

$$d = R_0^{1/2} \sin^2\theta + R_0^{-1/2} \cos^2\theta$$

Note that the order of matrix multiplication is important. If you reverse the order you will solve another problem, namely the effect of superimposing a homogeneous finite strain obliquely across a pre-existing simple flexural flow fold. When the initial ellipse axes are parallel to the x and y coordinate directions Equation 21.16 simplifies to that of Equation 21.12 ($\sin\theta = \sin 90° = 1$, $\cos\theta = 0$). The orientations of the principal finite strains from the folded bedding surfaces are given by

$$\tan 2\theta' = 2(ac + bd)/(a^2 + b^2 - c^2 - d^2) \quad (21.17)$$

and the orientations with respect to the x, y coordinate system

$$\theta'' = \theta' + \delta \quad (21.18)$$

The aspect ratio R of the finite ellipses is

$$R = \tfrac{1}{2}\{a^2 + b^2 + c^2 + d^2$$
$$+ [(a^2 + b^2 + c^2 + d^2)^2 - 4]^{1/2}\} \quad (21.19)$$

The functions 21.18 and 21.19 are plotted in Figure 21.35A and B respectively and show a number of significant differences from those curves of Question 21.5★ where the initial strain axes were symmetric with the layer surfaces. The angle α_δ between the strain ellipse long axes and the layering (dip δ) no longer shows a symmetric relationship across the fold (see variations of α_δ between any of the $\theta = 30°$, 45° or 60° curves and the dash–dot line of Figure 21.35A), a feature which had been initially noted in our card deck model Experiment D. This asymmetric relationship of the strain trajectories and potential cleavage to the main geometrical features of the fold is therefore a general one.

The variations of aspect ratio R of the finite strains for the curves of initial obliquity of 30°, 45° and 60° are also markedly asymmetric across the fold. Much higher strains occur on limbs with positive dip than on those with negative

dip. On the limb with negative dip each of the three curves reaches a minimum value. This is the position in the structure where the cleavage fabric intensity will be lowest (see the right-hand and left-hand sides of Figure 21.17). The minimum value of R on the curve for $\theta = 30°$ is unity, implying that the original strain has been completely effaced. This possibility occurs because an initial ellipse with

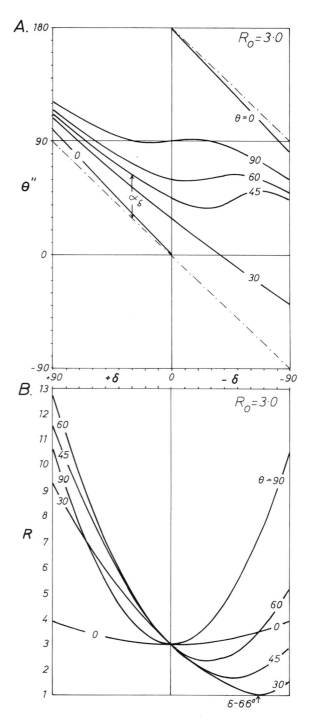

Figure 21.35. A: *Variations in the orientations θ'' of long axes of finite strain ellipses with angle of layer dip δ in layers with initial oblique (θ) strains with initial aspect ratio $R_0 = 3.0$. For any particular θ' curve the angle between layering and long axis at dip δ is α_δ. B: Variations in aspect ratio R of finite strain ellipses.*

aspect ratio $R_0 = 3 \cdot 0$ and orientation $\theta = 30°$ could have been formed by a right-hand sense simple shear parallel to the layering with shear strain $\gamma = -1 \cdot 15$ (see Figure 2.10). Where the left-hand shear in the flexural fold attains a shear value of $\gamma = +1 \cdot 15$ (at an angle of dip $\delta = -66°$) the original displacements are completely counterbalanced by the displacement in the fold, and so the total strain returns to an undeformed state.

KEYWORDS AND DEFINITIONS

Finite neutral surface   A surface joining points of isotropic strain ($R_{xy} = R_{xz} = R_{yz} = 1$) in a folded layer. In many folds in the process of development an **incremental neutral surface** can be found joining points where the incremental strains are isotropic (Figure 21.21).

Flexural fold   Fold showing parallel fold style formed by simple shear parallel to a planar structure (e.g. bedding, laminations). if the simple shear displacement is continuous throughout the structure a **flexural flow fold** will be formed (Figure 21.1A), whereas if the shear is discontinuous the fold is termed a **flexural slip fold** (Figure 21.1B).

Parasitic fold   Folds of small wavelength and amplitude located within folds of larger wavelength and amplitude in situations of **polyharmonic folding**. Parasitic folds normally show S or Z shaped asymmetric forms in the limbs of the larger structure (sometimes termed **drag folds**) and symmetric M forms in the hinge zones (Figure 15.21).

Rodding   A linear rock fabric generally found at the hinge zones of major folds produced by the parallel orientation of parasitic fold hinges (**fold rodding**) (Figure 21.16). **Quartz rods** are produced by the folding of quartz veins or by the separation of quartz veins during boudinage processes. Rod-like structural fabrics can also be produced by **intersection lineation** (see p. 468) and by the development of intense finite **stretching fabrics** (where $R_{xy} \approx R_{xz} \gg R_{yz}$).

Tangential longitudinal strain   A strain pattern developed in a folded and generally competent rock layer by layer parallel stretching in the fold arcs, (Figure 21.18). The reverse type of strain variation, generally characteristic of incompetent layers, is known as **inverse tangentional longitudinal strain**. (Layer Q, Figure 21.22.)

KEY REFERENCES

Cloos, E. (1946). Lineation. *Geol. Soc. Am. Mem.* **18**, 122 pp.

This is one of the classic texts describing the great range of linear structures found in deformed rocks and should not be missed.

De Sitter, L. U. (1958). Boudins and parasitic folds in relationship to cleavage and folding. *Geol. en Mijnb.* **20**, 277–286.

One of the best early attempts to relate the geometry of small scale and large scale tectonic structures and to account for the variations of form of small scale structures in terms of different progressive strain history in different parts of the fold.

Dieterich, J. H. and Carter, N. L. (1969). Stress history of folding. *Am. J. Sci.* **267**, 129–154.
Dieterich, J. H. (1970). Computer experiments on mechanics of finite amplitude folds. *Can. J. Earth Sci.* **7**, 467–476.

These papers marked a new era in structure geology techniques, the use of the large computer to carry out finite element analysis on viscous models. They showed how the predicted patterns of finite strain (and stress states evolving during the folding process) had remarkable geometrical correspondences with many of the small scale structural features and rock fabrics we see in naturally deformed rocks.

Kligfield, R., Carmignani, L. and Owens, W. H. (1981). Strain analysis of a Northern Apennine shear zone using deformed marble breccias. *J. Str. Geol.* **3**, 421–436.

Smith, R. B. (1975). Unified theory of the onset of folding, boudinage and mullion structure. *Geol. Soc. Am. Bull.* **86**, 1601–1609.
Smith, R. B. (1977). Formation of folds, boudinage and mullions in non-Newtonian materials. *Geol. Soc. Am. Bull.* **88**, 312–320.

Wilson, G. (1953). Mullion and rodding structures in the Moine Series of Scotland. *Proc. Geol. Assoc. Lond.* **64**, 118–151.
Wilson, G. and Cosgrove, J. W. (1982). "Introduction to Small-scale Geological Structures". George Allen and Unwin, London. 128 pp.

This paper shows how strain determinations over a large region of deformed rocks can be valuable in interpreting the geometry of large scale and small scale structures.

These two publications should be read by keen students wanting to keep up with exciting new ideas. They are both somewhat mathematical in their approach, but what is so useful to the field geologist is that some rather unexpected results of non-linear flow behaviour offer a completely new way of looking at the forms of certain types of structures in deformed rocks.

The development of modern structural geology owes very much to the work of Gilbert Wilson. His practical experience in industry led him to understand the importance of mapping small scale structures which he termed "tectonic weather cocks" for evaluating the movements that had taken place. This book is a slightly modified version of a paper published in 1961 (*Ann. Soc. géol. Belgique*) which, in its time, was a great eye-opener to the many geologists who were of the opinion that small scale structures were put into rocks by some devilish agent wanting to confuse the real geological picture. The long reference list contains many useful and often off-the-beaten-track references.

SESSION 22

Superposed Folding

The various different types of geological strain history which can lead to the development of folds with several different orientations at the same locality or in the same region are described. Initial interpretations of superposed folding sequences are usually based on an analysis of the two-dimensional forms of fold interference patterns. A more detailed three-dimensional analysis of fold geometry shows how the initial angular relationships of surfaces and linear features of early folds can be modified during a later fold-producing deformation and how many of the geometric characteristics of the superposed folds are controlled by the morphological features of the early folds.

INTRODUCTION

The fold structures found in most orogenic zones usually show considerable geometric complexity, and it is common to find that fold forms show complications in three dimensions that are the result of the superposition of folding instabilities on pre-existing sets of folds. In many continental terrains the crust may be subdivided into a basement, often containing folds and fractures acquired during earlier orogenic deformation, unconformably overlain by a sequence of flat lying or gently dipping sediments. When such a combined rock grouping is subjected to a compression, although the cover sediments may develop regular and relatively simple fold structures, the contraction of the basement may not be able to be accommodated by reactivation of the pre-existing structural forms. New structures are therefore developed which are superposed on the earlier folds, and these new structures are controlled both in orientation and in style by the anisotropy existing before their formation.

A second variety of superposed folding arises when the principal stress directions change during the history of development in an orogen. Research over the past two decades has indicated that this is a very common phenomenon in all orogenic zones, particularly in those parts of the orogen where the total crustal compression has been large. This type of superposed deformation may arise either as a result of the activity of discrete pulses of shortening separated by periods of quiescence and no crustal displacement, or as the result of separation of a regional deformation into zones where shortening is being accommodated by body translations and body rotations, and other zones where the shortening is taking place by internal deformation. At any one locality the periods of translation without strain and those periods of active deformation can vary with time, and the geometrical effect is to produce local discrete and superposed shortening and folding events. Local deformation events which can be separated on geometric criteria are generally designated $D_1, D_2 \ldots D_n$.

Although the local sequences may be regionally related, it is usually advisable to seek evidence of their absolute timing using radiometric age determinations of stratigraphic relationships. It is generally found that the time scale of such superimposed deformations is of the order of tens of millions of years. With such a time period between the principal deformation phases, the temperature and pressure conditions at any one locality generally change. The deformation style and mineral stability of the rocks are usually governed by the metamorphic conditions existing at the time of the individual deformation phases, and the successive structural forms are usually characterized by particular types of metamorphic fabric fingerprint and particular mineral assemblages.

A third type of superimposed folding occurs during a single progressive deformation as a result of smooth and systematic changes of stress and incremental strain during deformation. In Session 12 we indicated how a regular sequence of incremental strain variations can lead to complex shortening and lengthenings in a layer and how a spectrum of differently oriented folds and boudinage features might arise.

Structural forms which resemble those of truly superimposed folds can also be produced when a layer of competent rock is subjected to a constrictive type of deformation (ellipsoid Type 4 or 5, Session 10, p. 171) and where the layer is situated in a position within the ellipsoid so that it undergoes shortening in all directions (ellipse Field 3, Session 4, p. 66). With this type of deformation folding can take place synchronously in several directions and mutually intersecting fold forms of considerable complexity can develop. These folds generally show some geometric differences from those arising by successive superposition (Figure 4.11) and, because of their synchroneity, they should show relationships to the deformation fabric produced during metamorphic crystallization that are independent of their

Figure 22.1. *Folded discordant pegmatite dyke cutting metasandstone, Moine Series, Bettyhill, N. Scotland. See Question 22.1.*

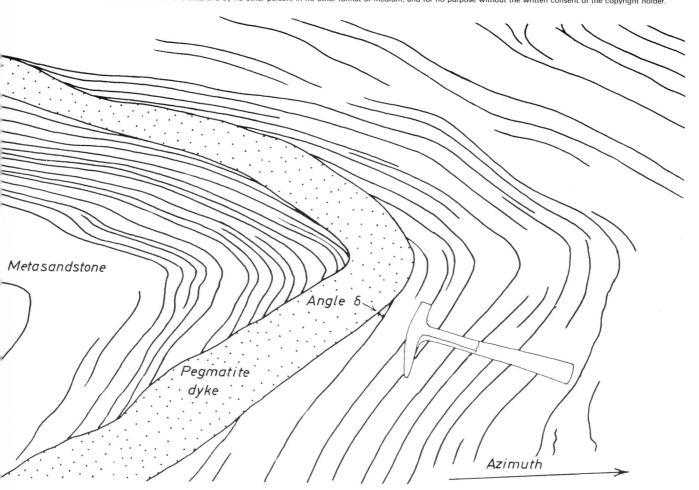

Metasandstone

Angle δ

Pegmatite dyke

Azimuth

orientation. These folds are also likely to be associated with prominent linear- or l-fabrics oriented at a moderate to high angle to the average orientation of the folded surfaces.

Although the existence of superimposed folds in many terrains has long been realized, the real understanding of the geometric details was only clearly established in the mid 1950s (Reynolds and Holmes, 1954; Weiss and McIntyre, 1957; Clifford *et al.*, 1957; Ramsay, 1958; Weiss, 1959). Although the geometric features of such systems are usually very complex, the complexities obey well defined rules. To make a detailed analysis it is usually necessary to make a careful study of a large number of field observations of the orientations of planar and linear features in the rocks. The analysis of the angular relationships of these data is most readily carried out using projection techniques (Sessions 9 and 16). However, we must emphasize that, because several fold axial directions are likely to exist in a terrain of superposed folding, a very thorough study of the geological map must be carried out before the data are plotted on an equal area projection, with special consideration given to the geograpical location of each datum.

QUESTIONS AND STARRED (★) QUESTIONS

The folding of non-parallel surfaces

Question 22.1

To begin our study we will first investigate some of the properties of folds which are developed in surfaces which are not initially parallel. Many of the special geometric complications found in regions of superimposed folding arise because there is a range of orientation of lithological and structurally induced surfaces as a result of the development of the first phase folds (variably oriented fold limbs with especially strong local variations around the hinge zone, intersecting lithological and cleavage surfaces). Our first problem will be to investigate particularly simple sets of intersecting surfaces, namely what happens when a fold is developed in a folded sedimentary layered sequence containing an initially cross cutting igneous dyke. Figure 22.1 is an outcrop of folded metasandstones which include a discordant pegmatite dyke. Outside the vicinity of this fold the angles made by the dyke with the bedding planes of the sediment are rather constant. An initial inspection of Figure 22.1 will show that the angles between the bedding and the dyke vary with position in the fold. An azimuth line has been constructed parallel to the axial trace of the fold in the bedding surfaces. Determine at as many points as possible the angle δ between the trace of the dyke contact and the sedimentary bedding and make a graphical plot of δ (ordinate) against orientation of the folded bedding plane trace (abscissa). Suggest models which might account for this variation. Plot the location of the traces of axial surfaces of the folded dyke and of the folded bedding. Do they have the same orientation? Do they have the same location? Now proceed to the Answers and Comments section and then continue with Question 22.2.

Superposed shear folds

Question 22.2

Certain features of superposed fold geometry can be easily studied in the laboratory using card deck models. We have already emphasized that folding by differential simple shear is only one of several types of displacement plan which can lead to fold development, but it is a geologically realistic displacement in certain environments, for example where folds form in certain types of ductile shear zones. In Session 3 we saw that differential simple shear across a set of initially planar surfaces will induce *similar folds*. In our study of superimposed folding we propose to use the card deck to show how simple shear affects the geometry of converging planes such as are found in the limbs of folds which predated the simple shear. Construct on the surface of a card deck a series of chevron fold shapes (Figure 22.2A) with limb orientations of surfaces p and q given by $\theta_p = +21°$, $\theta_q = -21°$. In order to compare the results of this experiment with some of the features of the folded dyke of Question 22.1 we have also drawn a cross cutting dyke d through the constructed chevron fold. Shear the card deck sideways using the sinusoidal shaped wooden endpieces that we used for the experiments of Question 3.1 and have shown in Figure 1.2. Figure 22.2B shows the results of this experiment.

Draw the traces of the axial surfaces of all sets of folds. Do the axial traces of the superposed folds run continuously across the folded axial traces of the first folds? Determine the interlimb angle α' of the deformed folds at different positions in the structure and make a graphical plot of α' (ordinate) against the inclination (θ'_p measured from the x-direction) of the first fold limb p. Discuss the pattern and significance of this variation.

What would be the orientation of the superposed fold axes (1) if the surfaces p and q indicated in Figure 22.2A were initially perpendicular to the surface of the card model (i.e. perpendicular to the diagram surface of the figure), and (2) if the surfaces p and q were inclined at some angle to the model surface? Make three-dimensional sketches to illustrate the resulting geometric forms.

What would be the effect on the overall geometry of the deformed first folds if the shear folding had been followed by a homogeneous plane strain with maximum compression perpendicular to the shear planes and maximum stretching within the shear planes?

Although it is recommended that this question should be carried out graphically, mathematical functions can be found to describe the geometrical features. If you have the time or mathematical interest proceed to Question 22.3★ below. If not go to the Answers and Comment section and return to Question 22.4.

Mathematical analysis of superposed shear folding

Question 22.3★

The displacement field of the experiment illustrated in Figure 22.2B which relates the coordinate of an initial point (x, y) to its final position (x', y') is given by

$$x' = x$$

$$y' = y + 2 \sin x$$

Find expressions which give the location of the axial surface traces of the new folds developed on the first fold limbs and which were initially inclined at angles of θ_p and θ_q to the

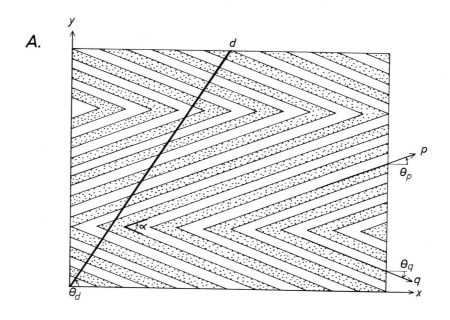

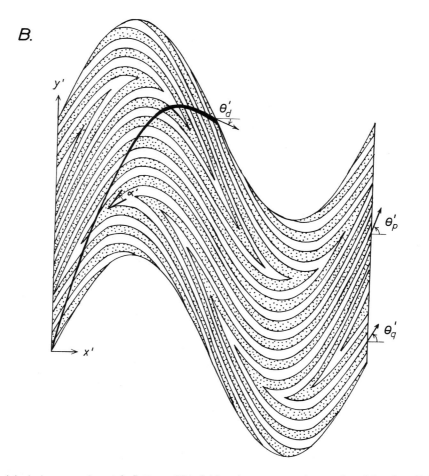

Figure 22.2. *Card deck shear experiment.* A: *Pattern of first folds to be constructed on card model surface.* B: *The resulting geometry after shearing the card deck with a sinusoidal endpiece. See Question 22.2.*

Figure 22.3. *Superimposed fold interference patterns in gneisses, Cristallina, Lepontine nappes, Switzerland. See Question 23.4.*

as to come to lie on the fold surface in a curvilinear form. The different parts of the curvilinear trace of the lineation lie on an imaginary plane (great circle on a projection) which is defined by the initial lineation direction and the direction of motion (a-direction in Figure 22.7D). This plane is "imaginary" in that no truly planar surface cutting the fold surface actually exists in the rock. This geometry sets up angular relations (α') between the deformed lineation and the fold hinge line direction which vary through the fold (Figure 22.7D). In the field the location of this lineation plane can often be seen directly in the outcrop by changing the observer's viewpoint of the deformed lineation (Figure 22.8), and it is usually a relatively simple procedure to make a direct recording of the deformed lineation plane. When this plane is intersected with the axial surface of the fold, the line of intersection directly determines the differential transport direction (a-direction) of the fold forming movements. It is fairly common to find that the slip direction is not perpendicular to the fold hinge line.

Where a shear fold is associated with a homogeneous strain the geometric features of Figure 22.7D are modified to those of Figure 22.7E. The plane on which any deformed lineation is situated comes to lie closer to the XY plane of the imposed strain and the slip direction of simple shear moves towards the X-axis (a'-direction). The intersection of the lineation plane with the axial plane of the fold now gives a direction which records the overall differential movement (a') which relates to the total fold form.

Interpretation of deformed lineation geometry in a superposed fold system

Question 22.6★

At locality X in Figure 22.4 observations of the orientations of first generation linear structures (quartz–feldspar rods parallel to F_1) in a second fold were made and tabulated (Table 22.1). The hinge line of the fold showed a plunge of 35° towards 213° and the axial surface had a strike of 45° and dipped 75° to the SE. Discuss the kinematic significance of these data.

Table 22.1.
First phase lineation orientations in a second phase fold.

1.	4° to 263°
2.	29° to 253°
3.	42° to 256°
4.	49° to 250°
5.	63° to 239°
6.	76° to 223°
7.	39° to 97°
8.	26° to 89°
9.	20° to 85°
10.	5° to 89°

ANSWERS AND COMMENTS

The folding of non-parallel surfaces

Answer 22.1

The angle δ between the contact of the dyke and the bedding surface has been plotted against the orientation α of the bedding surface in Figure 22.9. The data spread is uneven because of the rather sharp hinge zone form of the fold, and most of the data points fall in two groups related to the two fold limbs. The highest values of δ occur near the fold hinge but do not appear to coincide exactly with the actual hinge position. On both limbs the angle δ attains low values, particularly on the limb with α values of 150°–170°. The distribution of the variable δ-angles in the graph appears to be systematic and is clearly asymmetric with respect to the axial surface of the fold. When the δ-values in the fold are compared with those seen in dykes outside the folded environment, it is found that the high values exceed those of normal values, whereas the low values on the fold limbs are considerably less than normal. Such geometrical variation is characteristic of folded discordances on a wide variety of scales from major stratigraphic unconformities to small scale cross bedding. The reduction of angular discordances on fold limbs is a widely observed feature in many orogenic belts and leads to the "obliteration" of original discordances. For example Figure 22.10 illustrates an antiformal fold developed in cross bedded sandstone. The cross bedding relationships are quite clear in the hinge zone of the structure and indicate that the fold is an anticline; however, on the fold limbs it is exceedingly difficult to recognize the discordant bedding of the primary sedimentary structure because of the reduction of angle between the forset beds and the truncating bottomset of the stratigraphically overlying unit. It is important to realize that this effacement of the discordance is not just a locally produced deformation feature along one layer, but that the angular changes affect the whole body of rock in the fold limb. The angular changes are the result of the bulk strain state in the fold limb. Another especially important feature of this geometry is that initial angles are often *increased* near the fold hinge zone. For example, the maximum angle of repose of sand in the forset bed of a cross bedded unit can never exceed an initial value of 32°. However, in folded sandstones anomalously high angles sometimes exceeding 60° can be found in fold hinge zones (Figure 22.11). The application of these principles to a study of superposed folds should be clear. This discussion implies that the cross sectional shape of any initial fold (measured by the value of the interlimb angle and hinge curvature) will be changed as a result of re-folding. It therefore follows that correlation of fold phases using similarity of fold style alone is unsound, or, at best, should be used only where some knowledge of the range of variation of the initial foldforms in the superposed folds is available.

The axial surfaces of the folds in the bedding planes and in the dyke are defined by joining the positions of maximum curvature of the surfaces. Figure 22.9A indicates these axial surfaces: they are parallel, but they do not coincide, the axial surface of the folded dyke is shifted sideways relative to that of the folded bedding planes. Question 22.2 will look into the reasons for this lack of coincidence. Now return to Question 22.2.

Superposed shear folds

Answer 22.2

The axial surface traces of the various folds form two sets. Those of the first phase folds have folded forms because the

Figure 22.8. A linear structure produced by the intersection of a schistosity–lithological banding (parallel to the axes of a first set of folds) deformed when the schistosity surface was folded by a second deformation. A is a general oblique view of the structure and B illustrates the same fold viewed from a position where a particular curvilinear form can be aligned into a plane. Micaceous quartzite from Central Sahara, Algeria.

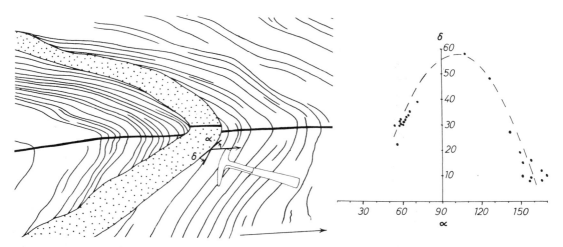

Figure 22.9. Answer 22.1. A shows the positions (heavy lines) of the axial surfaces in the pegmatite dyke and country rock. B is a graph of the angle between dyke contact and bedding planes δ plotted against the orientations of the dyke contact α.

Figure 22.10. Folded cross bedded sandstone showing variations of foreset cut-off angle with position in the fold. Moine Series, Garve, N. Scotland. Photograph by Jack Soper.

axial surfaces are displaced in essentially the same way as the layering in the limbs of the first folds. Those of the second phase folds are all parallel to the y-direction of the cards but show important en-echelon discontinuities across the traces of the first folds (Figure 22.12A). These offsets arise because the maximum curvature of each layer which defines the hinge of each fold is influenced by the initial orientation of the layer (θ_p or θ_q). Although the hinge lines appear to be located where the layers p and q are horizontal this is not quite correct, as will be shown mathematically in Question 22.3★. This non-horizontal relationship of the hinge position of the fold can perhaps be best appreciated in the folded dyke d. The dip of the dyke at the position of its maximum curvature is clearly towards the left of the diagram (θ'_d positive).

Plots of the interlimb angle α' of the first folds with variation in angle of dip of surface p are shown in Figure 22.13B as open circles. The variation is asymmetric with respect to the symmetry of the dip of the bedding planes and to the axial surfaces trace of the fold (located at $\theta'_p = +2\cdot1°$, see Answer 22.3★). The highest value of α' occurs at $\theta'_p = 21°$ and at this position the interlimb angle $\alpha' = 42°$, which is the same as that of the constant original angle before the second deformation. On both limbs of the second fold the simple shear deformation leads to a reduction in interlimb angle of the first fold. In fact with very strong shear deformation the angle decreases so much that the first folds become almost isoclinal in form. It is important to realize that the initially constant first fold profiles look very different at different localities in the second folds. This confirms the comments we made to Answer 22.1, that fold profile shape is not always a reliable correlation guide for folds of the same phase.

If the layers p and q are oriented perpendicular to the model surface of Figure 22.2 the fold axes in both surfaces will be oriented perpendicular to this surface (Figure 22.12B). If these surfaces are inclined at some other angle, then the axial directions of the folds will be governed by the initial orientations of the surfaces. In the case of shear folding, investigated in this question, the intersection of the shear plane (cards in the card deck) with the surface undergoing folding will have the same orientation throughout the p-surfaces and a constant, but different orientation through the q-surfaces. This geometry implies that the p-surfaces will all be cylindrically folded about their intersection line with the shear plane, and that the q-surfaces will be cylindrically folded about another line formed by their intersection with the shear plane (Figure 22.12C). This conclusion that the axial directions of the second folds generally varies from locality to locality in the deformed first fold structure and that these directions are, in part, inherited from the surface dips of the first folds is an extremely important one. These results mean that, in general, there will be more than two principal directions of folds in a two-phase fold system. Figure 22.12C shows that the first fold hinge lines vary from place to place as a result of deformation and that, because the first folds had a chevron fold, the new folds have two dominant orientations controlled by the initially constantly oriented fold limbs. In fact, the *minimum number of fold axial directions in a general two phase superposed system is four*, arising when both first and second generation folds have chevron forms. This regular, complex but systematic variation of axial direction leads us to make the suggestion that the apparent simplicity implied by the descriptive term **cross folds** for these structures is misleading and that the term should be dropped from current usage. It should be

Figure 22.11. *Highly deformed cross bedded sandstone (inverted) in the hinge zone of a fold. Moine Series, Cluanie, N. Scotland.*

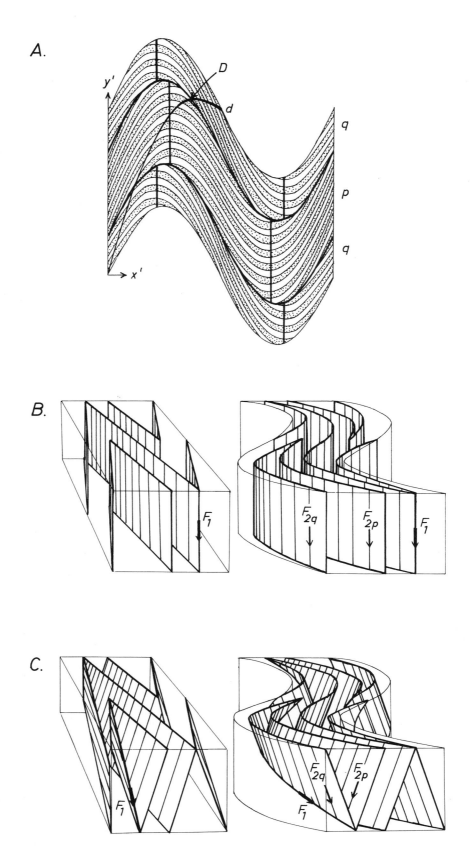

Figure 22.12. A gives an analysis of the axial traces of first and second folds in the deformed card deck of Figure 22.2. Note the continually curving forms of the first fold traces and the en-echelon offsets of the planar second fold traces. D marks the position of the axial trace of the folded dyke. B and C show the three-dimensional form of the deformed first folds where the initial fold axes F_1 are perpendicular and inclined to the card deck surface respectively. In B the hinge lines of the second folds F_{2q} and F_{2p} are parallel to the early fold hinges, whereas in C they have different orientations, being governed by the variable inclinations of the first fold limbs.

SUPERPOSED FOLDING 489

apparent from the geometry shown in Figure 22.12C that, where the p- and q-surfaces are not perpendicular to the shear direction, the angle α′ measured on the top surface of the model will be the **apparent interlimb angle** of the first folds and not the **true interlimb angle**. The true interlimb angle will be the dihedral angle between the p- and q-surfaces measured in a surface perpendicular to the first fold axis, and can easily be obtained by stereogram projection methods.

If the model second phase shear folding were later subjected to a homogeneous strain, the geometry we have described above would be modified depending upon the values of the principal plane strain ratios and orientations of the strain axes. If the surface of our model was the XZ plane of the later deformation with Z perpendicular to the shear plane (cards of the card deck), the following effects would be observed:

1. The curvatures of p- and q-surfaces would change and the positions of the second fold hinge lines and axial surfaces traces would not be the same as those for simple shear alone. The axial surface traces, however, would still show an en-echelon offset when followed across the first fold axial traces.
2. The finite wavelength W of the second folds would decrease and the amplitude A increase to values W' and A' such that $A'/W' = AR_{XZ}/W$ (where R_{XZ} is the ellipticity of the strain ellipse in the XZ plane).
3. The two directions of second fold axes would be modified: both would come to lie closer to the X-axis depending upon the strain ratio R_{XY} of the later deformation.
4. The apparent interlimb angle α′ of the first folds in the XZ plane would be modified; so would the true interlimb angle if the p- and q-surfaces were not oriented perpendicular to the XZ plane. Although the overall pattern of Figure 22.12B would be retained, the absolute maximum of α′ would be greater than that produced by simple shear alone, and the minimum values of α′ on the fold limbs would be decreased. The amount of change would depend upon the value of the ratio R_{XZ} of the later strain.

We have noted that the kinematics of the second fold formation of this experiment obey very special geometrical principles. In Sessions 19 and 20 we have already seen that many other displacement and strain plans are mechanically possible to account for fold formation. Why have we therefore gone into detail in investigating the geometry of this special model? The reason is that, in our experience, many of the geometrical features of naturally formed superposed folds do accord rather well in their broad features with this particular model. Undoubtedly buckling mechanisms do play some part in producing additional geometric complexities to those of the simple shear model (see Question 22.4), particularly in the selection of fold wavelength and in controlling the variations in form of the interfaces between layers of different competence. If a regularly oriented series of planar parallel multilayers is compressed, the axes of the initial folds so formed as a result of buckling instability in the layers should be perpendicular to the first principal shortening strain in the plane of the layering. The mechanical problem with superposed folding is that we have differently oriented groups of planar parallel

multilayers with each group being mechanically attached to the next across the hinge zone of the first folds. No one has yet investigated, on any exact theoretical basis, the likely instabilities which would form in such a complex combined system of multilayers. Indeed, it seems to us to be an extremely formidable analytical task to find a solution where the differently oriented layer groups are arranged in a completely general way to the bulk strain axes and imposed boundary conditions. It is not difficult to analyse the behaviour of each layer set separately, determining the maximum contraction directions for each set acting with mechanical isolation from the other sets. However, we are of the opinion that such predicted buckle fold directions would be unlikely to form. Our reasons are based on the compatibility problems of developing such differently oriented groups of folds across the first fold axial surfaces. The sideways deflections of one group of layers must be integrated with those of the differently oriented adjacent group of layers. The mechanical problems of the two groups cannot be treated in isolation and must be linked in some way. It can be shown that independent buckling by simple flexural slip or flexural flow of the two limbs of an initial first fold produce very complex compatibility problems across the first fold axial surface (Ramsay, 1963, 1967). In fact, these mechanical problems are the reason why corrugated steel plates are much stronger than a single unfolded sheet of steel. In geological situations our experience shows that superposed folding *is* of quite common occurrence, and that a previously folded rock multilayer in certain environments *can be refolded* across its initial fold axes. We suggest, therefore, that to produce such superposed folds the axial planes of the new folds in one set of folding surfaces have to interconnect in some simple way with those of an adjacent and differently oriented set of surfaces. For such folding to proceed in a compatible way there must be a geometric link of axial surfaces and fold axes across the first structure, and the directions of differential flow in the different first fold limbs setting up the second folds are unlikely to be perpendicular to any of the differently oriented second fold axes in these limbs. Folds with such complex buckle components have been termed **oblique flexural slip folds** (Ramsay, 1967, p. 396). Probably a better general terminology would be **oblique flow buckle folds**, in contrast to **normal flow buckle folds** where the overall fold forming displacement in the layer are perpendicular to the layer. In many respects oblique flow buckle folds have displacement plans rather like those of shear folds. In fact, the simplest compatible solution to our problem of second fold linkage across the axial surface of the first fold is to arrange the differential fold-forming flow direction to be parallel in the differently oriented sets of multilayers, a geometrical arrangement which is rather close to the simple shear model we have investigated in this question. It should be clear from this discussion that many mechanical problems in superposed fold systems need to be investigated further. We think there is much scope for future research, and that it would be most fruitful to combine analytical modelling with laboratory model experiments and further studies of the geometry of superimposed folds in naturally deformed rocks.

Answer 22.3★

The equations for the traces of the surfaces forming the

limbs p and q of the initial first folds are given by

$$y = m_p x + k_1 \qquad (22.1)$$

$$y = m_q x + k_2 \qquad (22.2)$$

where m_p and m_q are the slopes ($\tan \theta$ and $-\tan \theta$ respectively in Figure 22.2) and k_1 and k_2 are independent constant terms which depend upon the intersection of each surface with the y-axis. After deformation by simple shear parallel to the y-axis given by $x' = x$ and $y' = y + 2 \sin x$ the bedding traces of the two first fold limbs become curved:

$$y' = m_p x' + 2 \sin x' + k_1 \qquad (22.3)$$

$$y' = m_q x' + 2 \sin x' + k_2 \qquad (22.4)$$

The curvature c of any line whose equation is $y = f(x)$ is given by $c = d^2 y/dx^2 (1 + (dy/dx)^2)^{-3/2}$. The expressions for the second fold curvatures of the sets of lines p and q are therefore independent of their location (k_1 and k_2 values) and are

$$c_p = -2 \sin x'/(1 + (m_p - 2 \cos x')^2)^{3/2} \qquad (22.5)$$

$$c_q = -2 \sin x'/(1 + (m_q - 2 \cos x')^2)^{3/2} \qquad (22.6)$$

These functions are plotted in Figure 22.13A for the example shown in Figure 22.2B (with $m_p = \tan 21°$ and $m_q = -\tan 21°$). The hinge lines are found at positions of maximum and minimum curvature and the axial plane traces joining these points are parallel to the y'-axis but do not coincide. When passing from a p-orientation to a q-orientation in the y'-direction the axial traces of antiforms are shifted in a left-hand en-echelon sense, whereas synforms are shifted in a right-hand en-echelon sense. In Figure 22.13A the p-surface axial surfaces are found where $x' = 100°$ and $260°$, and the q-surface axial surfaces where $x' = 80°$ and $280°$. At these positions the dips of the surfaces θ'_p and θ'_q are almost parallel to the x-axis and can be obtained from the slopes of the lines of Equations 22.3 and 22.4.

$$\theta'_p = \tan^{-1}(m_p + 2 \cos x') \qquad (22.7)$$

$$\theta'_q = \tan^{-1}(m_q + 2 \cos x') \qquad (22.8)$$

which give $q'_p = 2·1°$ and $\theta'_q = -2·1°$.

The hinge position of the folded dyke is displaced even

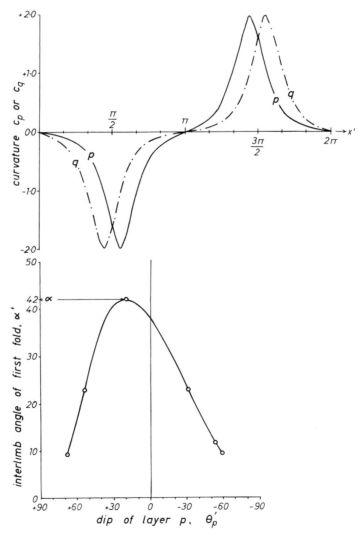

Figure 22.13. A: *Graphical representation of Equations 22.5 and 22.6 of Answer 22.3★ showing that the positions of the axial traces (maximum and minimum curvature) of the second folds in layer sets p and q do not coincide. B: Plot of the function of Equation 22.10 showing the variation of first fold interlimb angle with dip of layer p. The circles show the data derived by direct measurement in the laboratory model (Answer 22.2).*

more than are the hinges of the first fold limbs being found at $x' = 132.5°$. At this position the dip of the dyke contacts is $+12.5°$, not perpendicular to the axial trace of the fold in the dyke. In Session 17 we noted that such a relationship was geometrically possible and could lead to complications when classifying fold forms using the bedding thickness variation graph method. Here is, therefore, one possible kinematic model to account for this relationship.

The interlimb angle α' of the deformed first folds can be obtained from the difference between θ'_p and θ'_q along the axial trace of the first fold. This angle can be expressed in terms of the x'-axis position of the data

$$\alpha' = \theta'_p - \theta'_q = \tan^{-1}(m_p + 2\cos x')$$
$$- \tan^{-1}(m_q + 2\cos x') \qquad (22.9)$$

or it can be expressed in terms of the dips of surfaces p or q using Equations 22.7 or 22.8 in 22.9.

$$\alpha' = \theta'_p - \tan^{-1}(m_q + \tan\theta'_p - m_p) \quad (22.10)$$

$$\alpha' = \tan^{-1}(m_p + \tan\theta'_q - m_q) - \theta'_q \quad (22.11)$$

The function 22.10 is plotted in Figure 22.13B. The interlimb angle variation is asymmetric about the axial trace of the fold in layer p. The angle attains a maximum value at a position located on one side of the second fold trace, and the values are strongly reduced on the fold limbs. In the model

investigated here this maximum value is *equal* to the original value, all other values being reduced. This feature is a special property of the model we chose to investigate and, with other initial orientations of the p- and q-surfaces (where the first fold axial surface is not perpendicular to the second fold shear planes and where $|\theta_p| \neq |\theta_q|$), the value of the maximum can *exceed* that of the original interlimb angle.

Using similar mathematical techniques to those set out above it is not difficult to apply other displacement and strain systems to the simple shear geometry we have investigated and to determine exactly the effects of such modifications on the conclusions set out here.

Fold interference patterns

Answer 22.4

The locations of the fold traces are shown in Figure 22.14. The overall S-form of the pale bands of the whole outcrop arises from a pair of open folds whose traces run from upper left to lower right. These traces are the most constantly oriented of all those in the outcrop and they are interpreted as being the last (third) set to form. They are superimposed on an earlier (second) set of folds whose axial traces (marked with dash–double dot lines) also show a broad S-form. Where the two third fold axial traces cross the axial surface of the obvious second fold in the lower right of the outcrop

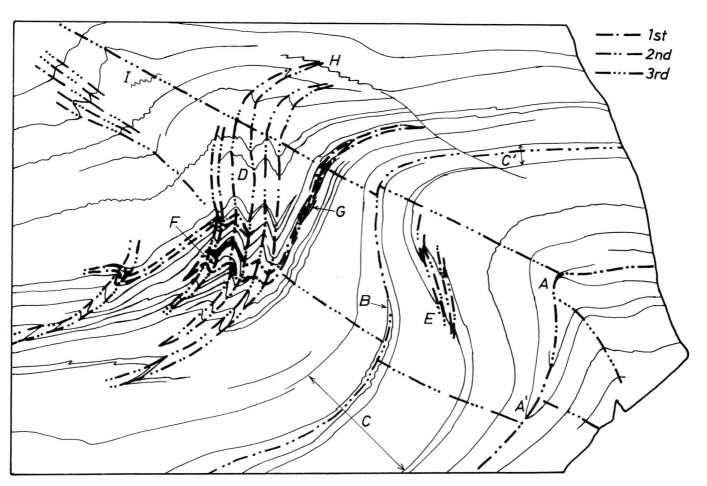

Figure 22.14. Answer 22.4 showing the positions of the axial traces of three sets of folds. For further details see discussion in text.

(localities A and A') they show a side step (cf. Answer 22.2 and Figure 22.12A). The second fold trace at locality B is based on one interpreted hinge zone. However, the convergence of the layering towards the trace at the top right of the outcrop (cf. distances C and C') together with an interpretation of the shapes of second phase parasitic folds (Z-shaped at D and S-shaped at E) support this conclusion. Six second fold traces can be determined around locality D and these re-fold earlier folds which appear to be the earliest (first) in the outcrop. The four axial traces of these first folds are clearly oblique to, and refolded by, the second set. The first fold pair at locality F shows an overall S-shape (the second fold superposition gives a complex Z-on-S pattern) while, what appears to be the same fold pair at G shows an original Z-shape. This suggests that the overall form of the layers in these first folds is closed, and it implies that the first

fold axes must pass in and out of the outcrop surface between localities F and G.

The ptygmatic forms of the small scale folds at H and I indicate that the pale pegmatite and aplite layers were more competent than the mica bearing granodiotitic gneiss during the second and third deformation phases. The competence contrast relationships existing at the time of formation of the first folding are not so clear, but alternate cuspate inner arcs and lobate outer arcs of the fold forms around the pale bands of the lower fold at locality F suggest that these bands were more competent than the surrounding gneiss.

This interpretation of fold forms based on interference patterns is completely consistent with the large scale regional fold pattern seen over an area of several hundred square kilometres. A close study of the relationship of the megamorphic minerals which grew as a result of Alpine thermal

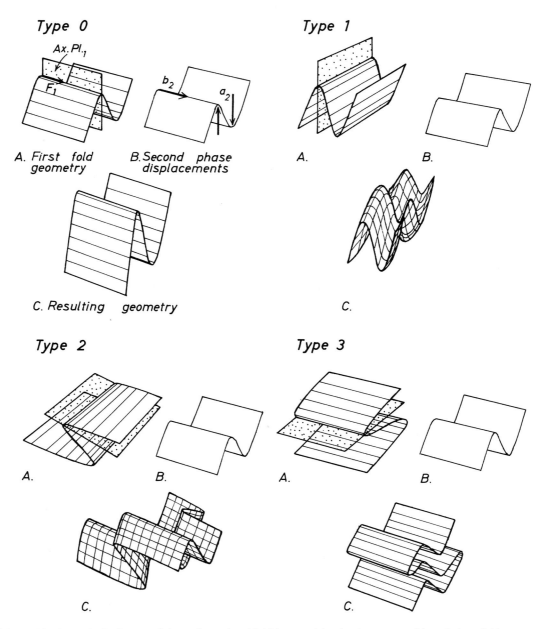

Figure 22.15. *The four principal types of three-dimensional fold forms arising by the superposition of shear folds on pre-existing fold forms.*

activity can be linked to the fold geometry and enables a broad correlation to be made between the thermal history and deformation history. The platy minerals are generally aligned sub-parallel to the axial surfaces of the second phase folds and cross cut the axial surfaces of the first fold. They are deformed in the hinges of certain third phase folds to give rise to a crenulation cleavage. These observations suggest that the main Alpine recrystallization reached a peak at about the same time as the formation of the second phase folds and that cooling of the rocks was taking place during the third folding period.

General discussion of interference patterns

Answers 22.2 and 22.4 have indicated that one of the most important features which enable two fold systems of different generations to be identified is from the special patterns of the traces of folded layering as seen in a model and on a rock outcrop surface. The kinematic superposition of two phases of fold forming deformation produce a variety of three-dimensional geometric forms on any initial set of planar layers, and differently oriented two-dimensional sections through the different types of layer morphology lead to different types of two-dimensional traces of the layers termed **fold interference patterns**. The card deck model of Figure 22.2 produced one type of pattern where the axial traces of the first folds were strongly curved and where the adjacent limbs of each fold either consistently opened out or closed together across the model surface. In contrast, some of the patterns produced by the intersections of first and second phase folds in Figure 23.3 were characterized by layer traces which formed closed or partially closed shapes which indicated that the hinge lines of the early folds emerged from the outcrop surface at one place and re-entered the surface at some other place.

The identification of the different types of interference pattern forms not only an extremely important method of recognizing the presence of superposed fold phases, but it also offers a simple way of determining the principal kinematic features of each fold forming deformation. Although the spectrum of possible interference pattern geometry is very wide, many of the observed differences are modifications of a few characteristic basic models. There seem to be four major **types of interference patterns**, the main features of their three-dimensional geometry are indicated in Figure 22.15. In each of the four types of pattern the initial fold geometry is indicated in the parts of the diagram labelled A, and this geometry varies with each of the four types. The displacement pattern of the superimposed second system is indicated in the B diagrams and remains a geometrically fixed reference frame for all of the types. The displacements of this second phase are considered in terms of a heterogeneous simple shear with the shear plane orientation given by a_2b_2, and the fold shapes which would be formed on any surface which was initially arranged perpendicular to the a_2 displacement direction are indicated in the B diagrams. Although this displacement system is almost certainly an over simplified scheme in terms of naturally produced superposed fold systems, we are of the opinion that it does provide an excellent *first approximation* to describe the dominant displacement of real systems (see discussion to Answer 22.2). Even in conditions of simple buckle folding of the layers the *dominant* fold generating displacements will

Table 22.2.
Geometric conditions leading to the four main types of interference fold patterns.

		Angle between F_1 and b_2	
		High	Low
Angle between AxPl$_1$ and a_2	Low	1	0
	High	2	3

have their largest differential sense in the axial surface directions of the buckle folds. In Figure 22.15 the various diagrams labelled C indicate the final result of superposing the second phase displacement scheme B on the morphology of the surfaces in the first phase folds. The three-dimensional form of the final structure depends primarily upon two factors: first, the angle between the first fold axial plane (e.g. Figure 22.15, the stippled surfaces Ax.Pl.$_1$) and the direction of transport a_2 of the second phase deformation; and second, the angle between the first fold hinge lines F_1 and the b_2 direction of the second phase deformation. The simplest relationships of these features which set up the four basic types of pattern are given in Table 22.2. Clearly there will be intermediate or mixed types where the various angular relationships of the two main controlling factors are no longer 0° or 90° and it should be obvious that the observed two-dimensional patterns will depend on the relationship of the two-dimensional cut to the three-dimensional layer forms. Figure 22.16 indicates some of these variables (C, A, G and I show what might be termed symmetric "pure" examples of Types 0, 1, 2 and 3 respectively, and B, D, E, F, and H are "mixed" types).

Type 0. Redundant superposition

This type has been termed "zero" Type because the interaction of the two sets of folds produces none of the geometric features that are generally held to be characteristic of fold superposition, and the resulting three-dimensional geometry is practically identical to that of fold structures produced during a single phase of deformation. All sections of the combined geometry resemble those of normal single phase folds (Figures 22.15 Type 0 and 22.16C). If the two phases of folding have the same wavelength, the resultant geometry would depend upon the in- or out-of-phase relationship of the two wave forms. The resulting form could vary from mutual amplification (Figure 22.15, Type 0,C) to produce symmetric in-phase folds, through out-of-phase relationships giving rise to asymmetric folds, to the mathematically possible but geologically unlikely situation of cancellation of the first displacement pattern by the second, leading to complete unfolding of the earlier structures. If the two phases of folding have differing characteristic wavelengths, various types of *polyharmonic folding* could be set up. As we discussed in Session 20, one way of producing the so-called *parasitic folds* on the limbs of large folds is to initially develop short wavelength folds in certain units of a multilayered complex (thin layers of high competence contrast), and then to modify these structures by developing later folds of a larger wavelength on the initial waves train as a result of late stage amplification of folds generated in other parts of the multilayer (thicker layers of low competence contrast). Although we generally think of this situation in terms of one phase of folding, the sequential development of the folds fits precisely into this zero Type superposed fold scheme.

Type 1. Dome–basin pattern

Because of the small angle between the first fold axial planes and the movement direction a_2 of the second folds, the differential shear of the second deformation does not strongly deflect these axial planes from their initial planar form. In contrast, the limbs of the first folds are deflected and develop new folds plunging in directions away from or towards the first fold axial surfaces. Although the planar form of the first axial surfaces may not be severely disturbed, the second generation movements will lead to strong undulations of the hinges of the first folds, and the production of culminations and depressions of the first folds. This effect is combined with culmination and depression forms of the second fold hinges (arising from the forms of the fold limbs on which they are developed) to produce culmination domes and depression basins. These domes and basins form an interlocking network and any one marker surface takes up a form like that of a cardboard carton used for transporting eggs: each dome is surrounded by four basins, and each basin is surrounded by four domes. Most sections through this three-dimensional structure show a regularly interknit series of closed forms of the layer traces (compare the models Figures 22.15 Type 1,C and 22.16A with Figures 4.13A and 22.17), but variations in the section plane can produce patterns which are not so symmetric, or which do not differ significantly from those of single phase systems (e.g. the two edge sections of Figure 22.15, Type 1,C).

One problem that may arise from the interpretation of dome–basin interference patterns concerns the relative timing of the two phases. In Answer 22.3 we saw how the axial surface relationships of the first and second folds (continuous-curved and offset-planar, respectively) can be used to indicate the relative ages of the two fold sets. With the Type 1 pattern, however, the axial surface relationships can be very similar and so the age relationship may not be so easy to deduce from geometric criteria alone. Problems such as this, although difficult to resolve from map patterns, are generally best tackled in the field by establishing the interference sense of associated small scale structures. For example, it may be found that one of the axial plane sets is related to a penetrative cleavage or schistosity, whereas the other is linked geometrically with the production of a crenulation cleavage fabric deforming the schistosity and hence formed after the schistosity.

The student should be aware that folds with changing axial directions leading to the development of domes and basins can be formed in many different ways. The existence of domes and basins does not always imply superposed fold events. In Session 4 we saw how certain types of this structure could arise in constrictional types of deformation (with

angle between first fold axis and b_2

Figure 22.16. *Summary of the main types of two-dimensional interference patterns resulting from horizontal sectioning of the forms shown in Figure 22.15. The four principal types in their simplest expression occur at the corners of the nine component boxes (A, C, G and I). Intermediate types are shown in boxes B, D, E, F and H (after Ramsay, 1967). The type numbers are indicated in the top left hand side of each box.*

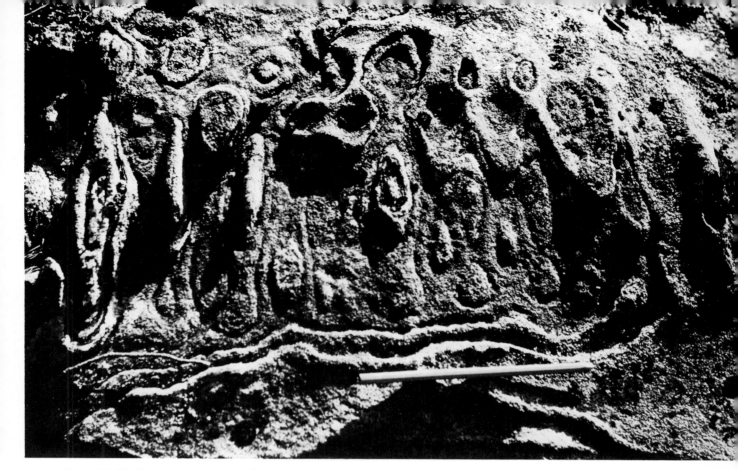

Figure 22.17. *Type 1 dome–basin pattern developed in metasandstones of the Moine Series. Monar, N. Scotland.*

Field 3 strain ellipses). Gravitational forces imposed on geological situations where low density material underlies denser material can also lead to dome-like instabilities known as **diapir structures** (salt domes, some granitic plutons) with attendant **rim synclines** or basins developed as a geometric compensation for the adjacent rising diapir. Any geologist who has worked in regions of high finite strain will also be aware of how such strains can lead to the change of shape of pre-existing morphological forms, and how such effects on folds with changing hinge line orientation lead to the formulation of very pointed domes and basins, "eyed"-folds and sheath folds. The critical features which distinguish two-phase interference domes and basins from these other types are first, that the pattern of alternate dome–basin forms shows a very high degree of geometric organization (Figures 4.11D, 4.13A, 22.16A and 22.17); and second, that in two-phase systems the folds related to the two separate generations generally show systematic differences in style, amplitude and wavelength and in the nature of their accompanying linear and planar fabrics. Today, there seems to be a current fashion to reinterpret many two phase interference patterns as single phase events. Although we agree that some of these patterns might have been falsely interpreted in the past, we are of the opinion that any universal reinterpretation of these forms as single phase sheath folds should be treated with circumspection and in the light of a careful evaluation of other geometric and deformational features of the rocks.

Type 2. Dome–crescent–mushroom pattern

With this type of pattern, because the angle between the differential movement a_2 of the second phase and the first fold axial planes is high, the first fold axial surfaces become strongly folded and because of the high angle between F_1

and b_2, the first fold hinges become strongly bowed. The arching of the layering of the first fold limbs and axial surfaces, in any one second fold, has the same sense relative to the first axial surface (cf. Figure 22.15, Type 2,C, and Type 1,C). The upbowing and downbowing of the first fold hinges leads to dome- and basin-like forms, but these structures are overturned in the same sense as that of the first folds. The basic two-dimensional interference patterns are best considered by considering what would happen to the layer traces as we progressively unroof the model of Figure 22.15, Type 2,C by horizontal sectioning (Figure 22.18A–F illustrate the sequence seen in unroofing the antiformal culmination). The first cut through the structure gives a *circular form* (A) representing a cut through the outward-dipping dome culmination at the apex of the first and second fold intersection. As we cut deeper this circular shape becomes modified into a *rounded triangular form* (B) as one side of the dome steepens and eventually becomes vertically inclined. At deeper levels this vertical sector becomes overturned (it represents the initially overturned first fold limb) and the two-dimensional section takes on a *crescent form* (C). The size of this crescent increases with depth (D) and at some critical depth where the section level is nearing the hinge zone of the underlying first fold synforms the dip on the inner arc steepens, becomes normal, and a *crescent with a protuberance* develops (E). At this level the edge of any adjacent crescent approaches the protuberance. The next deeper section shows a link-up of the layer traces of the two adjacent crescents and the development of a *mushroom form* (F). Where this occurs the section has dissected the first fold synform within the particular layer into two parts. In a multilayer all these different forms are present in any one section and lie within each other (Figure 22.18G). The reason should be obvious: it is because each successive level in the multilayer presents a different erosion level relative to the

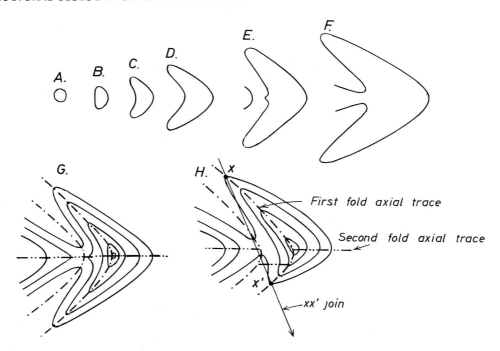

Figure 22.18. A–F: *Sequence of changing two-dimensional forms of Type 2 interference patterns in an asymmetric dome as a single surface is eroded to greater depth.* G *shows the type of pattern combining these forms resulting from unroofing a mutlilayer and where the interference structure has arisen from a perpendicular superposition of the two separate phases (cf. Figure 22.16G).* H *indicates how the basic pattern is modified where the angle between the first fold hinge lines and the b_2 direction lies between 0° and 90°: note the en-echelon arrangement of the second fold axial traces. The general regional trend of the first fold hinge lines is obtained from the xx' join.*

overall section. We have considered only the section pattern through a single upfolded first fold antiform, and clearly very similar deductions can be made about sections through adjacent downfolded first fold synforms. When these effects are integrated we obtain an interlocking series of forms (Figure 22.16G). Although this strange patterning might look somewhat improbable in geological terms, we must emphasize that such features are quite common in regions where recumbent fold nappes have suffered refolding (Figure 22.19).

The situation explored in Figures 22.15, Type 2,C and Figures 22.18A–G has been chosen so that the a_2b_2 plane is perpendicular to the first fold axial plane with b_2 at 90° to F_1. Our unroofing procedure was chosen to be symmetrically oriented to the overall structure and, as a consequence, the interference pattern formed also showed a high degree of symmetry. The first fold components show a mirror symmetry about the axial trace of the second fold: on one side of the second fold axial trace the layer traces represent an oblique cross section through the first structures while on the other side the first folds plunge in the opposite sense and therefore lead to a mirror image of the same oblique cross section (Figure 22.19A). The significance of the components in terms of axial traces of first and second folds is illustrated in Figure 22.18G. Had we chosen to dissect a fold structure where the two components were not so symmetrically related, the two-dimensional interference pattern would also show an asymmetry (Figures 22.18H, 22.19B). The axial traces of the first folds are continuously curved, but the axial traces of the second folds show en-echelon offsets where they cross the first fold traces (Figure 22.18H). In broad geometrical aspect, however, every point in the interference

pattern (e.g. the hinge point x) will have a geometric counterpart on the opposite side of some en-echelon second fold pair (hinge point x'). The join of the points x and x' will give the *average regional trend of the first fold axes.*

It will be obvious that sections which are differently oriented from those we have investigated here will show differences in interference pattern form. The patterns can vary from Type 0 to the next to be described (Type 3). We think that when the implications of the three-dimensional geometry of Figure 22.15, Type 2,C are fully understood it is not at all a difficult matter to determine the possible two-dimensional variations that are possible, and we therefore leave this investigation to the student.

Type 3. Convergent–divergent pattern

This type of fold morphology arises where the differential movement direction, a_2, of the second phase lies at a high angle to the first fold axial planes, like that of Type 2 described above. It differs in that the b_2 direction lies very close to the first fold hinge lines F_1. This relationship means that although the first fold axial surfaces become curved, the first fold hinges are not markedly deflected. The three-dimensional model of the purest form of this pattern is exactly that investigated in our card deck model (Figure 22.12B). All the fold axial directions, both first and second phase, tend to be sub-parallel. This relationship implies that in most two-dimensional sections the first fold hinges will be unlikely to pass into and out of the section plane and that layer trace forms of closed type will not occur. The limbs of first folds which originally converged together or diverged away from the first fold axial surfaces will

Figure 22.19. Type 2 dome–crescent–mushroom interference patterns. A is a perfectly symmetric form, the mirror line axis of symmetry representing the axial trace of the second fold (cf. Figures 22.16G and 22.18G), developed in Moine Series metasandstones. Loch Hourn, NW Scotland. B is an asymmetric pattern (cf. Figures 22.16H and 22.18H) in Alpinised pre-Triassic hornblende–biotite gneisses from the Lepontine nappes of the Swiss Alps.

Figure 22.20. *Type 3 convergent–divergent interference pattern. A: Two superposed sets of Alpine age folds in hornblende–biotite gneisses from the Pennine nappes of the Lepontine region of the Swiss Alps. B: refolded biotite schists and pale quartz–feldspar bands. Cristallina, Swiss Pennine Alps. The forms of the second phase folds in the pale bands indicate shortening of the layers by buckling mechanisms. The primary schistosity is parallel to the axial surfaces of the first phase folds, and this fabric has been refolded by the second structures to form a steeply inclined crenulation cleavage.*

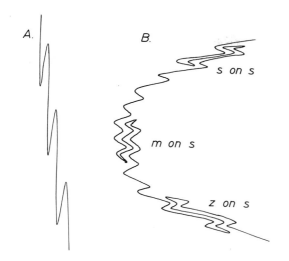

Figure 22.21. A: *S-shaped first folds.* B: *Effects of refolding the S-shaped initial folds into various forms of Type 3 interference patterns.*

therefore continue to do so even after refolding (Figures 22.16I, 22.20).

In this discussion we have investigated the four simplest relationships between the geometrical interactions of the two kinematic events and the types described above might be considered as the pure end members of a wide range of possible interactions. Clearly nature generally provides relationships which do not exactly accord with these special geometric conditions and most naturally formed interference patterns show more a complex asymmetry than those of the pure end members. If the student has time and interest to continue broadening his understanding of this geometrically fascinating and geologically important topic we would advise him to build a number of models of refolding in some easily handled nursery plastic material such as clay, wax or plasticene, and section the models in different ways with a knife (see for example Reynolds and Holmes, 1954; Ramsay, 1967, Figure 10.9). For the computer oriented geologist we would also recommend a study of the work of Thiessen and Means, where the effect of another orientation parameter, in addition to the two simple ones used in the above discussion, is examined.

Before concluding this discussion we would like to emphasize that not only are interference patterns of immediate use of the practical field geologist in helping him to understand the interaction of geometric factors in controlling the patterns of strata at the Earth's surface, but that the different types can be mapped in the field to help identify and locate the major fold components in a region. For example, Figure 22.21A illustrates the shapes (S) of folds which might be formed on one limb of a major structure. Figure 22.21B shows a possible geometric form of the Type 3 interference patterns that might arise as a result of the development of a later large scale fold across the earlier structure. The new small scale folds have either asymmetric S- or Z-shape or symmetric M-shape depending upon their location in the large second fold. These are combined with the early S-shapes to produce patterns which could be described as S-on-S, Z-on-S, or M-on-S, and by recording such variations it is possible to identify the location of the

observations point without two major fold systems (for further details of the possibilities of this methodology see Ramsay, 1967, pp. 477–479, 535–537).

Geometric analysis of superposed folds using projection techniques

Answer 22.5

Before any data plot can be made of a region of superposed folding it is essential to locate sub-regions, or domains, which are likely to be geometrically homogeneous in the sense that the domains are likely to contain cylindrical folds with fairly constant axial orientation. In Answer 22.2 we saw that surfaces which are differently oriented before the development of the second phase folds are likely to produce second folds with differently oriented hinge lines. A key method for subdividing a region into homogeneous domains necessitates construction of the axial traces of the first folds. Between any two of these traces the later folding is *likely to be* homogeneous. Figure 22.22A is a synoptic map of the terrain of Figure 22.4 which locates the traces of the major folds (first phase with dash–dot lines, second phase with dash–double dot lines). The first fold traces are recognized by their strongly curving form, whereas the second phase fold traces are generally oriented NE–SW and are only slightly curved. The overall map pattern is dominated by a large synformal structure (the Loch Monar synform), the axial plane of which dips SSE in the eastern part of the map and which is refolded to ESE or SE dips in the western part. The closure of the layering at the hinge of this structure implies a general WSW axial azimuth in the east, refolded to NNE and NW orientations in the west. The S-shaped folds with E–W axial traces in the southeast of the map have an orientation and asymmetry which suggests that they are best interpreted as WSW plunging parasitic folds on the southern limb of the Loch Monar synform.

The second fold axial traces can be followed across the trace of the Loch Monar synform without any definite en-echelon offsets. This is probably because both sets of folds are rather tight in cross sectional shape and cross each other almost at right angles. From a preliminary inspection of the strikes of vertical beds and the lowest dip readings on each limb of the Loch Monar synform it is possible to deduce that on the northern limb of the synform the second folds plunge at moderate angles (30–40°) to the southwest, while those on the southern limb plunge steeply (65–75°) to the south.

As a result of these preliminary deductions, it was decided to subdivide the region into three domainal areas indicated 1, 2 and 3 in Figure 22.22A. The data were analysed in three π-pole equal area projections in Figure 22.22B, C and D. Each of these plots shows distinctive π-pole great circle girdles indicating that each area is one of sub-cylindrical folding. Domain area 1 shows that the mean fold axis plunges at 15° towards 260°. This is interpreted as being representative of the first fold axial direction where the folds have suffered little effects of the refolding. The map indicates that the axial planes are inclined to the south, like that of the main Loch Monar synform in the eastern half of the region. Combining both map and projection data the best fit orientations for the moderately and steeply dipping first fold limbs are 37° towards 190° and 72° towards 173°, respectively.

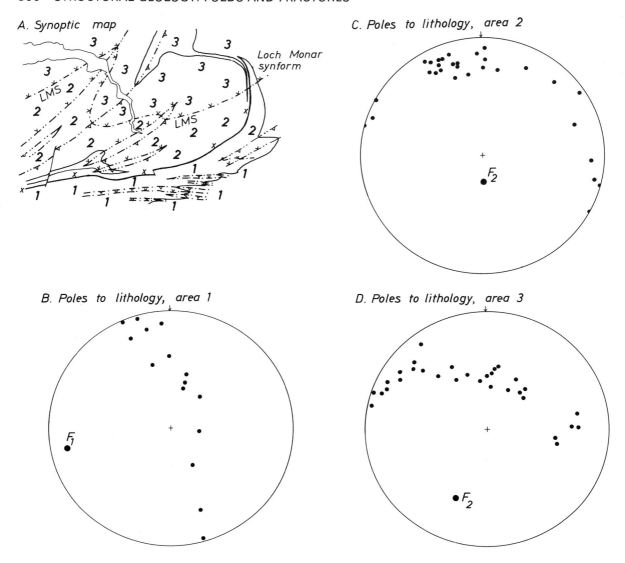

A. Synoptic map

C. Poles to lithology, area 2

B. Poles to lithology, area 1

D. Poles to lithology, area 3

Figure 22.22. *Answer 22.5. A shows the three main fold domains (1, 2 and 3) used in the analysis and the control of these domains by the axial traces of the first folds (dot–dash lines) and second folds (double dot–dash lines). LMS is the trace of the major first fold Loch Monar synform. The boundary marked x is probably a shear zone contact marking the initial base of the tectonically emplaced Lewisian gneisses. B, C and D show the π-pole plots of the three domainal areas.*

Domain areas 2 and 3 show sub-cylindrical folds with axes plunging 72° to 180° and 37° to 204° respectively. These two dominant directions appear to represent the major axes of second folding overprinted on the south-southeastward dipping steeply inclined (overturned) limb of the Loch Monar synform and on the south-southwestward moderately inclined (normal) limb of the synform. An analysis of the interrelationships of the various second fold axial directions and the first fold geometry is shown in Figure 22.23A and a schematic block diagram of the three dimensional structure is illustrated in Figure 22.23B.

Before leaving this discussion, one further aspect of the geology of this region should be mentioned. The region is comprised of both Lewisian basement gneisses and Moine cover sediments. The disposition of these two major rock divisions on the map of Figure 22.4 suggests that the Lewisian basement forms a downward closing wedge-shaped mass lying like a discontinuous stratum within the Loch Monar synform and surrounded on all sides by cover

rocks. This basement wedge has been folded like a stratum during the formation of the synform. How did the basement arrive in this position? To account for its presence we have to invoke some tectonic event that took place before the development of the Loch Monar synform. The basement wedge must either represent a fold core of a fold generation pre-dating that of the synform, or it was emplaced in the Moine Series as a thrust sheet. When the northern and western contact of the Lewisian gneisses and Moine sediments are examined in detail (Figure 22.22A, x-zone marked with a heavy line) the rocks of both groups are seen to be characterized by the development of a very strong imposed banding and platy foliation. The most likely solution to this problem is that this contact represents an early shear zone and that the Lewisian mass was emplaced by thrusting.

To conclude this Answer we should emphasize that, although the projection technique is an excellent one for defining exactly the various angular relationships of the collected field data (and in this region over 10 000 measure-

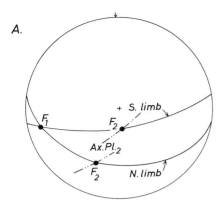

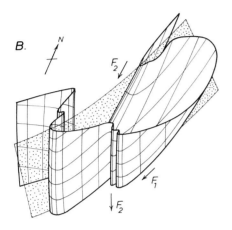

Figure 22.23. A: *synoptic equal area projection indicating the relationships of the F_2 folds to the inclinations of the north and south limbs of the Loch Monar synform. B is a diagrammatic representation of the structural geometry of Figure 22.4, the stippled plane showing the orientation of the axial surface of one of the major second fold antiforms.*

ments of planar and linear features were made and the original analysis was made in 22 separate domains) and for developing a proper understanding of the geometric aspects of the larger structural forms, there is often no good reason for reproducing all the projections in a descriptive publication. We have to admit that we have been guilty of overloading publications with such methods of data analysis

(Ramsay, 1958a and b). The youthful enthusiast who has assembled thousands of observations is often loath to let these data languish in his original field maps. Perhaps the best way of enabling the few who are interested in having these primary data is to deposit copies of the field maps with some responsible geological society or survey, and many national organizations will be found to be quite willing to act as source data agents for this valuable scientific information. In publications certain critical data presented in projection form might be used to strengthen some particular point, but for the general reader it is best to abstract the critical geometric data in the form of synoptic maps, synoptic projections (e.g. Figure 22.23A) and three-dimensional diagrams (Figure 22.23B).

Interpretation of deformed lineation geometry in superposed fold systems

Answer 22.6★

Figure 22.24 shows the data plot of the deformed lineations. The lineation poles plot about a great circle and this lineation locus best satisfies the kinematic model we evolved for shear folding, or shear folding together with a homogeneous strain (Figure 22.7D or E). From the lineation geometry and the class 2 (similar) profile shape of the fold there is no way to differentiate the two models. The a'-(or a-)direction of the differential movements which set up the fold form is obtained by intersecting the lineation locus with the axial plane of the fold: it plunges $74°$ towards $126°$ and is not perpendicular to the fold axis direction F_2. Returning to Figure 22.23A you will see that this a'-direction crosses the initial orientations of the limbs of the Loch Monar synform so as to fold them, together with the first fold axial surface, in the same directional sense. This is the reason why individual second fold antiforms and synforms can be traced across the Loch Monar synform in a coherent way. The overall interference pattern is that of a transitional variety between Types 2 and 3.

It has been noted that each second fold changes its axial orientation where it crosses the axial trace of the Loch Monar synform. This change of orientation is accompanied by a general change in fold amplitude and interlimb angle. The second folds are always tighter and have a higher amplitude on the northern limb than their equivalents on the southern limb. Although these differences might be corre-

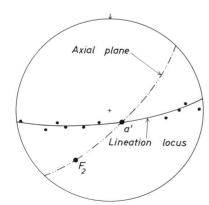

Figure 22.24. *Answer 22.6.★ The intersection of the lineation and the axial plane of the second fold gives the main differential movement direction a' forming the second folds.*

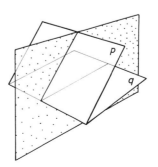
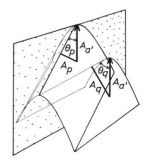

Figure 22.25 *Relationships of the amplitudes A_p and A_q of second folds superimposed on limbs p and q of an initial first fold to the overall differential movement vector $A_{a'}$ forming the folds.*

lated with real differences in the amount of displacement giving rise to each second fold, the same geometric effects are seen in hand specimens of refolded first folds. In these hand specimens it can be seen, from the way that the first fold axis parallel lineations are deflected around the second folds, that the second fold differential movement direction a' is parallel throughout the first fold and lies closer to the second fold axis developed on the steeply inclined first fold limb than it does on the moderately inclined limb. It therefore appears likely that the changes in second fold amplitude and interlimb angle are controlled by the angular relationships between the a'-direction and the surface undergoing folding. Figure 22.25 shows the geometric scheme proposed, and the relationships between the actual amplitude $A_{a'}$ of the fold-generating movements and the amplitudes A_p and A_q of folds developed in the initially differently inclined limbs p and q of a first fold. These fold amplitudes, measured in profile sections of each second fold (normal to F_p and F_q, respectively), will always be less than the value of A_a, by an amount given by

$$A_p \ = \ A_{a'} \sin \theta_p \qquad (22.12)$$

where θ_p is the angle between the second fold axis F_p and the a'-direction. It follows from this geometry that, as the layer undergoing displacement comes to lie closer to the a'-direction the fold amplitude and interlimb angle will decrease, and that where θ_p is zero, no folds can be formed. Evidence for this limiting case is suggested from observations of deformed early linear features showing regular wave-like forms on unfolded surfaces, and where the wavelengths and amplitudes of these undulating lineations are similar to those of folded lineations in the same general region (Ramsay, 1960; 1967, pp. 471–474).

In drawing our final conclusions to this session we would emphasize that, in our experience, superposed folding phenomena are exceedingly common in all orogenic zones. The complex fold geometry must be evaluated and analysed with extreme care if the various generations of movement are to be separated correctly in a way that will usefully lead to an understanding of the regional kinematics and the behaviour of the rock material located in plate collision zones.

KEYWORDS AND DEFINITIONS

Apparent interlimb angle	The angle between fold limbs seen on a randomly oriented section plane through a fold. This angle is the projection of the true dihedral interlimb angle (definition p. 313) on the section plane and can be greater or less than the true interlimb angle.
Deformed lineation locus	The geometric pattern of an initially rectilinear structure resulting from the imposition of a superposed heterogeneous displacement field. Initially rectilinar structures become spatially curved as the plane on which they lie becomes folded (Figure 22.7).
Diapir structure	A dome-like form produced by the uprising (and generally by a structural penetration) of a material of low density through an overlying cover of higher density as a result of gravitational instability.
Fold interference patterns	The special types of three-dimensional morphology of surfaces resulting from the superposition of two phases of folding and the two dimensional patterns of layer traces seen on sections through these forms. Four main types can be recognized. Their geometric constraints are indicated in Table 22.2, and their main forms indicated as follows:

> *Type 0. Redundant superposition.* p. 493, Figures 22.15, 22.16.
>
> *Type 1. Dome–basin pattern.* p. 494, Figures 22.15, 22.16, 22.17
>
> *Type 2. Dome–crescent–mushroom pattern.* p. 495, Figures 22.15, 22.16, 22.18, 22.19
>
> *Tye 3. Convergent–divergent pattern.* p. 496, Figures 22.15, 22.16, 22.20, 22.21

Normal flow buckle folds	Buckle folds (definition p. 309) formed where the main fold generating displacements of the fold hinge in the fold axial surface are perpendicular to the hinge line.
Oblique flow buckle folds	Buckle folds formed where the main fold generating displacements of the fold hinge in the fold axial surface are not perpendicular to the hinge line.

Rim syncline (or synform)   A synformal structure formed around an up-rising diapiric structure as a result of compensatory downward motion of the material around the dome.

KEY REFERENCES

Although there are very many publications describing superposed fold systems recognized throughout the world, most apply well known concepts to local and regional problems. We have selected below publications which historically relate to the sudden surge of new ideas in the late 1950s and early 1960s and which were the first to set out fundamental concepts and others which have marked subsequent rethinking or which have described new theoretical and practical concepts in this topic.

Ghosh, S. K. and Ramberg, H. (1968). Buckling experiments on intersecting fold patterns. *Tectonophysics* **5**, 89–105.

Well controlled laboratory model experiments are described, the main mechanism of the folding deformations being buckling, the experiments being performed with a free upper surface. This paper illustrates well the mechanical problems posed in the buckling of previously buckled layers and would form excellent material for widening the scope of this session.

O'Driscoll, E. S. (1962). Experimental patterns in superposed similar folding. *J. Alberta Soc. Pet. Geol.* **10**, 145–167.

This well illustrated paper sets out the results of experiments using card deck models to evaluate the mutual interference forms of two sets of inclined shear systems. It considers in detail the three and two dimensional aspects of the variety of superposition classified in this book as Type 1.

Ramsay, J. G. (1958). Superimposed folding at Loch Monar, Inverness-shire and Ross-shire. *Qt. J. Geol. Soc. Lond.* **113**, 271–308.

This sets out in detail the analysis we carried out in Question 22.5. The large plate showing the orientations of linear and planar structures would form useful additional data to that presented in this question.

Ramsay, J. G. (1960). The deformation of early lineation structure in areas of repeated folding. *J. Geol.* **68**, 75–93.

This paper analyses examples of early lineations formed in naturally deformed rocks over a wide range of scale, and the geometric analysis led to the recognition of the importance of the planar orientation of some types of deformed lineations.

Ramsay, J. G. (1962). Interference patterns produced by the superposition of folds of similar types. *J. Geol.* **70**, 466–481.

This first attempt at an overall classification of interference patterns is that used in this book. It extends the discussion presented above by considering in more detail the geometric effects of high straining proceeding with shear, and also indicates how high straining is the key to understanding the geometry of eyed-folds or sheath folds.

Ramsay, J. G. (1967). "Folding and Fracturing of Rocks." McGraw-Hill.

Chapter 10 (pp. 518–555) summarizes the state of geometric understanding at the time of publication and considers in more detail than we have space for here the concept of direction stability of the axial directions of folds, the effect of buckling components (active layering) and the reactivation of old folds. Pages 461–490 consider the geometry of deformed lineations, considerably extending the simple models discussed here, and pp. 491–517 discuss in detail the geometry of obliquely inclined surfaces.

Skjernaa, L. (1975). Experiments on superimposed buckle folding. *Tectonophysics* **27**, 255–270.

This paper analyses with laboratory models and real examples the mechanical input of first generation folds showing that the conclusions based on the shear fold model are not appropriate to situations where layer buckling is active.

Thiessen, R. L. and Means, W. D. (1980). Classification of fold interference patterns: a re-examination. *J. Str. Geol.* **2**, 311–326.

This paper elaborates and modifies Ramsay's original (1962) classification of superposed interference patterns showing that a third parameter is necessary for a complete scheme of reference.

Watkinson, A. J. (1981). Patterns of fold interference: influence of early fold shapes. *J. Str. Geol.* **3**, 19–23.

This gives a detailed examination of the mechanical effects of buckling anisotropy and shows, using laboratory models, that the early fold profile shape can exert a marked control on the interference pattern.

Weiss, L. E. and McIntyre, D. B. (1957). Structural geometry of Dalradian rocks at Loch Leven, Scottish Highlands. *J. Geol.* **65**, 575–602.

This paper was one of the first to describe, in rigorous geometric terms, the complex geometry of a naturally formed superposed fold system.

Weiss, L. E. (1959). Geometry of superposed folding. *Geol. Soc. Am. Bull.* **70**, 91–106.

This classic paper sets out many of the fundamental concepts of superposed folding and would provide excellent additional material for this session.

SESSION 23

Fault Geometry and Morphology

Terms are defined to describe the geometric features of single and groups of faults. Methods are established for determining the movement vector and vector components across a sector of a fault. A simple scheme of fault classification is given which leads to the recognition of the three main types of faults; normal, reverse and strike-slip. The detail of these three models can be complex, depending upon the way individual fault surfaces change orientation or link together into groups. The session concludes with a brief review of the special types of land surface morphology found in regions of present day fault activity.

INTRODUCTION

Faults are fracture discontinuities in a rock along which a significant differential displacement (> 0·5 mm and generally much larger than this minimum value) has taken place. Faults transect and displace lithological layers, and the intersection between the transected surface and the fault plane is known as the **cut off line** of that marker horizon. Although faults are normally formed during **brittle failure** of a rock under stress, transitions exist between true faults, where displacement has taken place on a sharply defined **fault plane**, and **ductile shear zones** (introduced in Session 3 and which will be further discussed in Session 26). Truly brittle faults are characteristic of rock failure taking place at relatively high crustal levels (upper 10 km of crust), and these shear surfaces pass downwards through a transitional zone of brittle–ductile shear zones into the deeper levels where displacement differences change smoothly and continuously through a zone of ductile flow (Figure 23.1).

Because faults are associated with breaking and mechanical disintegration of the rock on or around the fault plane, fault surfaces at the Earth's surface are generally subject to more rapid erosion than the surrounding rock and **fault traces** are therefore often seen as local topographic lows, depressions, gullies or river valleys (Figure 23.2). Sometimes a fault trace is marked by a significant change of height and character of the land surface on either side of the trace and the presence of a **fault scarp**. Because of these topographic expressions of the presence of a fault trace, faults are often spotted very quickly on vertical air photographs or satellite images as marked linear surface features or **lineaments**. However, not all linear erosion features crossing bedding traces are faults; many lineaments are formed along other types of fracture such as **master joints** along which little or no differential movement has taken place.

Natural exposures of fault plane surfaces are usually rare. Exposed faults are best sought in regions of high topographic

relief, in regions where erosional processes are especially active (e.g. coastal cliffs) or in regions of present day tectonic instability. Although uncommon, surface exposures of fault planes should be carefully sought out by the field geologist because much valuable information concerning the geological conditions during fault formation can be obtained from these exposures, together with critical data which may enable the movement sense and the orientations of the principal stresses at the time of fracture formation to be computed.

Nomenclature of faults

1. Orientation

The current nomenclature used for fault geometry is often complex, sometimes unnecessarily so. This complexity arises

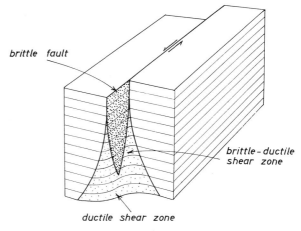

Figure 23.1. *A ductile shear zone at depth passing upwards through a brittle–ductile shear zone into a brittle fault.*

505

Figure 23.2. *An inclined fault cutting through banded gneisses. Aar Massif, Upper Urbachtal, Central Switzerland.*

in part from historical reasons. In mining operations it is often of key interest to discover the continuation of a faulted ore body, and in the past a number of practical terms were developed to describe the features of faults. Unfortunately each mining district had its own local terms which were often of local dialect origin, sometimes not self-explanatory and possibly even ambiguous as descriptive terms. Although you will find these terms in many text books we are of the opinion that many should be regarded as obsolete (e.g. hade, throw, heave to describe the orientation and movement on a fault plane). Second, and perhaps less defensible, are complexities of nomenclature that have arisen over the past two decades because of a resurgence of interest in fault geometry. Some of these new terms are undoubtedly useful and do help us to focus on important fault features which have been neglected or not understood in the past. However, we do feel that some of these new terms come into the

category of technical jargon and cannot be wholeheartedly recommended.

Fault surfaces can be planar or curved. If a fault has a statistically planar surface (to within $\pm 5°$) it can be termed a **vertical fault** or an **inclined fault**. The strike and angle of dip (obsolescent hade) of a fault are important geometric characteristics. If the fault is inclined the block overlying the fault plane is known as the **hanging wall**, the underlying block as the **foot wall**. These old miners' terms are descriptively accurate and convey useful geometric information. Curved faults, known as **listric faults**, are shovel (Greek *listron*) or spoon shaped, the fault surfaces usually flattening downwards to produced an overall upward facing concavity. Practically all fault surfaces change their orientation when they pass through layers of different rock lithology, and often show a *staircase-like form* in cross section. The part of a fault which passes with a

A. Fault reaching surface

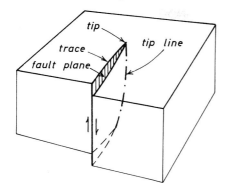

B. Blind fault

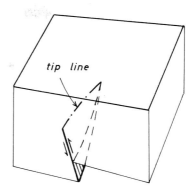

Figure 23.3. *Principal geometric features of a fault plane. With a blind fault the tip line does not break through the ground surface.*

relatively high angle through a layer is known as a **steep** or **ramp** and the section which is sub-parallel or parallel to the lithology is known as a **flat**. The terms ramp and flat are mostly used in descriptions of low angle overthrust faults where the hanging wall has moved upwards over the foot wall but, as descriptions, they can be applied to differently inclined sectors of any inclined fault surface. They are generally used irrespective of the actual geometric orientation of the lithological layering and, for example, a flat can be steeply inclined where the lithology is steeply inclined. Ramps and flats are connected to rock competence: ramps are found in the more competent layers in a succession (massive limestone, conglomerate, sandstone) whereas flats are found in the incompetent layers (evaporites, shales, marls, brittle coal horizons).

Individual faults are always of limited spatial extent. The boundary of a fault can take several forms. The fault displacement may die away to zero (Figure 23.3); the line bounding the physically displaced walls at the end of a fault

is then known as the **tip line**. The line where a fault plane cuts the ground surface is known as the **fault trace**, and this terminates at the **tip** or **tip point** (Figure 23.2A). In certain geological situations the tip line does not reach the Earth's surface and the fault is then termed **blind** (Figure 23.3B). A fault may also terminate abruptly against another fault, either because it has been truncated by a late fault, or because the two differently oriented systems acted together in a way which is geometrically related to give an overall compatible movement plan. Individual faults sometimes branch into a number of diverging **termination splay faults** at their ends. The term splay is also given to a fault which asymptotically branches off from another fault. Figure 23.4 shows the current nomenclature for **isolated-**, **diverging-**, **rejoining-** and **connecting-splay faults**. Where rejoining splays isolate a lens shaped mass of rock bounded on all sides by faults the structure is sometimes termed a **horse** (after an old Cornish miners's term). Where faults meet, the line of intersection of any two fault surfaces

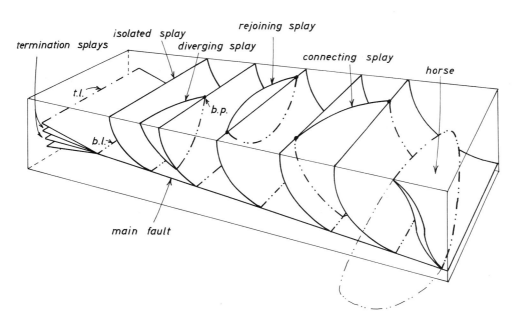

Figure 23.4. *Splay faults bifurcating from an underlying main fault. The fault tip line (t.l.) marks the limit of fault movement. The splays join the main fault along a branch line (b.l.), and this line can be exposed at the ground surface as a point or branch point (b.p.)*

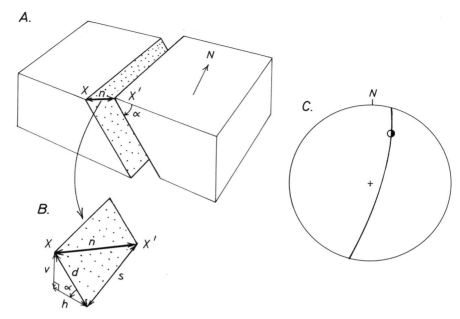

Figure 23.5. *The total or net slip n of a fault is the vector connecting points X and X' which were once in contact. In B this vector has been divided into strike-slip (s) and dip-slip (d) components, and the dip-slip component has been divided into a vertical component (v = d sin α) and a horizontal component (h = d cos α), where α is the angle of dip of the fault plane. C illustrates the projection method used to record the orientation of the fault surface and the split symbol pole to indicate the relative movements of the two sides.*

is known as the **branch line** and, if the faults crop out at the surface, this line will appear as the **branch point** (Figure 23.4, b.l. and b.p. respectively).

2. Movement

By matching points which were once in contact and which are now separated by the fault plane the *total displacement* or **net slip, n** (Figure 23.5) across that sector of the fault surface can be determined. This is a vector quantity requiring three parameters for its definition. In general the movement sense of the two sides can only be defined in relative terms (see Session 9, Figure 9.7 and Figure 23.5C for the projection method of representing the differential displacements across a fault plane). In some special environments, where fault movement is the result of recent tectonic activity, it may be possible to determine the polarity of the movement. For example the absolute movement can be determined along recent earthquake faults constrained by accurate topographic surveys. Where recent sediments with well marked environmental characteristics (e.g. tidal deposits) are displaced to levels outside their depositional environment, it may also be possible to determine if one side of a fault has suffered real uplift or depression relative to the sea level datum.

At a locality where the orientation of a fault plane is known the components of the displacement vector can be given in terms of two displacement components within the fault surface: a **strike-slip component, s**, and a **dip-slip component, d**. The dip-slip component may itself be subdivided into a **vertical component, v**, and a **horizontal component of dip slip, h** (Figure 23.5).

It should always be anticipated that any fault has components of strike and dip slip. To determine the movement vector in any sector of a fault surface is, in practice, not easy.

The geologist generally confirms the presence of a fault by the shift seen of planar markers—generally lithological layers—across the fault plane. The distance between points on the **cut off lines** of a faulted horizon is known as the **separation distance**. But the separation distance varies depending on the direction in which it is measured (Figure 23.6, distances XX', XX'', XX''', etc.) and only one

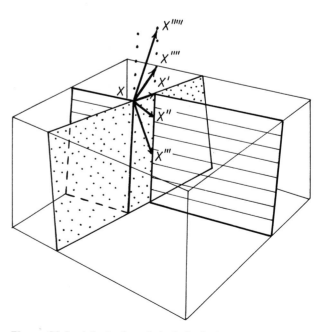

Figure 23.6. *A fault plane (stippled) displaces a marker horizon (lined). The intersections of fault and marker surface are the cut off lines of the horizon. The separation distance is measured on the fault plane between the cut off lines. The special separation distance XX' is known as the horizontal separation.*

of these measured distances correlates with the true value of the displacement across the fault. Because of this variation, the term **horizontal separation** is usually restricted to the separation measured in a horizontal plane, i.e. XX'). Note that *this is not the equivalent* of the strike slip component s. They will only be equal in faults which have no dip slip component or where a fault cuts a vertically oriented marker horizon striking normal to the fault plane. If the movement sense of the two sides can be constrained by observations of subsidiary structures on or around the fault surface, then the distance between the cut off lines in the direction of movement sense will give the movement vector. The movements of one block relative to another might be obtained from various structures. Fault planes are rarely absolutely plane surfaces and the movement of the blocks on such surfaces often leads to the formation of **fault grooves** and **striae** in the direction of displacement (Figure 23.7). These structures should be used with discretion, however, because sometimes they record only the last relative motion between the walls, and this movement might not necessarily be parallel to the total movement vector. The smoothed or polished surface on which the striae lie is termed a **slickenside** surface. Many faults are formed in regions of fluid flow and any small spaces which open up along the fault plane are sites of fibrous crystal growth. **Fibre vein systems** along fault surfaces may give a good indication of the movement vector direction and any change in the vector during the total fault history (Session 13, pp. 257–261). The walls of a fault sometimes undergo ductile deformation, and this can lead to the development of en-echelon vein systems and local cleavage formation. In these zones of flow the layering is sometimes displaced into

fold forms or is sometimes subjected to a microshear structure in which the integrated effect of the shear planes is to produce a kind of stepped shear fold in the layers. Such folds (Figure 23.8) are sometimes termed **drag folds**. The orientations of the axes of these folds are not controlled by the direction of movement but lie parallel to the intersection of the layer and the fault plane, that is the cut off line (Figure 23.9). In Sessions 2 and 3 we saw how shearing along a movement zone can set up extension fissures. The relationships of the shear zone (or fault plane) to these fissures might be used to determine the direction of the movement vector on the fault, and this technique forms the basis of our first question.

QUESTIONS

Determination of the movement direction on a fault surface

Question 23.1

Figure 23.10 shows a series of slates cut by a quartz-filled fault surface F with orientation strike 116°, dip 35° to SW. The walls of the fault show well developed quartz-filled extension veins V at a high angle to the fault with orientation strike 44° dip 85° to SE. These cut earlier veins running parallel to the cleavage, and these early veins are slightly folded near to the fault surface with fold axes plunging 15° to 273°. The fault surface is coated with fibrous quartz crystals with orientation plunge 25° towards azimuth 145°. Discuss the geometrical relationships of the small scale structures and the likely movement along the fault surface.

Figure 23.7. Fault striae on a polished fault surface (slickenside) in granitic rocks. Almora, Central Himalayas, India.

Figure 23.8. *Drag fold produced by movement along an inclined fault plane. Red Sea Hills, Sudan.*

Calculation of the movement vector on a fault surface

Question 23.2

To be able to compute unambiguously the differential movement between the walls in a particular sector of a fault it is necessary to be able to locate a point on one wall which was originally joined with a point on the other wall. Such information is rare in nature, and the most practical way to solve this problem is to try and discover a line element in one block which was originally continuous with a line which is now displaced in the other block. We therefore try and establish the geometry of such a line and find the two point intersections it makes with the two sides of the fault. One possibility (Figure 23.11A) is to discover the shift in a fold with regularly plunging hinge line which has been truncated by the fault. To establish the geometrical relationships it is necessary either to have underground mine workings or

rather strong natural typography. A second possibility is shown in Figure 23.11B. Two surfaces S_1 and S_2 intersect in a line L and, if such features can be established on either side of a fault surface, we have a way of determining the point intersection X and its counterpart X', and therefore of computing the movement vector. The types of structural features that might be of practical use for this construction are dykes, vein systems, pre-existing faults, or perhaps one of these features combined with a marker lithological surface.

Figure 23.12 shows a horizontal surface exposure of sandstone cut by a quartz-filled fault surface. This fault displaces pre-existing quartz veins with different orientations. The data are as follows:

Fault surface	strike 50°,	dip 64° SE
Vein S_1	strike 114°,	dip 78° NE
Vein S_2	strike 152°,	dip 54° W

Determine the movement vector across the fault plane.

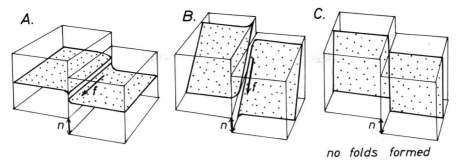

Figure 23.9. *The axes of drag folds (f) are controlled by the intersection relationships between the folded marker horizon and the fault plane and the movement vector n cannot be uniquely determined from the axial directions. Each of the fault situations illustrated has the same movement vector but the geometry of the stippled marker horizon depends upon the marker surface orientation.*

Now proceed to the Answers and Comments section and then return to the discussion below.

Combining the orientation of a fault plane with knowledge of the differential movement across the fault surface enables a fault to be classified and the fault type to be named. Three types of faults seem to occur frequently in natural fault systems.

A. Normal faults

These are inclined faults, usually with a dip exceeding about 50° where the dip-slip component is large relative to the strike-slip component, and where the hanging wall is moved downwards relative to the foot wall (Figure 23.13A). The name "normal" does not imply that this type of fault is any

more frequent than other types of fault. The name normal arises because it was this type of fault that was especially common in the main British coalfields, and if a coal seam was cut off by a fault during a mining operation, the normal mining procedure was to continue to drive the mine cut through the fault plane for a certain distance and then to sink a shaft to find the displaced coal seam. The geometrical reasons for this course of action should be clear from Figure 23.13A. Normal faults, like many other types of faults, often occur in conjugate systems, and the overall geometric features of the conjugate movements on faults inclined towards or away from each other lead to the relative sinking of a trough-like block or **Graben**, and the relative uplift of a linear elevated block or **Horst** respectively (Figures 23.14, 23.15 and session frontispiece). Horst and graben structures are very characteristic of zones of the

Figure 23.10. *Determination of the movement direction on a fault surface. See Question 23.1.*

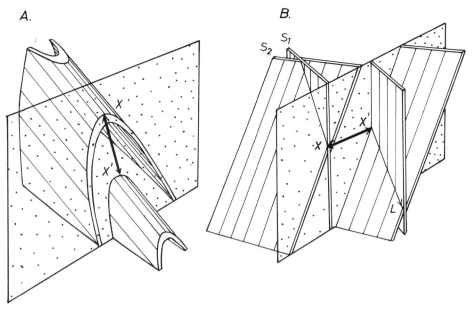

Figure 23.11. *Geometrical situations where it is possible to compute the total movement vector X–X' across a fault surface:* A *with a displaced fold hinge line;* B *with displaced intersecting surfaces* S_1 *and* S_2.

Earth's surface which have been extended predominantly in one direction and are often associated with the crestal zones of elongated domes (e.g. Rhine graben, Vosges and Schwarzwald horsts) or form during the development of separating motion in continental and oceanic crustal plates (e.g. East African rift system, Baikal rift, central Atlantic rift and graben system). It should be emphasized, however, that not all extensional environments give rise to symmetrically oriented normal fault systems, and that in some terrains only one set of normal faults is developed and the structure on the relatively down-thrown side is then termed a **half**

graben. The effect of a symmetrically paired set of conjugate normal faults is to allow the horizontal direction to extend at the same time as shortenings take place in the vertical direction (Figure 23.14). Such conjugate systems imply an extension of the upper crustal levels and, if developed on a regional scale (e.g. the Basin and Range of the western United States), almost certainly imply that the whole crustal thickness and perhaps the whole lithosphere has suffered horizontal stretching and vertical thinning. In such large scale thinning of the lithosphere the upper crust stretches by faulting, the lower crust by the development of

Figure 23.12. *Calculation of movement vector on a fault. See Question 23.2.*

A. Normal fault

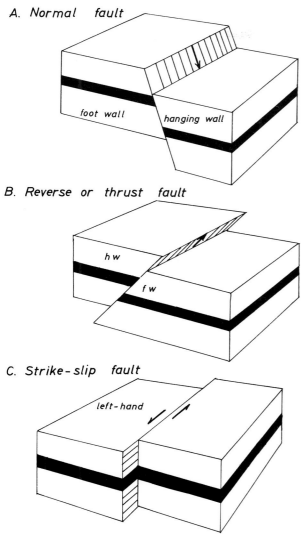

foot wall   hanging wall

B. Reverse or thrust fault

h w

f w

C. Strike-slip fault

left-hand

Figure 23.13. *The geometric features of the three main types of faults found near the Earth's surface.*

localized ductile shear zones (see Session 26) or by more homogeneous ductile strain.

The dihedral angle between a pair of conjugate normal faults is generally about 40–60° at the time of fault initiation. The bisector of this acute angle is sub-vertical and is generally parallel to the direction of maximum compressive stress σ_3 (see Appendix E for an analysis of stress). The relationships of the principal axes of stress to the fracture patterns they produce will be analysed in more detail in Session 25. Let us consider now the implications of movement on a conjugate system of normal faults. If the movement on a crossing pair of faults is synchronous it is only possible to allow the wedge-shaped graben to sink and the constraining horst blocks to move apart if large openings develop somewhere between the moving blocks (Figure 23.16 and Freund, 1974). In geological problems such geometric openings are not generally allowable. We have met this problem, the so-called compatibility problem, earlier in our discussion of ductile strain, and we will find that the geometric constraint during the movement of fault blocks, so that the connecting walls retain a continuity of the adjacent rock masses, is also an extremely important factor in controlling the geometric forms of the moving blocks and of the types of subsidiary structures that may develop in them. In laboratory experiments with rocks and model materials it is found that the individual inclined faults in conjugate systems act in sequence. Figure 23.17 illustrates the type of sequential development that leads to no compatibility problems at the cross over positions of the individual faults. Fault 1 is activated and then becomes stabilized: fault 2, with a conjugate relationship to fault 1, then becomes active and displaces the earlier fault plane. Fault 2 then becomes dead and movement proceeds on fault 3 parallel to the initial direction of fault 1. In such a sequential fault activity large overall displacements may be developed without any tendency for the fault blocks to separate and open up. The new developing faults might be formed in positions which are completely independent of previously active but now stabilized early fault planes (Figure 29.17C). However, when a

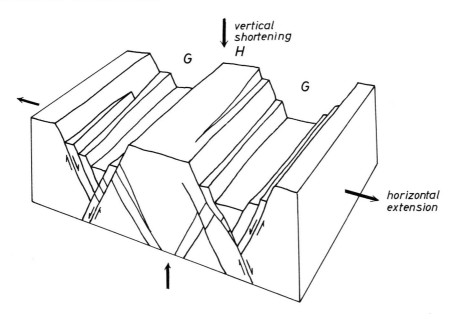

vertical shortening

H

G   G

horizontal extension

Figure 23.14. *Typical geometric features of horst (H) and graben (G) structures produced by the activity of conjugate normal fault systems.*

Figure 23.15. *Horst and graben features in Cretaceous limestones of the Säntis nappe, East Switzerland. The height of the vertical cliff is about 150 m.*

rock contains pre-existing planar weak zones oriented in the correct position to take up the necessary geometric displacement to continue the overall shape change, it is most likely that these weaknesses will be reactivated. It is possible for a truncated lower fault surface to be reactivated (point X in Figure 23.17B) and generate new faults in the overlying block, or for the truncated upper part of an old fault to be reactivated to cut the underlying block (point Y). Such a process of fault regeneration might be geometrically haphazard in that the upper and lower truncated parts of old faults might have an equal opportunity for propagating into the unfaulted adjacent block. However, because the properties of rock materials change significantly with depth from the Earth's surface, it might be easier to reactivate preferentially either the lower or the upper parts of truncated old faults. As a result of such preferential rejuvenation

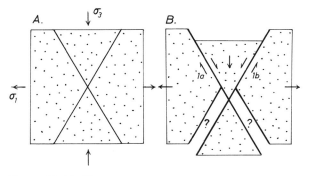

Figure 23.16. *The synchronous development of two crossing conjugate faults leading to the development of voids at certain parts, shown by question marks, of the fault system.*

the overall distribution of the rocks in the structure show markedly different relationships. Figure 23.18A shows the type of horst–graben structure produced when the lower fault surfaces are always the sites of later rejuvenation, and Figure 23.18B shows the structure resulting from preferential reactivation of the uppermost parts of truncated faults. Rejuvenation of the lower faults leads to the marked uplift of a central block of basement into the centre of the graben and the graben edge shows numerous complex fault splinters. In contrast, rejuvenation of the upper parts of faults leads to the sinking of a well defined simple graben compensated by the relative uplift of two blocks of basement along the side walls of the horst. These two uplifted blocks of basement show faults which dip in the opposite direction to that of the main faults bounding the simple graben structure. Their uplift, relative to the floor of the main graben structure and for the same regional extension, is less strongly developed than that shown by the central block in Figure 23.18A.

Special compatibility problems often arise during the formation of single sets of normal faults and the formation of **half graben** structures. One model that has been proposed to account for the geometry of half graben is shown in Figure 23.19A and is known as the **domino model** or **bookshelf model**. The model shows the sideways collapse of a number of fault bounded blocks as the two retaining walls of the model are moved apart. The fault blocks move against each other by shearing along a normal fault and each block undergoes a rigid body rotation (Figures 23.20, 23.21). In the simplest model no slip takes place inside the fault block and the angle of rotation of each block is given by the angle of dip of any initially horizontal

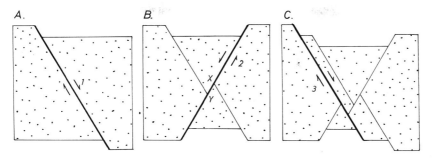

Figure 23.17. *Development of a conjugate fracture system by sequential movement along the faults. Fault 1 was displaced by fault 2 and is therefore truncated at positions X and Y. Fault 3, parallel to fault 1, truncates fault 2.*

bedding surface. Potential spaces occur in the triangular areas above and below each rotating fault slice and especially large potential spaces may develop where the rotating dominoes are in contact with the retaining unfaulted walls. In most natural occurrences of this type of structure the uppermost triangular basins are progressively filled with sediment derived from the eroded edges of the retaining walls and the uplifted edges of the rotating blocks. This sedimentary infilling generally keeps pace with the progressive block rotation, and individual stratal thicknesses generally show a marked increase into the down-rotated sector and develop a wedge-like form (Figure 23.19B) and may show interformational unconformities as a result of uneven rotation rates (Figure 23.22). Because there is a very precise relationship between the extent of the sedimentary infill and

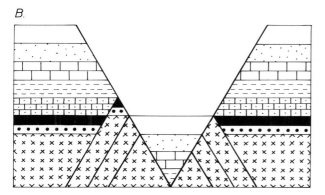

Figure 23.18. *Development of horst and graben structures: A, by reactivating truncated lower sections of pre-existing faults (e.g. Figure 23.17 at X); B, by reactivating truncated upper sections of pre-existing faults (e.g. Figure 23.17 at Y).*

the progressive rotation activity of the blocks such faults are sometimes termed **growth faults**. The space problems at depth may be accommodated by ductile flow of the underlying material into the potential spaces (cf. Figures 23.19A and 23.21) thinning the strata on the down-rotated block edges, and thickening the strata on the up-rotated edges. Full and half graben structures are often associated with large scale regional crustal thinning, a phenomenon which can also lead to the elevation of crustal isotherms and the upward migration of magma. The space problems that arise around rotating fault block sectors might then be accommodated by the development of magma infillings at depth and the outpouring of lava and pyroclastic material at the surface (Figure 23.19B). Clearly this simple domino model is somewhat naive but many of its basic geometric features arise in other related but more sophisticated models (e.g. the listric normal fault model described below). Strong crustal stretching leads to a very marked fault block rotation and to a corresponding decrease in the angle of dip of the fault planes. This effect may be less marked where surfaces other than those of the faults are activated during the rotation process. For example the bedding surfaces separating the different lithological units often become the site of localized shearing acting with the reverse shear sense from those of the main faults. The flattening of the main fault system reorients the fault planes in a way that they become less effective for taking up the overall extension: once the fault planes begin to rotate the rate of increase of slip on the fault surfaces compared with the rate of increase in overall extensional strain in the horizontal direction increases very rapidly. Frictional resistance along the fault surfaces will tend to inhibit fault slip and fault rotation and leads to a locking up of the domino block system. Eventually it becomes mechanically more efficient to set up new, more steeply inclined, normal faults which cross cut and displace the low angle early formed system (Figure 23.23A). The early faults and the bedding surfaces in the fault blocks between the new faults continue to rotate. The fault planes become progressively flatter (Figure 23.23B) and may pass through the horizontal to dip in the opposite direction from that in which they initiated. Such low angle faults can superficially resemble the next type of fault (thrust fault) to be described in this session. However, these rotated low angle normal faults still retain their characteristic geometry of layer extension and even when their dip is reversed, they always show younger rocks in the hanging wall riding over older rocks in the foot wall (Figure 23.23B). The regional positive extensions required for geometric forms of Figure 23.23B to be

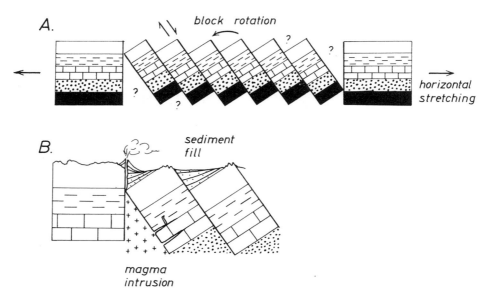

Figure 23.19. A: Development of a half graben structure on the principle of the domino- or bookshelf model. The model poses a number of compatibility space problems indicated by question marks. B shows filling of the upper block spaces by sediments and volcanic rocks and the possibility of magma intrusion in the lower block spaces.

developed have to be quite large. Structures of this type have been described in parts of the Basin and Range province, and an especially strong descriptive back up for such features has been made to account for the features of the Lower Colorado Region. Here, a whole series of differently oriented normal faults can be found at relatively high tectonic levels and these pass downwards and are cut by an extensional sole fault. The low angle major base to the complex normally faulted terrain runs close to, but not absolutely parallel with, the contact between a crystalline metamorphic basement complex and an overlying sedimentary cover. These large low angle normal or extensional faults have been termed **detachment faults**. Although this term has been connected with the historically well established French term

Figure 23.20. Fault blocks undergoing rotation as a result of development according to the domino model. Sanetsch Pass, Valais, Switzerland.

Figure 23.21. Rotation of a series of fault blocks in limestone according to the principles of the domino model with accommodation of the underlying incompetent shale by ductile flow. Gavarnie, W Pyrenees, France.

Figure 23.22. Intraformational unconformity in sediment filling a basin overlying a rotating fault block. Death Valley, California.

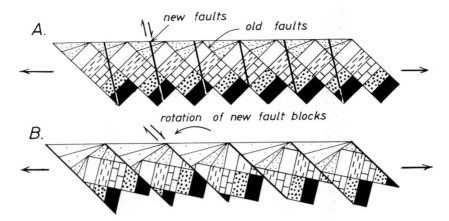

Figure 23.23. A: *Strong rotation of initial faults into positions of stability.* B: *the development and subsequent rotation of new, active fault blocks allowing further extension of the system.*

décollement fault (German *Abscherung*) the present usage of these terms is not quite synonymous. The term décollement (unsticking) is generally related to some special stratal control of the sole fault in a particularly soft or incompetent horizon (Figure 23.24A and B). In contrast, a detachment fault is not exactly parallel to any one incompetent horizon, even though the fault surface is controlled by the competencies and the overall orientation of the boundaries between the differing rock units (Figure 23.24C and D).

In many known examples of normal faults the fault surfaces are not planar. Two main types of change of orientation of the fault surface have been described: one where the dip change is abrupt (**steep and flat structure** generally related to a change of rock lithology), another where the change is more continuous to produce a smoothly curving fault surface or **listric fault**. Movement of the hanging wall relative to the foot wall in faults such as these leads to compatibility problems of a type which we have not yet encountered in our discussions of fault geometry.

Relative movement of the fault walls over an abrupt change of fault surface orientation often leads to the for-

mation of an open or partially open space around the more steeply inclined fault sector (Figure 23.25A). Such an opening generally becomes infilled by the fluid phase of the surrounding rock material and, because of the reduction of hydrostatic pressure (Appendix E) and space available for the growth of crystals, crystalline material is often deposited in this region (Figure 25.26). Because movement on the fault surface is generally discontinuous the cavity become progressively filled with new growing crystals in a periodic and often systematically sequential way (see Session 13, pp. 257–261). Such openings are well known to mining geologists, because they offer sites for the growth of ore minerals wherever the rock fluid (or fluid being introduced along the fault surface) contains metallic ions. Petroleum migration is also facilitated by the increased porosity produced in this

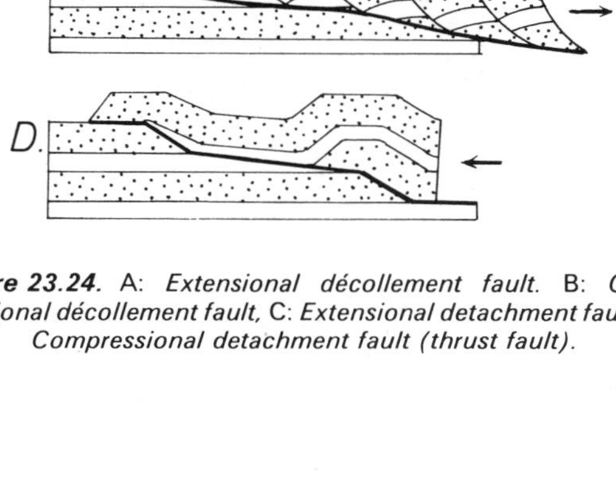

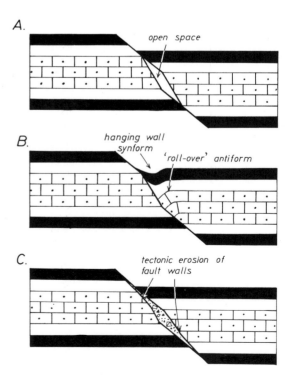

Figure 23.25. *Modifications of fault geometry as a result of movement along a fault with steep and flat structure.*

Figure 23.24. A: *Extensional décollement fault.* B: *Compressional décollement fault.* C: *Extensional detachment fault.* D: *Compressional detachment fault (thrust fault).*

Figure 23.26. *Development of carbonate-filled spaces along a normal fault cutting shale and sandstone in flat and steep sectors respectively. Tertiary flysch, Val d'Illiez, W Switzerland.*

way. A second possibility of resolving the compatibility problem of movement along irregular fault walls occurs where the rock walls are able to deform to take up the imposed constraints. Because normal faulting generally occurs at relatively high levels in the crust both walls do not input mirror symmetric solutions to the problem.

Gravitational forces are predominant in enforcing the overall rock displacements and the hanging wall usually adjusts its geometric form to the constraints of the foot wall. The geometric principles are basically simple, although in detail the actual geometric modifications are likely to be tightly constrained by the properties of the hanging wall rocks. The hanging wall steep adjusts by forming a "**roll over**" **monoclinal** or **antiformal fold** as the hanging wall is transferred into the less steeply dipping foot wall (Figure 23.25B). The hanging wall flat overlying this sector has to adjust by subsiding as a **hanging wall synform** (Figure 23.25B). The exact geometric forms of the antiformal roll over and hanging wall synform depend on the rock rheology. The forms may be governed by vertical gravitational sag into the underlying potential space, or they may be controlled by layer anisotropy and therefore obey the rules of kink band geometry (Suppe, 1983). A third possibility of solving the compatibility problem of a sudden change of initial fault plane dip is to allow the fault surface to take on a compromise orientation lying between the orientations of flat and steep orientations. Such a compromise requires a mechanical erosion of parts of the fault walls to provide the rock debris which will act as mechanical infill for the spaces developing at the separating parts of the faults surface. Generally those sectors located at the maximum initial change of dip receive maximum tectonic

erosion, whereas those sectors which are relatively planar become the sites of deposition of this debris (Figure 23.25C). This last geometric solution predominates when the differential slip across the fault walls is large compared with the spaces between the differential inclinations between flats and ramps. Any fault with large displacement difference across its walls solves its long term compatibility problem to best effect when the walls become as planar as possible.

Normal faults often show a curved or **listric** profile, particularly those developed in sediments where the rheological properties of each stratum do not differ greatly. The development of concave upward listric normal faults can be attributed to mechanical effects (Session 25) or to geometric constraints. One purely geometric reason for the formation of listric faults is that this geometry enables a much smoother integration of the local normal fault geometry with movements taking place on any underlying detachment fault (c.f. Figures 23.19A and 23.24C). If the listric geometry is such that the fault surfaces have profiles which are those of circular arcs, then it is possible to glide the two walls past each other without opening any spaces along the actual fault surface. However, if you make such a circular rotational motion between different blocks you will find that problems open up in other parts of the model. For example, if we slide the fault blocks between the circular arcs of Figure 23.27A we produce a situation whereby an increasingly large space opens up underneath the blocks (Figure 23.27B). Such a situation is clearly geologically untenable. One compatible solution to keep the uprising end of this model in contact with its underlying sole fault is shown in Figure 23.27C. This model allows the uplifted block of model B to subside vertically by simple shear on vertical shear surfaces. Such a

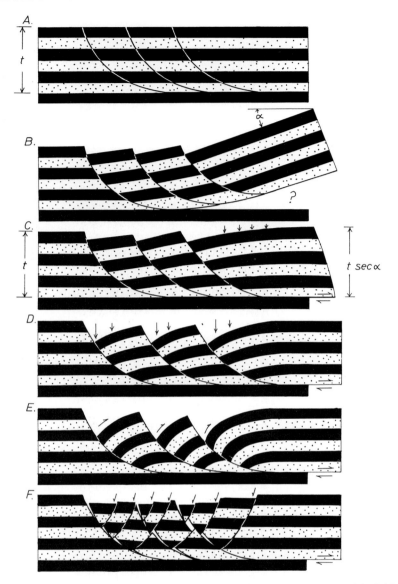

Figure 23.27. *Models of listric normal faults. A, initial position of fault planes; B, movement of rigid fault blocks setting up space problems (at question mark); C, modification of model B by vertical subsidence of fault blocks to keep contact with an underlying sole fault; D, modification of each listric block by allowing vertical subsidence and the formation of hanging wall "roll over" antiforms; E, modification of each listric block developing hanging wall "roll over" anticlines by flexural slip or flexural flow mechanisms; F, development of conjugate listric faults or counter faults to the main listric fault set.*

resolution produces a thickening of the horizontal strata on the right-hand side of the block to a new thickness of $t \sec \alpha$ (where t is the original stratal thickness). It should be apparent that such a solution is geometrically possible, but geologically naive, because clearly the excess space under the progressively rotating block faults would be *progressively* eliminated as the structure formed. However, the overall relationship between a strata thickened "toe" compensating for a composite listrically thinned "heel" is not uncommon in many geological situations where horizontal stretching has occurred.

Another type of resolution of this problem is shown in Figure 23.27D. In this model the internal rigidity of the listrically rotating fault blocks has not been retained. This model has been generated by allowing the appropriate amount of vertical subsidence to take place that is necessary to allow for the normal fault to develop. Such a geometrical

development gives rise to hanging wall "**roll over**" **antiforms** not formed in the model of Figure 23.27B, together with thinning of the strata to accommodate for the subsidence (in this example simple shear on vertically oriented shear surfaces). In this model the overall extension is accommodated without change in the thickness of the outward moving right-hand block.

A combination of situations like those shown in Figures 23.27B and C is often found in nature, particularly as a result of near surface movements of soil or of relatively unconsolidated weathered rock masses. The types of geometrical solution are shown in Figure 23.28. It is possible to rotate a sector of strata on a basal circular fault plane keeping the rotating block and its sub-stratum everywhere in contact. However, such a rotation is generally triggered in a situation of gravitational instability of the slope as a result of the downward motion of its underlying block support

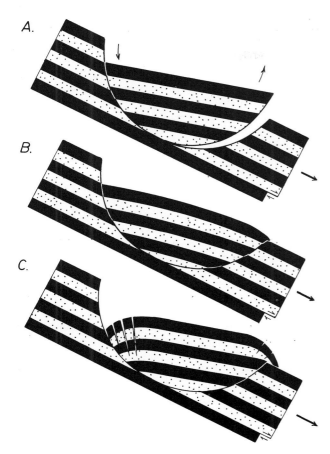

A.

B.

C.

Figure 23.28. *Listric faults found in situations of landslip fault-ing. A illustrates the basic compatibility problem, B shows a solution which keeps the upper listric fault walls in contact, while C keeps the lower listric fault walls in contact.*

(Figure 23.28A). In this situation a potential space is opened beneath the rotating block. Figure 23.28B illustrates how the rotating block might subside on its unsupported side by vertical gravitationally induced simple shear. A second type of solution can be found by keeping the lower part of the rotating block in contact with its sub-stratum and allowing open fissures to develop along the upper part of the listric normal fault (Figure 23.28C).

If listric normal faults are formed in well bedded rocks with contrasting lithology the geometric adjustments taking place in the upper fault block are often accomplished by gliding on the bedding surfaces and by the production of flexural slip (or flow) folds (Figure 23.27E). This type of displacement plan leads to the formation of "roll over" antiforms, but the intensity of folding measured by the change in layer dip in the constant layer thickness, parallel folds so formed is greater than that of the gravitational subsidence model of Figure 23.27D.

Another type of solution of fault wall compatibility is shown in Figure 23.27F. The progressive sliding on the main listric fault has been accommodated in the hanging wall of each fault by the development of a conjugate set of **counter listric faults**. These counter faults can partially or even completely counteract the normal rotation expected in the hanging wall but, if the blocks between the two inclined fault sets are completely rigid, voids have to open up along the main and conjugate fault sets. It is possible that these

geometrical problems arising during the development of counter listric faults could be accommodated by ductile flow in the hanging wall producing secondary "roll over" antiforms.

B. Reverse faults

These are inclined faults, with inclinations generally less than 45°, with a zero or small strike-slip component of displacement and with a marked dip-slip component which give an elevation of the hanging wall relative to the foot wall (Figure 23.13B). Low angle reversed faults of this type (dip less than 45°) are frequently termed **thrusts** or **thrust faults**. Thrust faults are especially abundant in the upper levels of the external parts of compressional orogenic zones and often apear to be the first formed major tectonic structure in these regions. Because an orogenic foreland region practically always consists of a normal polarity sequence of sub-horizontal cover sediments overlying a basement, thrust faults practically always bring older rocks to lie on younger rocks. Thrust faults only bring young rocks to overlie old rocks where they are developed in strata which have been previously folded or faulted, or where a thrust cuts through an angular unconformity dipping more steeply than the thrust. Thrust faults are often associated with large scale vertical duplication of regionally sub-horizontal strata. The sheet forming the "allochthonous" or moved hanging wall block is termed a **thrust sheet**. **Nappes** are thrust sheets which have moved more than about 10 km relative to the foot wall. Those sheets with a smaller relative displacement are sometimes called **parautochthonous nappes**. Although the term nappe is, on geometric principles, relatively clear, the nomenclature autochtonous, parauthochthonous and allochthonous as adjectival descriptions for unmoved, slightly transported and moderately to strongly transported, although generally in current usage, is fraught with difficulty. It is frequently almost impossible to decide if any block of rock has moved or not relative to its original site of deposition. Because a fault is located along a surface of relative displacement it is not always possible to determine whether the overlying thrust sheet has been displaced relative to an unmoved foot wall (an *overthrust*) or whether the foot wall has moved beneath the hanging wall as an *under-thrust*. It might be useful here to return briefly to Session 4 (pp. 61–62) and to check the concepts of absolute, relative and local displacements.

When sub-horizontal overthrust masses are subjected to erosion, elements of a thrust sheet may be isolated from the parent unit in much the same way that **outliers** of strata may be separated from their originally continous counterparts. Such an island-like, isolated fragment of an once continuous thrust mass is known as a **Klippe** (Figure 23.29). Erosion processes may also lead to a local unroofing of an overlying thrust sheet to expose the rocks beneath the thrust sheet. The exposed section of underlying rocks may be completely "framed" by the overlying thrust sheet, when it is known as a **window** (Figure 23.29). Where erosional processes have proceeded to a sufficiently deep level to allow for a partial connection of the localised window with the more extensively outcropping frontal area of sub-thrust geology, the structure is generally termed a **half window** or **breached window**. The region at the back of a thrust sheet passes downwards into the surface is classically known

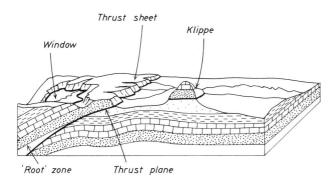

Figure 23.29. *The terms used to describe parts of thrust sheets exposed at the surface.*

as the **root zone** (Figure 23.29). Current terminology does not always favour this term but, because it is firmly entrenched in the excellent descriptive literature of the past and because we think it is still a very useful term to describe a particular locality or zone in a thrust sheet, we recommend the continuation of its usage. The main problem with the term root zone is that its genetic overtones suggest that it represents the *origin* of the particular rocks forming the thrust sheet. We suggest that the term is employed much in the same way of that of the position where a tree enters the ground through massive "roots". In this case we all are aware that the source of these roots may be many metres beneath the actual soil surface.

Although sets of reversed faults and thrusts are most often arranged with a uniform sense of overall translation of the overlying rock sheets relative to the underlying rock masses sometimes a conjugate set of thrust faults or **back thrusts** appears either at the same time or, more frequently, at a late stage of tectonic evolution of the main reversed fault structure.

Thrust faults often show variations in dip as they pass

from one lithology to another. **Flats** are those sections which lie close to the planar lithology of generally incompetent layers. **Ramps** are those sections which climb more steeply across the lithological layers and are generally found in more competent beds (Figures 23.30, 23.31). As the hanging wall of the thrust moves relative to the foot wall various compatibility problems arise if large spaces are not to be opened along the flat sectors. The geometric constraints are of a similar type to those arising with movements along any non-planar surface and introduced in our discussion of normal fault geometry. The compatibility constraints on the development of a thrust sheet in the uppermost parts of the crust are generally resolved by allowing the hanging wall to adjust its shape to the form of the underlying fault and, as a result, to undergo folding. Such fault plane induced folds have been known for many years (Rich, 1934), but a much more complete appreciation of the quite complex geometric implications of the interrelations between the fold and fault structures has mostly come about by the work of Boyer and Elliott (1982), Butler (1982) and Suppe (1983).

The progressive development of the hanging wall fold structure is complex in that there are several different fold forming and unfolding possibilities which depend on the location of the hanging wall with respect to the foot wall dip variations. Figure 23.30 shows a model developed by Suppe and Namson (1979) and Suppe (1983). In this model the foot wall is completely inert and remains undeformed. Folds are formed only in the hanging wall as it is forced over the underlying staircase-like steps. In this particular model these folds are kink-like with axial planes bisecting the fold limbs at angles which depend upon the changes of dip from the flat to ramp sectors. This geometry conserves cross sectional area (and therefore rock volume in three dimensions), bed length and layer thickness of all the strata involved in the fold. Because of the kink-like geometry of the folds the model postulates that the rocks are folded in a similar way from the lower part of the hanging wall to the top of the

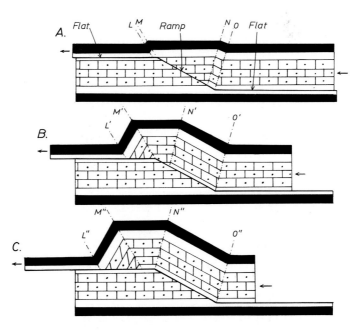

Figure 23.30. *The progressive change in geometry of an overthrust sheet sited on a flat–ramp–flat fault surface (after Suppe, 1983). A, B and C illustrate successive geometric modifications of the thrust sheet with increasing displacement.*

Figure 23.31. *Ramp and flat structure developed in Cretaceous shelf carbonates and marls in the Säntis region of the eastern Swiss Helvetic Alps.*

thrust sheet. The sequential development of the various kink folds formed during overthrusting is shown in Figure 23.30, and we will consider the sequential development of each of the four axial surfaces separately from left to right across the model. Synform *L* is initiated as soon as the hanging wall ramp passes on to the foot wall flat. Once it has been formed the axial surface remains fixed relative to the rocks through which it passes and, with further forward thrusting, the complete fold is transported passively with the thrust sheet (*L*, *L'*, *L''*). Antiform *M* is also initiated at the joining position of the higher level flat with the ramp. Its position is initially (and during the early thrust stages) fixed by this change in fault plane dip. As the thrust sheet advances the beds progressively pass through the axial surface *M'* to take up a position on the steeply inclined limb of the *hanging wall antiform*. The rocks between *L'* and *M'* acquire an internal strain as a result of the layer parallel simple shear in the fold limb. At a still later stage the rocks which were initially situated in a horizontal orientation above the lowermost flat (right-hand side of Figure 23.30) pass over the ramp to come to lie on the upper flat. When this first occurs the axial surface becomes fixed relative to the overthrust sheet and it is then bodily translated with the thrust sheet (*M''*), stabilising the limb size (separation distance between *L''* and *M''*) of the hanging wall antiform. At the lower end of the ramp a separate pair of kink folds is formed (*N* and *O*). Antiform *N*, initiated where the hanging wall flat moves on to the steeper ramp, is first fixed relative to the overthrust mass and is transported along the ramp (*N'*). When it reaches the top of the ramp its position relative to the fault surface becomes fixed, while the rocks of the thrust sheet migrate through it. On the right-hand side of the fold trace (*N''*) the rocks are internally strained by a positive simple shear strain parallel

to the bedding, but on passing through the axial surface they are unstrained. Finally, synform *O* also has a unique and simple history compared with the other folds. It is always spatially fixed at the change of slope from the lower flat into the ramp and as the thrust sheet moves the rocks pass from an unstrained state through the axial surface.

Now follows a starred question. If you do not have time to undertake it proceed to the discussion directly below.

STARRED QUESTION (★)

Internal geometry of ramp–flat thrust sheets

Question 23.3★

Using the model of Figure 23.30C, indicate which sectors of rock:

1. remain unstrained throughout the deformation history;
2. become strained during a negative sense shear on the bedding planes;
3. become strained during a positive sense shear on the bedding planes;
4. become strained by a negative bedding shear and unstrained by a positive bedding shear; and
5. become strained by a positive bedding shear and unstrained by a negative bedding shear.

The amount of shortening in the thrust sheet is less at the frontal part of the sheet than at the rear. Why is this so?

Develop mathematical expressions for (a) the angles of dip of the pair of axial surfaces *L* and *M*, and (b) the pair

Figure 23.32. *Imbricate or schuppen structure developed in Palaeozoic quartzites and carbonates. Central Appalachians, USA.*

N and *O* in terms of the angle *θ* between the ramp and the horizontal flat sector of the thrust. Determine the values of bedding plane parallel shear strains in the two inclined sectors of the hanging wall folds and the aspect ratios of the strain ellipses.

Proceed to the Answers and Comments on this question and then return to the discussion below.

The special features of this model clearly arise from the necessity of making geometric adjustments in the thrust sheet to conform with the morphology of the underlying foot wall. Although this model suggests that no folds should be found in the foot wall block, in fact folds are often found in this position. These folds can be developed by the formation of new thrust planes with ramp–flat geometry in the foot wall, either by the forward propagation of the lower thrust along the incompetent layer (see below), or by developing independent systems of thrusts with ramp–flat geometry in strata beneath the thrust plane shown in Figure 23.30. Both of these possibilities are known to occur, and the development of foot wall folds in this way will clearly modify the geometrical disposition of the initial hanging wall structures. Other possibilities for forming folds in the foot wall, which are not uncommon, arise by development of a zone of ductile simple shear strain along the base of, or on both sides of, the initial thrust. Such shearing would lead to the formation of a foot wall synform underlying the ramp section of the thrust, and to the oversteepening or reversal of the strata in the lower part of the dipping limb of the hanging wall anticline (Figure 23.33A). A further possibility for developing folds in both the hanging wall and footwall would be to initate the thrust after a certain amount of layer buckling of the competent and incompetent strata has taken place (Figure 23.33B).

The geometric adjustments necessary for a thrust sheet to be progressively transported according to the ramp–flat model requires a continuous energy input, part of which is taken up in the processes of straining and unstraining the rocks located in the sector overlying the ramp (Figure 23.55, between *M″* and *O″*). The amount of energy expended in this zone clearly depends on the distance between *M″* and *O″* (a function of ramp length and dip), the amount of strain involved (a function of change in fault plane dip) and the thickness of the thrust sheet—because the model assumes that the axial surfaces of the folds pass without modification to the surface of the thrust sheet. If the transport of the thrust sheet relative to the foot wall is large the expenditure of energy for this folding and unfolding process is clearly uneconomical. Nature is lazy in the sense that a process using less energy than another to accomplish more or less the same results will proceed with advantage over one consuming more energy. In our experience this is the case with large scale overthrusting; the geometry preferred for minimum work is that where any initially formed ramp and flat geometry becomes smoothed away during later stages of thrusting. The thrustplanes located where relative displacements in excess of about 10 km occur are all sub-planar or very gently bowed (except, of course, where they have ceased moving and been affected by some later tectonic disturbance). The wearing down of the ramp–flat treads in the thrust surface geometry is accomplished by the formation of various types of special fault rocks derived mechanically by wall rock abrasion, brecciation or flow (gouge, breccia, cataclasite, mylonite—see Session 25 for descriptions). Although the ramp–flat geometric models are undoubtedly applicable to certain thrust fault situations, they seem to be especially applicable where movement on the frontal part of some major overthrust is subdivided and dissipated into several subsidiary thrust splay faults which are practically always localized at a relatively high tectonic level. In such regions where ramp–flat folds are developed, an understanding of the geometric evolution often gives valuable insight into the solution of problems of economic interest. In particular, the recognition of fold traps for petroleum and the identification of those sectors of competent layers which might be located in parts of the fold where porosity increases may have developed as a result of the straining and unstraining processes have important implications in the search for new oilfields.

Before we leave this simple discussion of the relationship between folding and thrusting we should emphasize that folding may be associated with the formation of overthrust sheets and nappes as a result of quite different types of processes from those described above. We have already noted that the localization of overthrust faults is such that each fault generally tends to migrate upwards through the stratigraphic pile as it is followed from an internal to an external tectonic position. During nappe transport, particularly if the bulk displacement is large, the upper parts of the nappe may be displaced more than the lower part so that the overall nappe sheet undergoes a shear deformation. This shear will lead to a shortening of the layering disposed obliquely to the stratigraphically upward cutting thrust surface and consequently will have a tendency to form buckle fold instabilities within the more competent layers (Figures 23.33C, 18.16 and the commentary to Answer 18.3).

Thrust sheets and nappes often develop with sub-parallel orientation and lead to a geometric situation where thin

nappes (thin relative to their areal extent) are stacked one above the other almost like beds in a stratigraphic succession, but differing in that each is a tectonic unit separated from the adjacent sheets by a **thrust plane**. The nappe and its underlying thrust plane are generally given the same name; for example in northwest Scotland the Moine nappe is the tectonic sheet overlying the Moine thrust. Thrust complexes made up of several thrust sheets are sometimes bounded below by a well marked **sole** or **floor thrust**, and may also be bounded by an uppermost **roof thrust**. A floor thrust is often controlled by a particular incompetent lithology which allows for a geometric unsticking or **décollement** of the upper element from its underlying foundation. Often a series of reversed faults branch upwards from a floor thrust and produce a *splay fault fan structure* which is termed an **imbricate-** or **schuppen-structure** (Figures 23.32, 23.34). In imbricate faults the strata are stacked like a reversed overlapping series of roof tiles. Such imbricate fans have been subdivided on which of the subsidiary thrusts shows maximum displacement: that type where the maximum slip occurs on the frontal fault is known as a **leading imbricate fan**, whereas that where the maximum slip occurs on the most internal fault is known as a **trailing imbricate fan** (Figure 23.34). In an imbricate fan, although the displacement on any fault may be small, the aggregate displacement of a series of imbricate faults may be very large. Individual imbricate faults may terminate upwards, but often they curve asymptotically towards a roof thrust so that the imbricate zone is contained between a roof and floor thrust (Figure 23.34). Such a structure is termed a **duplex** after a particular type of living apartment in north

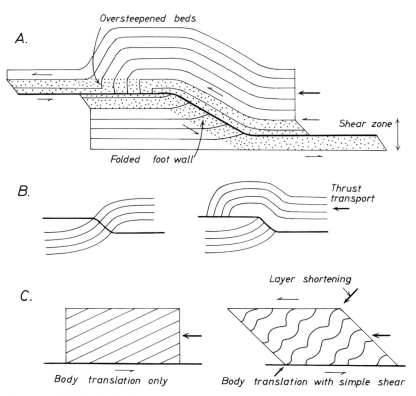

Figure 23.33. A: *Development of folds in the foot wall as a result of ductile shear in a zone (stippled) along the thrust.* B: *Development of ramp and flat structure in previously folded rocks.* C: *Folds developed by throughgoing simple shear in a transported thrust sheet.*

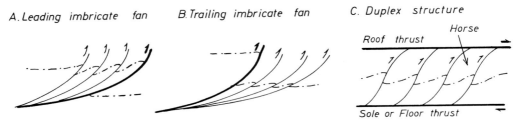

Figure 23.34. Types of imbricate fan and horse structure. A marker stratigraphic horizon is shown schematically with a dash–dot line.

America (self-contained, on two levels and with a floor and roof!). In a duplex, the individual imbricate units are now **horses**: lens-like in form and completely enclosed on all sides by faults. A duplex imbricate structure appears to develop as a result of sequential initiation and development of successive fault block elements under a major thrust sheet. Figure 23.35A shows the sequence of events suggested by Boyer and Elliott (1982). At an early stage a major thrust sheet, with total slip S_0 shows ramp and flat structure as it is guided by particular lithological levels with differing com-

petence. At the base of the foot wall ramp the original flat at P propagates along the lower glide horizon and generates a new thrust flat with frontal ramp rejoining the main thrust. The small horse so generated becomes transported forward by an amount S_1, and in so doing uplifts the overlying main thrust sheet. This horse then itself becomes inactive, and slip motion proceeds to propagate from point P_1 along the lower glide horizon. A new horse develops, transporting the earlier formed horse on its back over the underlying floor thrust through an amount S_2, and again lifting the overlying main

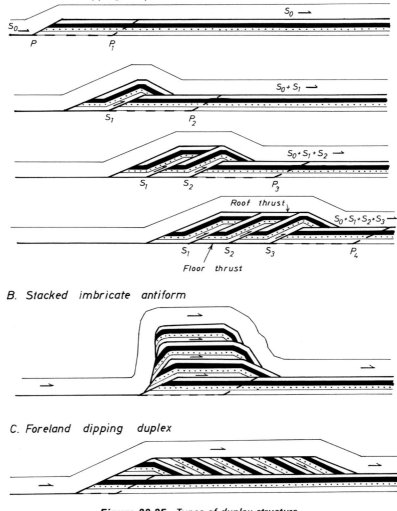

Figure 23.35. Types of duplex structure.

thrust sheet. The structure normally develops further by successive activations of horse units in the frontal part of the growing imbricate structure. Occasionally, reactivation of more internal units can lead to so-called **out of sequence thrusts**. The normal type of duplex activity leads to a transfer of slip along a glide surface at low tectonic level to a higher level. One especially important geometrical feature of this progression is that different parts of the main thrust are active at different times during the forward transport of the overlying thrust sheet, and in consequence the same thrust plane will have differing amounts of displacement at different positions in the overall structure. In this type of duplex development the overall dip of the imbricate splay faults and the internal bedding of the horses is away from the overall movement direction sense of the thrust sheet and the duplex is termed a **hinterland dipping duplex** (Figure 23.35A). If, however, the slip on individual horses becomes larger, and about the same as the length of the individual horses, each successively activating horse provides an uplift to the previously formed and deactivated horses and develops an overall antiformal structure in the imbricate zone and in the overlying thrust sheet (Figure 23.35B). Such **antiformal imbricate stacks** show a progressively decreasing downward amplitude of their componential elements, and the antiformal structure dies to zero at the décollement along the floor thrust. A further type of imbricate structure develops where the slip on individual horses is greater than the length of the horse. Where this occurs the successive uplifts of the roof thrust and previously formed horses occur *behind* the early formed and forward transported horses. This rearward uplift changes the dip sense of the imbricate faults and the overall dips of their contained strata and the imbricate structure now forms a **foreland dipping duplex** (Figure 23.35C).

It should be clear from the discussion above that the movements of fault blocks over irregularly oriented flats and ramps sets up folds in the moving block, and that the location of the axial surfaces of these folds will be controlled by the locations of the changes of orientation of the fault surface. Although most geometric analysis of these folds has been undertaken in studies of thrust faulting, such features are characteristic of all types of faults showing non planar fault surfaces. In the case of reverse or thrust faults a special nomenclature has been developed to describe different types of ramp and flat geometry, and each type has its own special effect in terms of fold development (Figure 23.36). Ramps which form perpendicular to the main direction of overthrust transport direction are known as **frontal ramps**. A frontal ramp can occur anywhere within a thrust sheet and need not occur in the most forward part of the sheet. Ramp structures can be inclined at other angles to the sheet movement: those that are approximately parallel to the transport direction are known as **lateral** or **sidewall ramps**, those inclined at other angles are **oblique ramps**. When a thrust sheet moves over a surface with a complex interconnecting array of different types of ramps the thrust sheet geometry takes on an "inside-out" form of the underlying thrust geometry mould (Figure 23.36B). Forward movements of an initially shaped trough-like thrust sheet with an advanced frontal ramp and two lateral ramps leads to the transposition of this thrust sheet into a flat roofed dome-like structure. This is just one way of forming a **culmination dome** in a thrust sheet. Another way of forming a culmina-

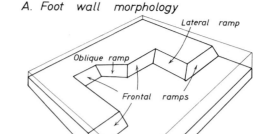

A. *Foot wall morphology*

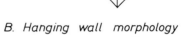

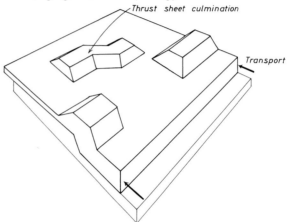

B. *Hanging wall morphology*

Figure 23.36. *Influence of the geometry of the thrust foot wall on the moving thrust sheet of the hanging wall.*

tion is to produce a duplex structure on splay faults which have forms of frontal and sidewall ramps. In this case not only do we have a possibility of producing an anticlinal stack with uplift controlled by the direction of the frontal ramps, but the antiformal stack will be bowed into a culmination by the movement on the sidewall ramps of the imbricate structure.

C. Strike-slip faults

These faults are usually steeply oriented, often vertical with differential displacement between the walls that is predominantly horizontal (Figure 23.13C). The movements on a strike-slip fault are termed either *right-hand* (dextral) or *left-hand* (sinistral), depending on the relative movement of the wall of the fault opposite to that of the observer. Because of its simple descriptive clarity we recommend the term strike-slip fault in preference to the many others that are in current usage (lateral-, wrench-, tear-, and transcurrent-fault). A **transform fault** is the name given to a type of strike-slip fault and which is commonly found at certain boundaries separating the major crustal plates making up the Earth's lithosphere. Transform faults link other major features along plate boundaries (subduction zones, extension ridges) and form an important and necessary structure whereby the relatively rigid plates are able to move past each other while conserving the surface area of the Earth (Wilson,

Table 23.1

Strike-slip fault	Transform fault
1. Terminate by splay faulting or the bending of the fault to its receding side	Terminate abruptly at special extensional or contractional features
2. Displacement varies, and decreases toward the fault termination	Equal displacement along the fault
3. Displacement smaller than 20% of fault length	Unlimited displacement
4. Adjacent parallel faults show similar displacement sense	Adjacent parallel faults can show opposite displacement
5. Formed as part of the internal strain pattern within continental plates.	Formed at plate contacts and found at ocean–ocean, ocean–continent, and continent–continent sites

1965). Although there are certain geometrical similarities between the normal type of strike-slip faults and transform faults Freund (1974) suggested that there are some significant differences, as listed in Table 23.1.

Although strike-slip faults are generally in sub-parallel sets, a second conjugate set is sometimes developed making an angle of about 60° with the main set. The second set generally shows the opposite relative displacement of the walls from that of the main set. En-echelon groupings of strike-slip faults are relatively common. These are described either as a **right-hand** or **left-hand en-echelon** pattern depending upon whether adjacent faults show a right or left sense of strike shift (Figure 23.37). Where strike-slip faults are grouped in an en-echelon set, or where individual strike-slip faults show curved surfaces, special compatibility problems arise during movement along the main fault surfaces. Because the dominant slip movement is horizontal the predominant subsidiary structures arising from geometric adjustment develop by vertical compensatory displacements. Two types of effects predominate.

1. The shear sense and either the en-echelon sense or fault strike change are opposed. Movement on en-echelon faults sets up special strains in the rock sector between the two faults and the development of a compression oriented obliquely to the main fault sets. In situations of curved major faults, the local shear displacement along the fault must be accompanied by shortening normal to the fault surface, a displacement type sometimes termed **trans-**

pression (Harland, 1971) (Figure 23.38A). This compression is generally relieved by vertical uplift of the sector accompanied by the formation of thrusts and folds. These features may range from small scale **pressure ridges** of near surface superficial material to extensive thrust faults and folds and the development or **rhomb-shaped horsts** and **uplift terrains** on a regional scale (Figure 23.39A). If the movements on the fault are predominantly those of simple shear the initially formed fold traces and thrust strike directions will be oriented at 40–45° to the main fault trace (Figure 23.37A, angle α). The displacement vector on the thrust surfaces may show components of strike slip as well as reversed dip slip. With further movement in the fault zone the axes of the initially formed folds are likely to be rotated towards the main fault trace and, as rotation proceeds, stretching is likely to occur sub-parallel to the fold axes. In regions of recently active strike-slip faulting the uplifted region is likely to be preserved as a topographic high and be the source of sediment during normal erosional and depositional processes. In certain instances strong transpression is known to result in a tectonic extrusion of rock material from the fault zone in a lateral sense. The structure that results consists of a series of convex upward reversed or thrust faults (Figure 23.40). The cross sectional appearance of the overall structure has led to the names **flower-** or **palm tree-structure**. The relative movements on the thrust faults in this structure are likely to be complicated by the shear displacements along the main strike-slip fault, and the thrusts are likely to have strike-slip movement components with the same sense of shear as that of the main fault.

2. The shear sense and either the en-echelon sense of fault strike change are the same. Movement on systems of this

A. l.h. shear, r.h. en-echelon   *B. l.h.shear. l.h. en-echelon*

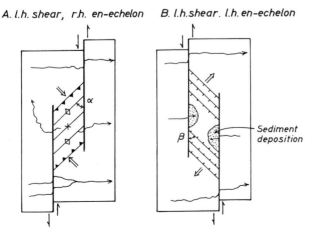

Figure 23.37. Types of secondary structure developed in the overlapping sectors between en-echelon strike-slip faults.

A. l.h.shear, r.h. bend   *B. l.h. shear, l.h. bend*

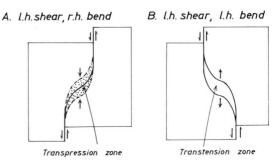

Figure 23.38. Transpression and transtension developed at bends in strike-slip faults.

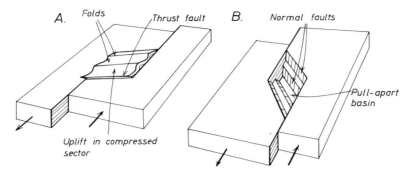

Figure 23.39. A: *Uplift region developed in a left-hand shear, right-hand en-echelon or transpression zone.* B: *Pull-apart basins or rhomb-shaped grabens developed in a left-hand shear, right-hand en-echelon or transtension zone.*

types leads to the extension of the sector, to stratal thinning and to the development of normal fault systems obliquely inclined to the main fault strike (Figures 23.37B, 23.39B). Such extension accompanying strike-slip motion is generally termed **transtension** (Figure 23.38B). The orientation of normal faults can be governed by pre-existing weaknesses. If, however, they are initiated in structureless rocks by the shear motion they should form obliquely to the main fault at angles of 45–50° (Figure 23.37B, angle β). Further simple shear movement would increase this angle by fault block rotation in the sense of shear of the main fault, and some compressional structure might be anticipated normal to the strike of the normal fault blocks. On a small scale this type of extension structure leads to the formation of land surface depressions, **sag ponds** (locally the sites of temporary or permanent lakes), while on a large scale major crustal depressions may be formed as **pull-apart basins, rhomb-shaped grabens or rhombo-chasms**. These basins will localize the accumulation and deposition of sediment derived from the topographically higher regions bounding the graben (Figure 23.37B). This sediment often shows marked thickness changes from the input locality at the edge to the centre of the basin and, because the strike-slip bounding side walls of the basin are usually moving horizontally relative to the basin floor the deposits of successive sedimentary levels may show a change with time. The maximum thickness of each deposit measured by isopach

lines will then show a change of position in the same sense as that of the horizontal displacement along the fault (Crowell, 1974). The development of large pull-apart basins may lead to such severe crustal thinning that the uprise of the overall isotherm structure may produce regions of abnormally high heat flow (e.g. Salton Trough, S California) which can lead to the rise of magma intrusions, perhaps with associated extrusive igneous material in the centre of the basin.

The surface traces of zones of strike-slip faulting frequently show a braided pattern, with the individual faults breaking into several splays or linking together. The isolated horse sectors arising from the interaction of strike-slip faults show combinations of uplift at the end of the horse where transpression occurs with depression at the end where transtension occurs. This combination often leads to a strong overall rotation of the horse block relative to its confining walls (Figure 23.41).

In concluding this section on strike-slip faults we should mention that these faults are often associated with a wide range of secondary shear features. The most important are known as **Riedel shears** after their recognition in laboratory experiments performed in a clay sheet overlying a pair of rigid blocks sliding past each other on a sharply defined vertical fault surface (Cloos, 1928; Riedel, 1929). The structures that developed in the clay consisted of a set of en-echelon shear fractures oriented at an angle of 10–15° to the direction of the underlying shear surface, followed by a second set oriented at an angle of 75–80° to the surface. These two sets are designated R_1 and R_2 respectively. Where the main shear plane has a left handed displacement the Riedel shears R_1 and R_2 show left-handed and right-handed shear sense respectively (Figure 23.42). The acute and

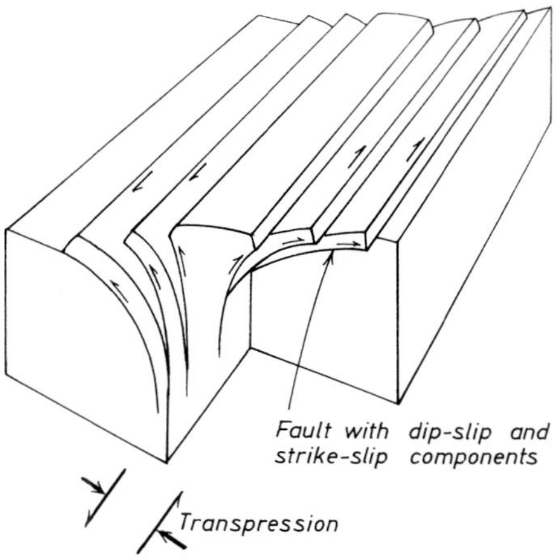

Figure 23.40. *Flower- or palm tree-structure with sideways oblique over-thrust faults developed along a left-hand transpressional strike-slip fault.*

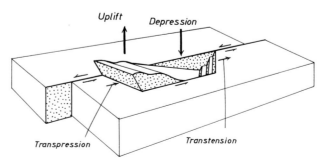

Figure 23.41. *Rotation of a horse contained between two left-hand strike-slip faults as a result of combined transpression with uplift and transtension with subsidence.*

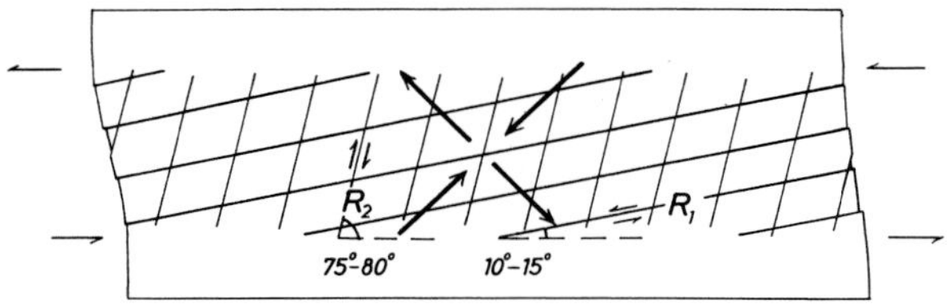

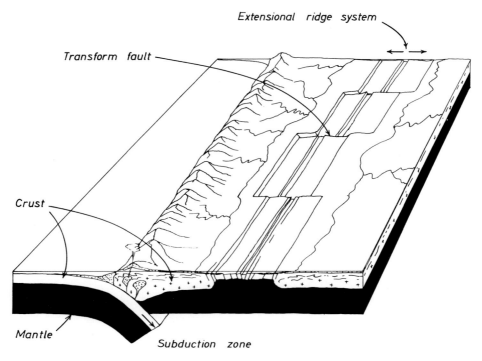

Figure 23.42. *Conjugate Riedel shears R_1 and R_2 resulting from secondary fault development in a zone of left-hand shear. The heavy arrows show the principal axes of incremental strain developed as a result of simple shear in the zone.*

obtuse bisectors of R_1 and R_2 lie in the directions of maximum and minimum incremental shortening respectively. We will discuss the significance of these structures in Session 25 when we have made a study of certain mechanical aspects of fracture formation.

Grouping of different types of faults

We have noted that faults often occur in conjugate sets which appear, from geometric considerations, to be broadly synchronous even though individual faults may operate sequentially. The actual fault pattern of a region may be more complex and characterized by more than two fault sets. Sometimes it is possible using geometrical criteria based on intersection relationships or on stratigraphic evidence (a young rock overlying unconformably one set of faults yet itself cut by another set) to subdivide the fault sets into groups of differing ages. It sometimes appears, however, that complex linked fault systems are broadly contemporaneous because the displacements implied by some of the sets could not have been geometrically possible without movement on some linked set. For example, a stratal extension developed by normal faults at one locality might be

compensated by stratal shortening due to thrust faulting at another. Such relationships are well known as a result of the formation of surface landslides (Figure 23.28) and similar features have been described as resulting from large scale tectonic processes (Coward, 1984). The concept of inter-linked movement zones is well known as being one of the basic geometric constraints of plate tectonics, whereby the semi-rigid plate-like lithospheric sheets can move relative to each other along certain well defined types of movement zone. Three main types of lithospheric plate boundary are recognized which effectively create, destroy or conserve the surface area of individual parts of a plate. These three types of plate contact namely the **extensional ridge system**, the **subduction zone** and the **transform fault** are, in broad terms, geometrically analogous to normal, reverse and strike-slip faults respectively (Figure 23.43). It is well known how these three types of plate contact can interconnect in various ways to accommodate the differently moving plates (Wilson, 1965). One such relay system is illustrated in Figure 23.44 and shows how a **transform fault** connecting two extensional zones allows separation of two plates or rock blocks. In a similar way individual faults or fault sets may be interconnected to produce different types of block movement with compatible displacement relationships. Figure 23.45 shows some of the possibilities that can develop across a strike-slip fault where the amount of horizontal displacement varies across the fault surface. In this figure the development of normal faults (A), thrust faults (B) and localized folds (C) require differential vertical motions between certain parts of the overall strike-slip system. These geometric relationships also necessitate other types of geometric adjustment in regions which lie outside the areas of the blocks shown in this figure. What we emphasize here, however, is the range of possibilities for fault linkage which enables the development of a coherent and compatible displacement pattern to occur.

Figure 23.43. *Interrelationships of the three major types of crustal plate contacts.*

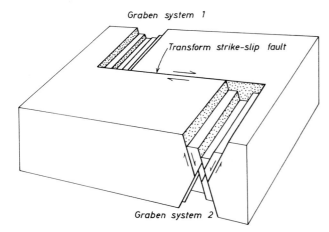

Figure 23.44. *Compatible displacement relationships linking two extensional graben zones with a strike-slip (transform) fault.*

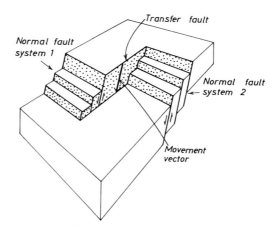

Figure 23.46. *Compatible displacement relationships linking two normal fault zones with a transfer fault. The movement vector on the transfer fault contains both horizontal and vertical components.*

Grouped and linked fault structures often necessitate rather complex movement vectors across some of the fault planes, and the faults may not always be fitted into the three groups of the simple classification scheme outlined above. Figure 23.46 shows how one set of normal faults can be linked with another by a steeply dipping fault termed a **transfer fault** (Gibbs, 1984). The two groups of normal faults could not exist in a compatible way without this transfer fault, and the transfer fault is analogous to the *sidewall ramps* that link two differently located frontal ramps in thrust fault systems (Figure 23.36). The movement vector on this transfer fault has both horizontal and vertical components: the strike-slip component allows for the horizontal stretch required by the normal fault sets and the vertical component is given by the dip-slip components of each of the normal fault groups.

Figure 23.47 illustrates the complex interlinked fault pattern arising as the result of the irregular uplift of horst form. The movement of the most strongly uplifted part of the principal horst requires an extension along the layers forming the graben floor. These secondary extensions can be accommodated if a secondary set of conjugate normal faults is formed sub-perpendicular to the main faults. Note how the main primary fault bounding the horst (movement direc-

tion s_1) acts as a transfer fault to the sets of differently inclined secondary normal faults on either side.

Rotation across fault planes

Many fault discontinuities terminate at a tip line. A particular fault may terminate by being truncated by another fault, or by having the movements relayed by another type of fault in the ways we have discussed in the previous section. However, it is quite common for a single fault to terminate quite simply without recourse to linking up with other faults. Figure 23.48 illustrates a planar horizontal surface of banded gneisses cut by a vertical fault plane. When the layers on either side of the fault are correlated it will be seen that, although there is a clear mismatch on the left-hand side of the outcrop, the horizontal separation distance in the outcrop surface becomes less and less towards the right-hand side. On the far right-hand side the fault ceases to exist. When this type of mismatch geometry can be studied in three dimensions (Figure 23.49) it is realized that this difference in the displacement vector implies a difference in the orientation of any faulted lithological surface, a feature which leads to a rotational effect across the

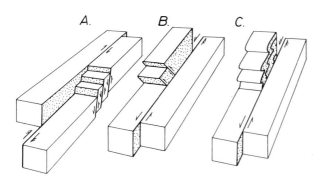

Figure 23.45. *Subsidiary structures allowing for compatibility across a strike slip fault with differing amounts of strike-slip movement component (indicated by one head or double head arrows). Each relationship sets up combined strike-slip and vertical components of movement at some localities on the main fault.*

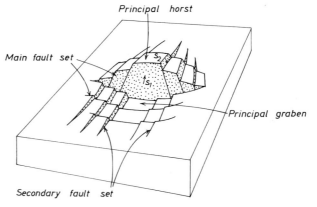

Figure 23.47. *Horst and graben structure developed on a main set of conjugate normal faults associated with sub-perpendicular set of secondary faults; s_1 and s_2 show the movement directions one of the main faults and one of the secondary faults respectively.*

Figure 23.48. *Horizontal outcrop surface of a banded amphibolitic gneiss showing variation in horizontal separation along a fault trace. Outer Hebrides, Scotland.*

Figure 23.49. *Small scale normal fault cutting horizontally layered gneisses. The displacement across the fault decreases to zero away from the observer implying a rotational component with axis of rotation normal to the fault surface. Note the movement striae on the exposed fault surface. Gotthard Massif, SE Blitzingen, central Swiss Alps.*

fault. In Session 9 (Question 9.6★) we discussed how the angle of rotation might be measured. The rotation axis is the *normal* to the fault plane. Generally the angles of rotation are in nature rather small and rotations of more than 10° are most unusual. Note that rotations of this type found in so called **scissor-** or **hinge-faults** are quite different from those associated with rotations on listric faults. Rotation of the fault walls in a listric fault occurs about an axis which is *sub-parallel* to the fault surface, and the rotation angles may be quite large.

The differential rotation of any planar element in a scissor fault varies along the fault plane, being minimal at the fault tip lines and at the centre of the fault at the location where maximum displacement has taken place. Faults commonly relay each other in an en-echelon manner and often the dying away of one individual fault is replaced by the development of an adjacent fault in the en-echelon set (Figure 23.50A). Such a system of faults can lead to a rather constant overall displacement across a region, even though the local movement on each individual fault surface may be quite variable. In Figure 23.50A it will be seen that the total vertical and horizontal components of dip slip from one side of the block to the other along three parallel traverses a, b and c is the same. The dip-slip component in traverses a and c (d_a and d_c, respectively) are equal, and these are equal to the sum of the dip-slip components in traverse b ($d_b' + d_b'' = d_b$). The sectors of unfaulted rock lying between the en-echelon fault relays are termed **fault bridges**, and these usually integrate the rotations on any two en-echelon faults.

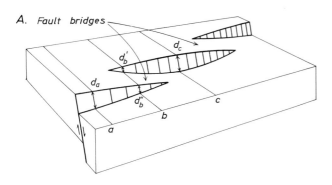

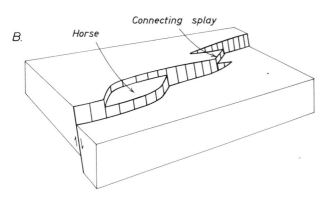

Figure 23.50. A: *Left-handed en-echelon group of normal faults with rotated fault bridges. The total displacement across the three cross sections a, b and c is equal, so there is no differential rotation across the main blocks on either side of the en-echelon zone.* B: *Modification of A where the fault bridges have been broken by connecting fault splays.*

so that they are systematically inclined to the main wall regions of the en-echelon zone. Because of the development of a strong differential bending of the fault bridges stresses may build up within the bridge and lead to a secondary failure across the bridge or to a sideways propagation of one or both tips of an en-echelon fault pair into the adjacent fault. The bridge zone is then either reduced to a wedge-shaped fault splinter joining the wall rocks separated by a connecting splay fault, or to an isolated fault horse between the linked initial faults (Figure 23.50B). If the development of the connecting faults between an en-echelon fault pair occurs fairly early in the faulting sequence the horse may show little differential rotation with adjacent walls, whereas if the curving tips of the two faults develop at a late stage the horse may retain the differential rotation of the original fault bridge.

Land surface morphology and near surface geological features in regions of active faulting

In regions of present day tectonic activity fault surfaces frequently cut the Earth's surface. Where they do so, characteristic types of surface geomorphology may be produced depending upon the relative movements of the two sides of the fault.

A. Normal faults

Normal faults are characteristic of regions of active crustal extension and the overall topographic pattern is controlled by the differential vertical movement components on the fault surfaces: elongated somewhat flat-topped ridges or **horsts** are separated by downthrown trough-like regions or **grabens**. In many regions of current crustal stretching the grabens have absolute displacement components which depress the original crustal surface, and in continential regions the graben surface level may even lie below sea level (e.g. Death Valley, California; Dead Sea, Israel). In continental regions the horst blocks are subjected to rapid erosion and the erosion debris is fed continuously into the adjacent grabens. The considerable crustal thinning that often takes place in regions of stretching often leads to the upwelling of lava and the development of volcanic activity along the faults or in the grabens between conjugate normal faults (e.g. Rhine graben of south Germany; East African rift system). In regions where oceanic crust is actively undergoing positive extension the horst and graben systems are usually completely submarine. Crustal extension here is not only connected with the formation of conjugate normal faults but is also accomplished by the upwelling of basaltic magma along steeply inclined composite or **sheeted dykes** which break surface either in the form of submarine basaltic **pillow lavas** or subaerial **flood basalts**. The regions undergoing depression in the submarine grabens are the site of the accumulation of the extensive lavas or their associated erosional debris (pillow lava breccias, palagonitic tuff and other pyroclastic deposits) and marine sediments with characteristics dependent on the climatic environment and the location of the calcite- and aragonite-compensation depths (C.C.D. and A.C.D.) with respect to the level of the depositional surface.

A normal fault which cuts the surface of a continent generally gives rise to the formation of a fairly regularly

A. Normal fault

perched terrace

main fault scarp

subsidiary fault scarp

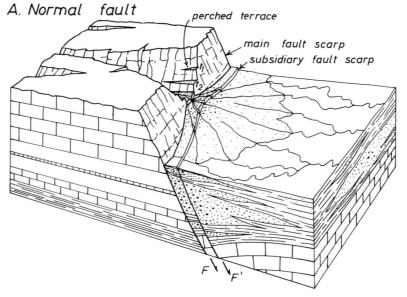

F   F'

B. Thrust fault

topographically irregular scarp

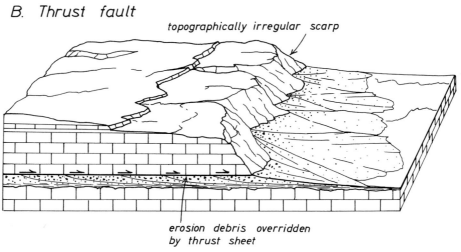

erosion debris overridden
by thrust sheet

C. Strike-slip fault

local topographic scarp

deflected river channels

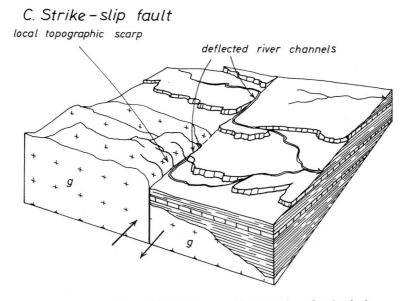

g

g

Figure 23.51. Topographic expression of active faults.

oriented **fault scarp** feature coinciding fairly closely with the trace of the fault surface (Figures 23.51A, 23.52). The uppermost part of the scarp on the relatively upthrown block retreats away from the fault surface as a result of normal local erosional processes, but the base of the scarp is always being reactivated as a result of successive rejuvenations of the fault plane (Figure 23.51A). The side valleys cutting the relatively uplifted block are continuously uplifted relative to the graben and erosional activity tends to be very marked leading to the production of valleys with V-shaped cross sectional forms. The debris derived from the erosion of the horst block is fed into the graben and the sudden change of slope at the position of the fault scarp normally leads to the building out of a marked alluvial fan. These debris fans may have their apex at the base of the fault scarp but during periods of relative stability the fan surfaces may build back into the valley to form local valley debris fill. Reactivation on the fault plane leads to the formation of a small scarp in these recent sediments and successive rejuvenations of fault movement may lead to the development of successive **perched alluvial terraces** on the edge of the horst block located high above the general level of the valley alluvium (Figure 23.51A, terraces t_1 and t_2, and Figure 23.52). Small fault scarps associated with movements on subsidiary faults or splays from the main faults (Figure 23.51A, F') often lead to the development of small fault scarps within the main mass of alluvium. Generally such surface scarps are rapidly obliterated, being formed in unconsolidated clastic debris and being constantly buried by the influx of new debris. However, at depth they do persist as marked fault features.

If extension takes place predominantly on one set of normal faults (e.g. in parts of the Basin and Range Province of Nevada, USA), relative rotation of the down faulted block generally associated with a curved or listric fault surface can give rise to very marked variations in the thickness and facies of the sedimentary fill of the basin developing as a result of block rotation. In such situations the displacement across correlatable marker beds is always least in the uppermost part of the succession, and increases with depth to approach that of the total bedrock displacement (see Figure 23.51A, fault F').

B. Reverse or thrust faults

Surface uplifts and depressions do occur in regions of active thrusting, but, in contrast to the activity on normal faults, it is the hanging wall side of the structure which shows relative uplift. Although a type of fault scarp is produced in such a situation which may regionally be very imposing (e.g. the Himalayan thrust front of north India and Pakistan), the actual scarp face is usually more irregular than that produced along a normal fault (cf. Figures 23.51A, B). The local features of the topographic scarp are irregular on account of the general low angle of inclination of the fault surface. The actively eroding part of the overthrust mass, generally known as the thrust sheet **toe**, provides debris which is fed into the region of lower topographic level in the front of the thrust. Movement on the thrust leads to an overriding of the thrust mass on to a land surface made up of debris of the thrust mass (Figure 23.51B). This is a classic situation first

Figure 23.52. Fault scarp along an active normal fault. Note the abruptly truncated spurs, the V-shaped valley sections, the alluvial cones and the uplifted alluvial terraces on the left. Death Valley, California, USA.

realized along the northern front of the European Alpine mountain chain and described as "the nappes riding over their own debris". Another well known example of this type of near surface tectonics is that of the Muddy Mountain and Keystone Thrust areas of southeastern Nevada. The overriding thrust mass here is at least 4 km in thickness, shows a minimum strike length of 210 km, and probably has a total movement in excess of 50 km. The mechanical problems for the movement of such a large rock mass are very considerable and Johnson (1981) has suggested that progressive erosion of the thrust toe might make the movement mechanisms easier to understand. If the thrust toe was being eroded back at a reasonable rate of 1 mm per year the frictional resistance to forward transport would be much reduced. In such a situation a total displacement of 50 km at the back of the thrust sheet would not necessarily imply that the frontal region of the thrust sheet overrode the foreland region by the same distance.

C. Strike-slip faults

Although the morphological trace of an active strike-slip fault is generally a well marked linear feature, the fault line is not always marked by the presence of a fault scarp. True strike-slip faults only show scarp morphology where the horizontal motion along the fault brings together rocks with marked differential resistance to erosion (Figure 23.51C). The most consistent topographic feature of active strike-slip faults is the systematic offset in the displacement sense of the fault movement shown by rivers and streams crossing the fault trace.

Strike-slip faults, like many other types of faults, are frequently arranged in en-echelon groupings with curved fault terminations or cross faulted regions linking the ends of the main en-echelon fault components. If the en-echelon sense of two major strike-slip faults is different from that of the relative movement sense on the faults the differential movements in the en-echelon overlap sector tend to produce a strong oblique contraction and the formation of a morphologically elevated thrust or fold block (Figure 23.39A). If the en-echelon sense of two adjacent faults is the same as that of the relative movement sense on the main faults, the movements taking place on the two major faults tend to open a space in the offset zone of the two faults. Such an extension may lead to the development of oblique conjugate normal faults and the formation of a near surface rhomb-shaped graben depression (Figure 23.39B).

ANSWERS AND COMMENTS

Determination of the movement direction on a fault surface

Answer 23.1

An equal area projection of the data is shown in Figure 23.53. The steeply inclined extension veins intersect the fault surface in a line plunging 35° towards 219°. If these veins were formed by extension processes in the fault zone the intersection line (i) should be positioned perpendicular to the shear movement direction producing them and oriented in the fault plane with azimuth 124° and plunge of 7° (Figure

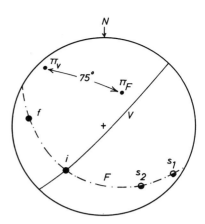

Figure 23.53. *Equal area projection of the data of Question 23.1.*

23.53, point s_1). The angle between these veins and the fault surface is 75°. This angle suggests either that the initial angle was around 45° and this initial angle was subsequently steepened by left-handed shear strain in the fault zone, or that the shear was accompanied by a strong negative volumetric dilation (see Session 3, Figure 3.21B). From the overall geometry of Figure 23.10 and the sigmoidal forms of some of the veins the first explanation looks most likely. The fibre structure of the quartz crystals in the fault surface is oriented parallel to the slide direction during the time of their formation. This direction (Figure 23.53, s_2) does not coincide with the slip direction s_1 deduced from the geometry of the en-echelon vein system, and we suggest that it implies a change in direction of movement during the development history of the fault. The relative movement sense of both s_1 and s_2 is that the top left-hand side of the fault moved towards the left relative to the lower right-hand side, and that the movement sense was left-handed from the viewpoint of this outcrop surface. This is shown in split symbol form in Figure 23.53. The fold axis direction f cannot be used to determine the direction of movement as the folds are drag folds formed by shear displacement of the early veins. The general displacement sense of the fold forms is, however, in accord with the displacement senses deduced from the vein and fibre orientations. A change in slip direction along a fault surface during fault development is of fairly frequent occurrence. In this example we do not have any clear assessment of the direction of the total displacement vector. We can suggest that it lies somewhere between the positions s_1 and s_2, but whether it lies nearer to s_1 or s_2 is not a simple decision. The extent of the shear deformation of the en-echelon veins suggests quite a large shear strain, whereas the length of the quartz fibres (e.g. see directly below the hammer head) suggests a small localized deformation. Probably the total displacement vector across the whole shear zone lies in the fault surface closer to s_1 than to s_2.

Calculation of the movement vector on a fault surface

Answer 23.2

By using either projection techniques (Figure 23.54B) or by drawing the sub-surface contours on the two veins and on the fault surface, it is possible to determine the intersection

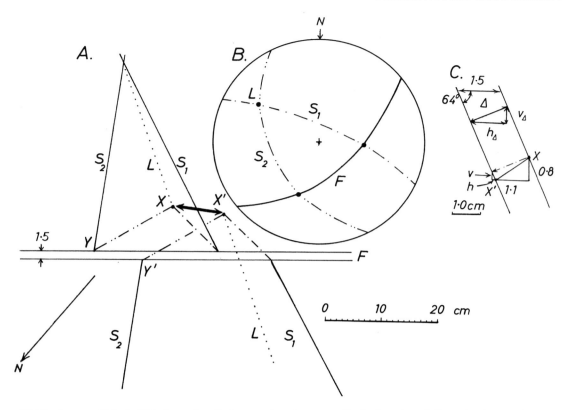

Figure 23.54. *Answer 23.2.* A: *XX′ shows the displacement sector across the fault surface F.* B: *projection used to compute the intersection lines.* C: *Cross section of the fault showing the relationships between the points X and X′ and the dilation vector* Δ. *For details, see text.*

lines of S_1 on S_2 (line L), S_1 on F and S_2 on F, and so determine the spatial position of the intersection points of line L on the SE wall of the fault (X) and on the NW wall of the fault (Y). The join $XX′$ gives the total components of the movement vector as projected on to a horizontal surface (Figure 23.54A). The vertical distance separating X and $X′$ is given by determining the lengths and plunge angles of YX (16·2 cm) and $Y′X′$ (17·0 cm). The plunge angles are 45°. The vertical separation of X and $X′$ is therefore 16·2 tan 45° − 17·0 tan 45° = 0·8 cm. The movement vector $XX′$ across the fault could therefore be described in terms of horizontal component 9·3 cm and vertical component 0·8 cm with azimuth 237°.

It is not possible to express this vector in terms of the components of strike-slip s, horizontal dip-slip h, and vertical dip-slip v, because the vector does not lie in the actual fault surface defined by either of the fault walls. This is because the fault movement has been accomplished together with a fault wall perpendicular dilation. If we describe the total movement vector in terms of s, h and v we must also include a wall perpendicular dilation Δ or its horizontal h_Δ and vertical v_Δ components in a fault profile. The width of the quartz zone filling the fault is 1·5 cm, so the dilation perpendicular to the walls is given by Δ = 1·5 sin 64 = 1·35 cm. This has components h_Δ = 1·35 sin 64 = 1·2 cm, and v_Δ = 1·35 cos 65 = 0·6 cm. The difference of vector $XX′$ direction from the fault strike is 238° − 230° = 8° and this gives the horizontal separation of X and $X′$ parallel to the fault as 9·3 cos 8 = 9·2 cm, and the horizontal separation normal to the fault as 9·3 sin 8 = 1·1 cm. The vertical

separation has been previously calculated as 0·8 cm, so now we have the total vector expressed by three components related to the orientation of the fault walls. The horizontal separation parallel to the fault is the strike-slip component s = 9·2 cm. The true dip-slip component must be obtained by removing the dilation component from the horizontal separation (1·1 cm) and vertical separation (0·8) respectively. What we have to do is to bring point X along the dilation vector so that it lies on the same fault wall as $X′$ (Figure 23.54C). We then find that h is given by the difference in total horizontal separation component and h_Δ, and v by the difference in total vertical separation component and v_Δ. We then can resolve the total movement vector into five components given by

$$h_\Delta = 1·2 \text{ cm}$$

$$v_\Delta = 0·6 \text{ cm}$$

$$s = 9·2 \text{ cm left-hand movement}$$

$$h = h_\Delta − 1·1 = 0·1 \text{ cm}$$

$$v = v_\Delta − 0·8 = −0·2 \text{ cm}$$

It should be clear from this list that the fault is predominantly a left-handed strike-slip fault with a significant dilation component, and insignificant dip-slip (h and v) components. You might have realized from inspection of Figure 23.11 that the motion is predominantly of strike-slip type but the effects of dilation across a fault can produce quite complex

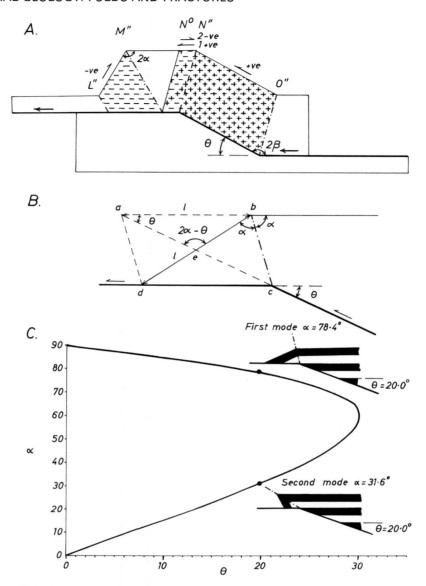

Figure 23.55. Answer 23.3★ A *shows the shear state of the various sectors of the structure.* B *illustrates the geometric features necessary to compute the interlimb angle 2α of the frontal fold pair, and* C *shows the various possible solutions relating ramp dip θ and interlimb angle 2α*

geometrical effects that make the dip-slip components not so easy to determine by simple observation. You will have noted that the traces of veins S_1 and S_2 on either side of the fault are not quite parallel. This might be the result of rotational components across the fault surface (see Session 9, pp. 161 and 163). In the example of Figure 23.11 this effect is the result of an undulating outcrop surface cutting regularly oriented structures.

Now return to the discussion on p. 511

Internal geometry of ramp-flat thrust sheets

Answer 23.3★

Figure 23.55A illustrates the strain relationships in the model of Figure 23.30C. Areas remaining unstrained throughout the forward movement of the thrust sheet are unornamented. The limb sector between the frontal hanging wall synform L''

and antiform M'' shows a negative sense of bedding parallel shear whereas the inclined sector overlying the ramp between antiform N'' and synform O'' shows a positive bedding parallel shear. The sector of horizontal rock lying between antiform N'' and a line drawn parallel to this axial trace through the meeting point of axial trace M'' and the thrust surface (N^0) originated on the rear hanging wall flat. It was subsequently passed on to the hanging wall ramp (where it acquired a positive shear deformation) and was then unfolded by a negative shear as it passed through the fixed position of fold trace N''. There are no sectors of rock which were initially strained by a negative shear and subsequently by a positive shear.

The amount of shortening at the front of the thrust sheet is less than that at the rear because part of the shortening has been taken up by the development of the folds. If the bedding limb length of the fold limb between L'' and M'' is x units, and that between N'' and O'' is y units, and if the interlimb angles of folds L'' and M'' is 2α and folds N'' and O'' is 2β, the difference is shortening of the frontal part of the thrust from that at the back is given by:

$$x(1 - \cos 2\alpha) + y(1 - \cos 2\beta) \qquad (23.1)$$

Figure 23.55B illustrates the geometric relationships that exist to control the orientation of axial surface M where θ is the inclination of the ramp relative to the flat, and 2α is the interlimb angle of antiform M. Using sine rule relationships in triangle abe:

$$\frac{be}{\sin \theta} = \frac{ab}{\sin(2\alpha - \theta)} \qquad (23.2)$$

Because the layer length $ab = 1$ remains unchanged in the fold limb ($bd = 1$) and because $be = 1/2$ it follows that Equation 23.2 may be simplified. In its simplest form this gives either:

$$\tan \theta = \frac{\sin 2\alpha}{2 + \cos 2\alpha} \qquad (23.3)$$

or

$$(\tan^2 \theta + 1)\cos^2 2\alpha + 4\tan^2 \theta \cos 2\alpha + 4\tan^2 \theta - 1 = 0 \qquad (23.4)$$

Equation 23.4 is a quadratic form in the variable $\cos 2\alpha$. For values of θ of more than $30°$ no real values of α exist, but for values of θ less than $30°$ two real solutions of α may be found. Graphically these solutions are shown in Figure 23.55C. The two possible geometric solutions have been termed by Suppe (1983) the first and second mode folds. Most folds found in natural ramping situations appear to be first mode structures. Second mode folds are clearly geometrically possible and may also be mechanically possible. For example, although the internal deformation by bedding plane shear in second mode folding is greater than that of first mode folding, the volume of rock over which this high deformation is developed is smaller and, in certain circumstances, the energy input may be smaller. With the simple model used here for the calculations no solutions for fold geometry can be found where the ramp dip exceeds $30°$

and, if ramps exceed this inclination, modifications must be made in the constraints of constancy in bed length and bed thickness. Suppe has shown that ramps with dips greater than $30°$ can be geometrically feasible keeping the geometric conditions of constant bed length and bed thickness if the flat sector above the ramp is also inclined and cross cuts the layers.

The angular relationships of folds N and O are simple functions of ramp inclination, θ, and the interlimb angles 2β are given directly from

$$2\beta = 180 - \theta \qquad (23.5)$$

The shear strains γ on the limbs between L and M and between N and O depend only on the interlimb angles 2α and 2β and are, respectively:

$$\gamma_{LM} = -2 \cot \alpha \qquad (23.6)$$

$$\gamma_{NO} = +2 \cot \beta \qquad (23.7)$$

The principal strains have orientations (angle ϕ measured from the bedding surfaces) of $\tan^{-1} (2/\gamma)/2$ and values of the principal quadratic extensions λ_1 and λ_2 of $[2 + \gamma^2 \pm \gamma(\gamma^2 + 4)^{1/2}]/2$. The ellipse aspect ratios are calculated from $(\lambda_1/\lambda_2)^{1/2}$.

In conclusion we should emphasize that, in this session, we have considered faulting only from a geometric viewpoint. We are of the opinion that geometric considerations, based on the principles of compatibility, offer an essential key to understanding fault systems. Over the past decade this approach to faulting has led to a major re-examination of theoretical concepts and has given extremely positive results in helping to solve practical problems of economic interest particularly in the search for petroleum traps. However, this geometric approach has its limitations. Problems related to the dynamic interpretation of fractures and the reasons for the initiation, propagation and mechanical development of faults cannot be solved using geometric concepts alone. Here we need to develop the concepts of stress analysis and the mechanical properties of rocks under stress, and we will investigate these aspects of fracturing in Sessions 25, 26 and 27.

KEYWORDS AND DEFINITIONS

There are so many new terms in this session that it is impractical to redefine them here. We give only a list below of those terms which the student should understand and where a description may be found in the forgoing text.

KEY REFERENCES

One of the special aims of this session has been to analyse the principal features of fault geometry and, in order to explain adequately the fundamental features, space has not permitted descriptions of many classic examples of known fault zones. The student is strongly advised to make up this deficiency by looking into as many as possible of the general references cited below and by reading the particular papers suggested as special references.

General references

Davies, G. H. (1984). "Structural Geology of Rocks and Regions", 492 pp. Wiley, New York.

This textbook is written in a very lively style, and the excellent illustrations will fill many gaps in the material of Session 23. Particularly recommended are the chapters on plate tectonics (pp. 163–200) and faults (especially pp. 261–305).

Hancock, P. L., Klaper, E. M., Mancktelow, N. S. and Ramsay, J. G. (ed.) (1984). Planar and linear fabrics of deformed rocks. *J. Str. Geol.* **6**, 1–215.

Several of the papers published from the proceedings of a special conference develop aspects of the relations of small and large scale structures, microscopic rock fabric and strain partitioning in thrust fault regions.

McClay, K. and Price, N. J. (1981). "Thrust and Nappe Tectonics." Special Publication 9. Geological Society London.

Perry, W. J., Roeder, D. H. and Lagerson, D. R. (eds). 1984. North American thrust-faulted terrains. *Am. Assoc. Petr. Geol.* Reprint series 27.

Sylvester, A. G. (ed.) (1984). Wrench Fault tectonics. *Am. Assoc. Petr. Geol.* Reprint series 28.

Williams, G. D. (ed.) (1982). Strain within thrust belts. *Tectonophysics* **88**, 201–362.

This collection of papers arose from an international conference, and it covers many aspects of fault geometry, fault mechanics and descriptions of regional geology.

An excellent collection of 25 previously published papers all of which are very informative and several of which are classics.

A companion volume to that of Perry *et al.* with a collection of 19 previously published papers.

This presents 10 papers, read at a Tectonic Studies Group Conference, concentrating on the geometric aspects of thrust faulting. This collection of papers contains many new and stimulating ideas.

Special papers

Aydin, A. and Nur, A. (1982). Evolution of pull-apart basins and their scale independence. *Tectonics* **1**, 91–105.

Boyer, S. E. and Elliott, D. (1982). Thrust systems. *Am. Assoc. Petrol. Geol.* **66**, 1196–1230.

Charlesworth, H. A. K. and Gagnon, (1985). Inter-cutaneous wedges, the triangle zone and structural thickening of the Mynheer coal seam at Coal Valley in the Rocky Mountains foothills of Central Alberta. *Can. Petrol. Geol. Bull.* **33**, 22–30.

Freund, R. (1974). Kinematics of transform and transcurrent faults. *Tectonophysics* **21**, 93–134.

Gibbs, A. (1984). Structural evolution of extensional basin margins. *J. Geol. Soc. Lond.* **141**, 609–620.

Horsfield, W. T. (1977). Contemporaneous movement along crossing conjugate normal faults. *J. Str. Geol.* **2**, 305–310.

Suppe, J. (1983). Geometry and kinematics of fault-bend folding. *Am. J. Sci.* **283**, 684–721.

Wernicke, B. and Burchfiel, B. C. (1982). Modes of extension tectonics. *J. Str. Geol.* **4**, 105–115.

This provides a very good summary of the geometric features of curved and en-echelon strike-slip faults.

This important paper reviews much of the previous literature on thrust fault geometry and extends very many of the classic ideas both in terms of nomenclature and basic perception of the problems.

In our discussion of thrust fault geometry we briefly described the possibility of producing thrust surfaces with displacements in an opposed sense to that of the main orogenic translation. This publication illustrates the kinematics of how such backthrusts might develop.

This discusses the geometric consequences of movement along strike-slip faults with particular emphasis on splaying and bending of faults. The analysis is supported by natural examples of faults from New Zealand, Iran and Israel together with the description of laboratory model experiments.

This paper reviews the geometric features found in normal faults and provides new data on fault geometry from seismic profiling of the oil basins of NW Europe.

This paper discusses the problems of crossing fault systems and gives the results of laboratory experiments suggesting a rejuvenation of truncated faults as a method of solving the compatibility problems arising at the fault intersections.

This paper describes the geometry of the folds developed by movement of thrust sheets over irregularly oriented thrust planes. It would form a valuable extension to the discussion of Question 23.3★.

This gives an account of the geometric features of extensional faulting and gives methods of calculating the amounts of extension from the fault geometry.

SESSION 24

Faults and the Construction of Balanced Cross Sections

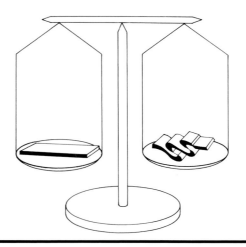

This session discusses the basic principles for constructing geological cross sections using realistic geometric constraints, how these constructed sections may be checked, and how such sections may be restored to an undeformed stratigraphic template or palinspastic reconstruction. Different types of geometric constraints are appropriate to different geological environments. Sections may be drawn or restored using conditions of no change in bed lengths, in cross sectional areas or in bed volume. In foreland fold and thrust belts, where thin-skinned tectonic styles predominate, the cover and basement are often decoupled. Such decoupling provides an especially simple lower boundary constraint and it may be possible to determine, from the geometric features of the major structures at the Earth's surface, the depth of the decoupling or décollement surface. The session concludes with a discussion of section balancing and section restoration techniques in regions where the individual layers have undergone complex strain variation, techniques which are more appropriate to tectonic levels characteristic of the internal and deeper parts of orogenic zones than those used in the thin-skinned external foreland regions.

INTRODUCTION

A geologist is often required to use the observations that he or others have made at or near the surface to interpret the structural forms likely to be found at deeper levels. Such constructions form a key to the understanding of regional tectonics and they often play a vital role in industry. The exploration for oil and gas in particular requires the best possible control on underground structure in order to site drill holes for stratigraphic investigation or for producing wells. The structural geologist whose predictions lead to many dry holes will not be a popular member of a petroleum exploration team! We have discussed some aspects of the problems of extrapolating surface structures to depth in Session 18, and we will now enlarge the scope of this discussion by considering what types of constructional guides we might use where faulting as well as folding is a feature of the sub-surface geological structure.

The techniques of section construction have taken a major step forward over the past two decades, especially as a result of the search for hydrocarbons in the frontal fold and fault belt of the Eastern Canadian Rocky Mountains (Bally *et al.*, 1966; Dahlstrom, 1969, 1970; Boyer and Elliott, 1982). In particular, a technique which has come to be known as **balanced section construction** has led to various types of geometric constraints being placed on section drawing methods, constraints which enable the geologist to become more aware of errors, or even of gross impossibilities, that might arise during the production of a geological cross section by other methods. These constraints also enable the geologist to adjust and redraw his first

attempts at reconstructing the sub-surface geology in a way which enables more realistic structural forms to be postulated.

The basic premise behind the concept of balanced section construction is not new to us: it is the fundamental idea of *compatibility* that we have discussed in several previous sessions. Compatibility implies that the body translations, rotations and strains developed in a deformed rock mass obey geometric rules (which may be formulated in mathematical terms) that are requisites for the rock mass to remain coherent after deformation. In most general terms, this coherence can be thought of as the property of the rock mass in which no large voids are developed, and no overlaps occur in the sense of two originally separate parts of the original mass coming to occupy the same physical space after deformation. Although this is probably most frequently used to constrain the deformations in heterogeneous ductile deformations, we have already seen in Session 23 that the broad principles can also be applied to discontinous deformations along faults to discover how the walls of a fault plane have to undergo folding or further fracturing as a result of fault displacements taking place on irregularly oriented fault surfaces.

The important constraints on section balancing are in principle very simple. The most important is that the geometric features of any constructed section *must be restorable* to a pre-deformational form without loss of material volume and in a way that the disposition of the strata, and the lengths and thicknesses of individual strata restore to produce a coherent picture or **palinspastic reconstruction**. The geometric features of the structures seen in the section must be chosen so that they accord with the actual structural forms seen at or near the surface and they should not infringe well established geometric rules (thrust faults should generally cut up-section from lower to higher levels in the transport direction; extension faults should cut down-section in the transport direction). If seismic sections are available along all or part of the section plane these are likely to provide an extremely valuable geometric input into the constructed section. From surface data alone it is sometimes possible to construct a number of cross sections which are valid balanced sections. Seismic data will then be of very great importance for the differentiation of these models into more likely and less likely geological solutions.

Many aspects of the balanced section methodology have been developed in geometric situations characteristic of the marginal regions of orogenic zones, an environment where a relatively thin (1–6 km) sequence of continental shelf sediments originally deposited on a continental type basement of crystalline gneisses and igneous rocks has been deformed by horizontal shortening as a result of plate collision. In such regions the basement is often very competent compared with the overlying sediments, and the cover is often mechanically detached from the basement along a **décollement fault** and involved in complex faulting and folding of a style not shared by the underlying crystalline basement. Situations such as this are known as **thin-skinned tectonics**. The presence of a planar detachment fault to the thin skin of cover rocks, generally located at the unconformable surface or at some level of particularly incompetent strata in the cover, provides a very important geometric constraint, or boundary condition, on the deformations that are possible in the cover strata. In regions of thin-skinned tectonics deformation by thrust faulting often tends to predominate over other styles, these thrusts usually associated with local accommodation of deformation by the development of kink folds and box folds formed when the layers are transported over undulating ramp–flat surfaces. These special types of folds have been termed **fault-bend folds** and **fault propagation folds** (Suppe and Chang, 1983). The finite strains within these folds are not usually very high, and generally arise by localized or diffuse layer parallel simple shear. In such deformation the bedding surfaces coincide with surfaces of no finite longitudinal strain in the strain ellipsoid. Profile sections through the folds will therefore show no change of bed length from that of the original bed length and, because simple shear is a plane strain deformation, the area of cross sections of original layers will remain unchanged. The first of these geometric constraints leads to a technique known as *line length balancing*; bed lengths of different layers between original fixed vertical reference lines, or **pin lines**, should be equal (Figure 24.1). Before the contraints of line length balancing are used in section construction it is strongly advised that a field check is made to confirm that there is little or no shape change in the bedding surfaces. For example, fossils which lie on the bedding surface should show no length changes or distortions. Simple shear strains parallel to the bedding can build up quite high finite strain states, particularly if a state of **overshear** (p. 448 and Figure 21.1C) is developed as a result of concentration of shear displacement along incompetent layers. One special feature of simple shear displace-

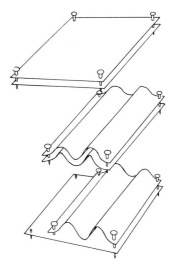

Figure 24.1. *The concept of consistency of bed lengths that forms the basis of line length balancing of cross sections. After Dahlstrom (1969).*

ment is that the X-direction of the finite strain ellipsoid and the XY plane always make angles (θ') of less than 45° with the shear (bedding) surface (Equation 2.4 and Figure 2.10). For line length balancing to be an appropriate technique there should be no truely penetrative cleavage (parallel to XY) at angles greater than about 40° to the bedding surfaces. The folds produced during bedding plane parallel slip should all show a parallel form (Class 1B). If data exist to show that there are variations of bed thickness which are systematically related to the position of the layer in the fold, the folds must have deformational components other than those of simple shear, and simple bed length balancing will be incorrect. The overburden pressure and rock temperatures in regions of thin-skin tectonics are generally insufficient to produce extensive metamorphic reconstructions of the original clastic or diagenetic minerals making up the rocks. Very local reconstructions may be found (anchimetamorphism) but these generally lead only to the reordering of the crystal lattices (e.g. change in Illite cristallinity index). In regions where the metamorphic grade exceeds low greenschist facies rock deformation usually occurs by ductile flow and tends to produce strain variations which often have considerable layer parallel extensions and, in these regions, line length balancing techniques are usually in error.

Certain other geometric constraints are also required for line length balancing to be correct. The deformation must be a plane strain; there should be no movement of material away from the plane of the section and there should be no loss of volume during deformation. Such features are generally present in zones affected by thin-skinned tectonics where the axial directions of fold hinge lines are more or less parallel. In arcuate fold zones, however, these constraints might be invalid. Arcuate fold zones can arise in a number of ways. In those zones where fold arcuation is essentially a single phase deformational event the geometric features of any profile section depend on whether the convex side of the fold arc has been active (as a result of convergent flow of the foot wall region, Figure 24.2A) or, in contrast, whether the concave side has been transported over a foreland (resulting

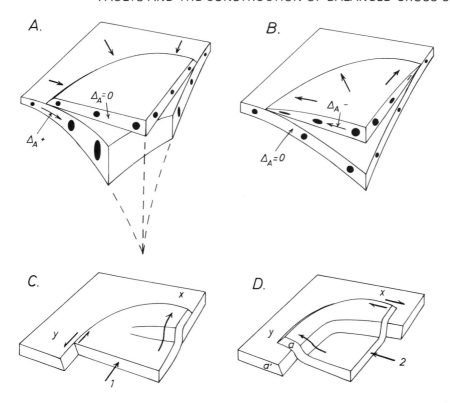

Figure 24.2. *Geometric effects of developing thrusts and folds with arcuate traces. A, convergent movement of the foot wall leading to increase in thickness of the foot wall; B, divergent movement of the hanging wall leading to thinning; C and D, two stage developments of an arc by successive simple shear displacement.*

in divergent flow of the hanging wall region, Figure 24.2B). These two situations lead to complex strain patterns in the foot wall and hanging wall respectively, resulting in area and length changes of layers in any profile section. A third model for the development of arcuate fold trends that avoids the non-plane strains of the previous two is shown in Figures 24.2C and D. In this model the overthrust sheet is developed along a curved fracture by two (or more) differently oriented transports. Figure 24.2C shows an early movement where one part of the sheet (at x) is simply overthrust whereas at another position (at y) the movement is taken up only by horizontal strike-slip movement. Between x and y the movement proceeds by a combination of overthrust and horizontal strike-slip motions. At a later stage (Figure 24.2D) a second movement with movement vector in a different direction to that of the first takes place in the thrust sheet, giving rise to horizontal motions only at x, and overthrust motions at y. Providing that each strain is accomplished by simple shear motion the overall strains in the overriding block will be close to that of plane strain simple shear (with total movement vector being the resultant of the sum of the separate vectors). It should be clear from Figure 24.2D why continuous sections drawn perpendicular to the thrust front (or normal to the local fold axes) will be potentially unbalancable. For example the material in the thrust sheet at position a was originally in the same line of section as position a' and is unlikely to be geometrically coherent with the profile section directly under position a. The only section lines likely to be matchable and balancable above and below thrust fault surfaces are those which *contain the direction of*

the total displacement vector. It is therefore very important that the section line be chosen with great care so as to contain this vector. If the traces of the folds and thrusts on a map are reasonably parallel it is usual to accept the movement vector direction as the perpendicular to these traces. In young orogenic zones this assumption is usually correct, but in old (Caledonian and Precambrian orogenies) zones later differential uplifts can produce complications. If the map traces of folds and thrusts are slightly arcuate it is best to choose a section near the areas of maximum displacement which lies near the centre of the bowed form (the "bow and arrow" rule of Elliott, 1976). Sections taken near the end points of a local arc are likely to contain differential rotations above and below thrust planes and balancing may then become difficult as a result of local movements out of the section plane.

It may be possible to use certain types of field information to determine the movement vector. Shear fibre structures on fault surfaces (Session 13, pp. 257–261) can sometimes be used in this way, but grooves and striae on slickenside surfaces are always suspect because their directions generally relate only to the last movement direction on the fault: each time the fault moves the earlier structures are generally erased and the newly smoothed surface overprinted by new striae. In regions of strong straining the orientation of linear stretching fabrics (X-directions) and mineral alignments might be used to determine the section plane because in simple shear displacement the XY plane does contain the slip direction (a-direction). However, if the strain is strong enough to develop a noticeable linear fabric it is likely that

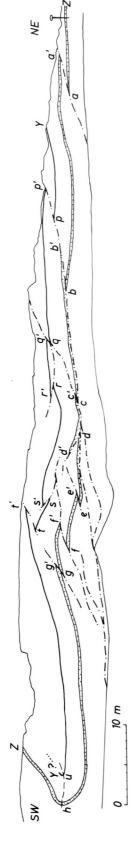

Figure 24.3. Exposure of Visean limestone, Hénaux Basse Normandie Quarry, north France, with marker horizons Y and Z (data from Mark Cooper). See Question 24.1.

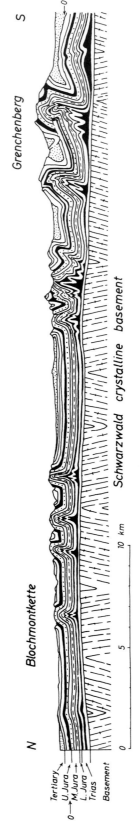

Figure 24.4. Profile through the north eastern part of the Jura Mountains, north Switzerland. After Buxtorf (1916). See Question 24.2.

line length balancing of the beds will be incorrect. The orientations of the axes of small scale folds developed in strained rocks, particularly those close to fault surfaces, should not be assumed to be perpendicular to the transport direction. These structures might be **drag folds** (p. 509) or be structures whose axes have been rotated by strain or body rotation towards the transport direction (Figure 10.28; **sheath folds** or **eyed folds** Figure 15.5).

QUESTIONS AND STARRED (★) QUESTIONS

Making an undeformed template for a cross section by simple line length balancing

Question 24.1

Figure 24.3 shows a simplified sketch of a railway cutting in Visean limestones of the Boulonnais region, north France (Cooper *et al.*, 1982, 1983). This was prepared from a mosaic of down-plunge directed overlapping photographs corrected for parallelax. Two marker bands Y and Z can be traced through the rock face, and are offset by low angle thrust faults (dot–dash lines) which dip both NE and SW. The region forms part of a thrust complex of Hercynian age. The section of Figure 24.3 lies beneath a major thrust plane and is, in part, underlain by a duplex and a basal or sole thrust.

A pin line has been selected on the right-hand side of the section. Measure and record the bed lengths of the faulted sectors of the two marker horizons (pin line to a, a' to b, b' to c, etc.). This can be done using a length of cotton and a scale or, faster and more accurately, using a map measuring wheel or linemeter. Make a reassembly of the faulted units into an undeformed template assuming that there have been no changes of length of the beds as a result of the thrust faulting. You will find that you will have to make some rotations of the faulted blocks together with some internal

shape changes inside these blocks. Compare the bed lengths in the actual rock face and in the reconstructed template and calculate the shortening in the section.

Now check the Answers and Comments section and return to Question 24.2.

Checking if a published section is line-length balanced: Example 1

Question 24.2

Figure 24.4 illustrates a classic cross section of the folded Mesozoic rocks of the Swiss Jura made by Buxtorf in 1916. This section was developed from careful surface mapping together with data derived from the development of a railway tunnel through the Grenchenberg. A key feature for the interpretation of this section is that nowhere in any of the anticlinal fold cores can the lowest Triassic arkosic sandstone (Buntsandstein) or underlying well lithified Palaeozoic rocks and crystalline metamorphic rocks be found. The stratigraphically lowest rocks to appear in the anticlinal cores are rocks of the middle Triassic Muschelkalk formation containing shales, marls, salt and anhydrite. Buxtorf interpreted these geometric features by postulating a décollement of the Mesozoic strata above the Buntsandstein, an interpretation that is now completely accepted and supported by seismic profiling. It is interesting to note that the horizon of shearing between the detachment nappe and underlying rigid sub-stratum does not occur along the actual line of unconformity between Mesozoic sediments and pre-Mesozoic basement. The mechanical separation of basement and cover does not coincide with the stratigraphic separation. Two factors are important in localizing the décollement surface. One is clearly the presence of the mechanically weak horizons in the middle Trias. The second is that the mineralogical similarity of the lower Triassic arkose and the underlying quartz–feldspathic

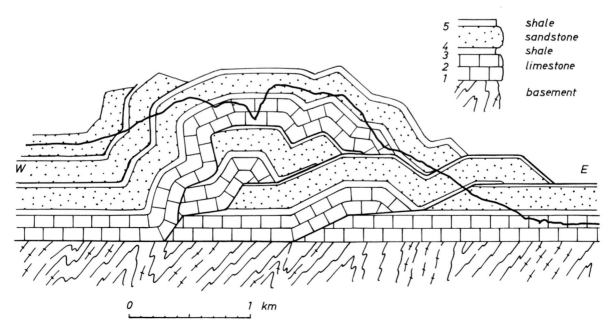

Figure 24.5. Cross section of folded and thrust sediments. See Question 24.3.

gneisses leads to a mechanically similar behaviour across the actual unconformity, the two units behaving as a single competent packet.

In Figure 24.4 measure the lengths of the individual strata. Some reconstructions can be made of certain strata eroded away in the anticlinal folds cores, but the highest stratigraphic units cannot be restored without making certain assumptions about the fold geometry which might not be completely justified. Check if the bed lengths are the same across the section. Determine the total shortening of the Jura cover "nappe" relative to the underlying structural basement beneath the décollement horizon.

Now check the Answers and Comments section and return either to Question 24.3★ or, if time is not available to Question 24.4.

Checking if a published section is line length balanced: Example 2

Question 24.3★

Figure 24.5 shows a cross section which is said to be balanced. Make a restored template of the strata and comment on whether or not the section is geologically realistic and balanced.

Check the Answers and Comments section and return to Question 24.4 below.

Calculation of the depth to detachment

Question 24.4

In Question 24.2 we examined some aspects typical of thin-skinned tectonics where cover sediments had been sheared off from an underlying basement. This type of detachment leads to the uplifting of certain regions of anticlinal fold development or overthrusting. These geometric features sometimes enable the geologist to calculate the depth to the detachment surface from a datum or from the level of some particular marker stratum in the succession. Figure 24.6 illustrates the simplified geometry of this type of situation. The initial bed length l has been shortened to a length l'. Providing that there is no loss of rock volume and no change of area in the cross section, then the volume of material removed by shortening of the sedimentary pile must be equal to that of the uplifted portion (stippled in Figure 24.6). In plane strain conditions the cross sectional area A must be equal, and

$$A = d(l - l') \tag{24.1}$$

where d is the depth of the décollement surface below a particular marker level x. To calculate this depth we must measure the uplifted area A (with a planimeter, or by using Simpson's rule, or simply by superposing the area over a rectangular grid and counting the number of unit squares) and the shortening $(l - l')$. The original length l is generally assumed to equal the total length of a marker bed in the structure. This bed is best chosen as the one which shows maximum competence, and any problems that may arise as a result of strain development during uplift are minimized by taking the measurements along the neutral surface of the layer located more or less midway between the boundaries of the competent layer.

Figure 24.7 illustrates a profile section through a part of the folded Jura mountain of north Switzerland where Mesozoic sediments overlie a more rigid basement composed mostly of crystalline metamorphic rocks. Using the technique described above in the two anticlinal and thrust-faulted uplifts of Mont Terri and Clos du Doubs above the rather flat-floored synclines between and on either side of these uplifts calculate the depth to the décollement horizon beneath each anticline. We recommend that the best horizon to use for the calculation of the shortening in each anticline is the mid-layer of the Hauptrogenstein Formation (x–x of

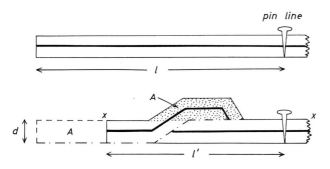

Figure 24.6. *Relationship between uplifted area A and initial geometry. d is the depth to décollement below level x.*

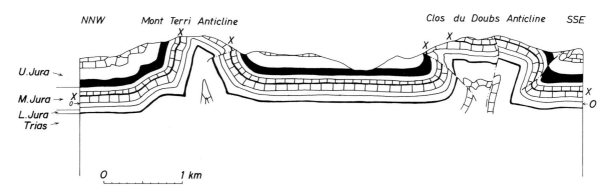

Figure 24.7. *A profile through folds in part of the Jura Mountains, north Switzerland (after Laubscher, 1962). See Question 24.4.*

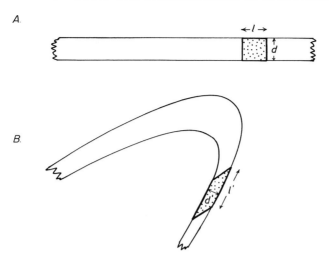

Figure 24.8. A: *Shape and length relations of an initial element bed length l and bed thickness d.* B *gives the modified element shape and modified dimensions l' and d' after folding.*

Middle Jurassic age) because the competence of this formation is high (it is an oolitic limestone) and the strain markers within it show that very little layer parallel shortening by strain has gone on. Complete the section to depth.

Line length measurements in strained rocks

Question 24.5

We have seen that the technique of line length balancing is only correct for concentric and kink folds where the finite strains are built up by heterogeneous simple shear parallel to the layering. In such folds the strains are plane and the layer surfaces always coincide with one of the two surfaces of no finite longitudinal strain in the various strain ellipsoids. Such types of strain distribution are not generally found in nature. Because the constituent rock layers have different rheological properties, each layer will have its own characteristic strain distribution which departs from that of the layer parallel simple shear model. For example, folded competent layers often show layer parallel stretching along their outer arcs and layer parallel shortening along their inner arcs. Although such variations in bed length might approximately compensate one for the other along a fold wave train containing several antiforms and synforms, it is unwise to assume that the total layer length of a competent–incompetent rock interface records the original length of that surface. We have already seen that it is best to make layer measurements as close to the neutral surface as possible but to locate this surface with precision is not always easy, especially where the fold arc length is small compared with the thickness of the competent layer (Figure 21.21) and the layer curvature is high. Another common feature that precludes simple line length balancing occurs where early layer parallel shortening has taken place during the early stages of folding (where competence contrasts are not high, see Figure 19.10B, Figure 20.11A, B and C). We have also seen that where the competent layers are widely spaced within incompetent material the layer lengths in the incom-

petent materials in the contact strain zones between the competent layers are always significantly shorter than their original lengths (Figures 17.7, 17.10).

In order to restore measured bed lengths to their original lengths we need to have information about the finite strain though the fold, and to establish a correction factor F_l at each locality in order to make the restoration. The geometry of a small sector of bed before and after deformation is illustrated in Figure 24.8. The basic equation relating the quadratic extension to the values of the principal reciprocal quadratic elongations λ_1' and λ_2' and the angle between the long axis of the maximum extension direction and the bedding trace (ϕ') is that developed in Appendix D (Equation D.7).

$$\lambda' = \lambda_1' \cos^2 \phi' + \lambda_2' \sin^2 \phi'$$

This equation may be rewritten in terms of the multiplying correction factor $F_l = \lambda'^{1/2}$ necessary to restore the bed length to its undeformed state, the ellipticity of the strain ellipse $R = (\lambda_1/\lambda_2)^{1/2}$ and the area change that has taken place in the section $\Delta_A = (\lambda_1 \lambda_2)^{1/2} - 1$

$$F_l = (1 + \Delta_A)^{1/2}[R^{-1} + (R^2 - 1)R^{-1} \sin^2\phi']^{1/2} \qquad (24.2)$$

This expression is in its most convenient form for the practical solution to our problem and graphical solutions of F_l for different values of R and ϕ' are given in Figure 24.9, with the only unknown being the area change Δ_A. If we can assume plane strain without volume change in the section ($\Delta_v = 0$, $e_2 = 0$ and $\Delta_A = 0$), these curves are very simple to use, but remember that Δ_A is a function of both volumetric dilatation and value of the intermediate strain e_2.

The curves of Figure 24.9 indicate that the correction factor may neccessitate an increase of measured bed length ($F_l > 1.0$, generally where the strain ellipse long axis makes a high angle with the bedding trace) or decrease in measured length ($F_l < 1.0$ where the ellipse axes are at a low angle to the bedding). Where $\Delta_A = 0$, one curve shows $F_l = 1$, and this is the curve which is appropriate to the simple folding by bedding parallel simple shear.

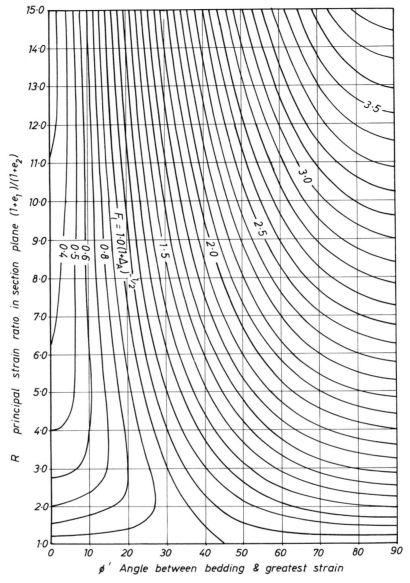

Figure 24.9. *Graphs for evaluating the bed length correction factor F_1 knowing the principal strain ratio R, and the angle ϕ' between the bedding trace and the long axis of the strain ellipse.*

The next step in the restoration of bed length throughout the folded layer requires us to put together the correction factor data from different positions in the fold (Figure 24.10). We are unlikely to have a continuous link of strain data at all positions along the layer because natural outcrops are usually separated by regions of unexposed terrain or because appropriate objects for strain measurement are unvailable. However, we can assume that the correction factor will change systematically and smoothly from outcrop to outcrop. This assumption is generally valid because it is based on the concept of strain compatibility. Discontinuities in the value of the correction factor F_l are only likely where there is a physical discontinuity in the layer (e.g. a fault). Sudden changes in dip of the layer will possibly be associated with sudden changes in the value of F_l, particularly where the axial surface of the kink does not bisect the limbs of the kink, but these changes will not be random

providing that no fault break occurs at the kink axial plane. In kink folds the search for good strain data should be intensified at these positions. The data for the correction factor F_l and measured length of a bed from some pin line l' is plotted graphically l' as abscissa and F_l as ordinate (Figure 24.11C). The data points are then joined with a continuous curve (except where faults are present which cut the bed). To make the total correction for the original length l though the complete structure we have to integrate the correction factors for each small sector length $\delta l'$. The original length l_0 is given by

$$l_0 = \lim \delta l' \to 0 \sum_0^{l'} F_l \delta l' = \int_0^{l'} F_l dl' \qquad (24.3)$$

This integral is the area under the correction factor curve, and it can be found by using Simpson's rule, by counting out

Table 24.1
Data for Question 24.5

Locality	Distance from A' (m)	Strain ratio R	Angle between long axis of strain ellipse and bedding trace, ϕ' (deg)	Correction factor F_l	Correction factor F_d	d' (m)	d (m)
A'	0·0	2·3	−47			14·9	
B'	31·0	2·8	−45			11·8	
C'	55·0	3·0	−43			10·0	
D'	61·0	3·0	−45			9·5	
E'	70·0	3·1	−56			9·5	
F'	76·0	4·3	−75			11·0	
G'	87·0	4·8	+42			8·0	
H'	96·0	5·0	+21			4·8	
I'	113·0	13·0	+12			4·0	
J'	123·0	6·3	+30			7·0	
K'	132·0	0·0	+64			11·0	
L'	135·0	0·8	90			12·5	
M'	144·0	4·3	−55			8·6	
N'	164·0	3·0	−52			7·0	
O'	179·0	2·5	−51			6·5	
P'	208·0	2·5	−52			7·0	

graph squares, by using a planimeter or using a digitizing table with appropriate computer input.

Figure 24.11 shows a cross section of a fold where strain data have collected at localities A' to P'. Table 24.1 records the information available. Using the correction factor curves of Figure 24.9 with the assumption that the area change on the section $\Delta_A = 0$ compute the correction factor F_l for each locality. Plot the correction factor curve and find the corrected lengths of each bed length sector between A' and B', . . . O' and P', etc. Calculate the total shortening in the fold along the layer between localities A' and P', and express this as a horizontal value $1 + e$ and as a percentage of the original length AP.

Correction of layer thickness in strained rocks

Question 24.6★

At the various localities A' to P' in Figure 24.11 the orthogonal thickness d' was measured, and these data are presented in Table 24.1. Make a graphical representation of the variation of d' through the fold, plotting l' as abscissa and d' as ordinate. Discuss the significance of this curve. The correction factor F_d relating original thickness d and final thickness d' can be expressed by

$$d = F_d d'$$

where

$$F_d = 1/F_l$$

(24.3)

Construct a new graph which shows how the original thickness of the layer varies with original position in the fold (l as abscissa and d as ordinate).

ANSWERS AND COMMENTS

Construction of an undeformed template

Answer 24.1

Figure 24.12 shows a line length balanced reconstruction of

the Hénaux quarry outcrops of Figure 24.3. In this duplex structure restoration of bed lengths the total shortening of layers Y and Z can be expressed as negative extensions. Between the pinning point and locality u the extension is given by $(495 − 592)/592 = −0.16$ and the percent shortening is 16%. Between the surface outcrop of layer Z and the fold hinge location h the extension is $(411 − 462)/462 = −0.11$ (or 11% shortening). The differences in these values of the two marker layers are to a large extent dependent upon the choice of pin line position and the positions of the end points u and h of the two layers.

Structurally below the duplex shown in Figure 24.3 is a lower duplex showing a number of interesting geometric features not encountered in the upper duplex zone. When this lower duplex is restored by line length balancing (Figure

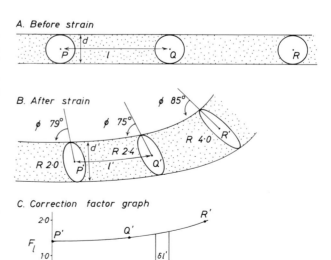

Figure 24.10. *The method of producing a bed length correction graph from three observation points P′, Q′ and R′ along a folded surface.*

24.13A and B) the region between two marker horizons shows a systematic thickening from NE to SW. Because there is no evidence here or elsewhere suggesting that these changes in bed thickness are primary, it would appear that the bulk strain in the overall duplex structure is not solely the result of the stacking of imbricate thrust faults. The layering in the fault restored section *B* is therefore less than that of the original length. This section can therefore be further corrected using the **area balance method** (Hossack, 1979; Elliott and Johnson, 1980). From the geometry shown in Figure 24.14 if the area *A* of a horse unit is divided by the initial stratigraphic thickness *d* the initial layer length can be obtained. The technique employed here (Cooper *et al.* 1983) for determining the areas of individual horses employed cutting up a paper model of the duplex and weighing each sector on a chemical balance. The total length of the restored section can be applied by using this method either on the original section or on the fault restored section. The advantage of using it on the partially restored section is that it enables the total shortening to be separated into fault and ductile strain components. From this construction (Figure 24.13B and C) the present section length of 61 m extends to a length of 76 m when the faults are restored and to a length of 99·5 m when the layer shortening strain is removed. The total longitudinal strain along the layering can be calculated as a percentage ($100(99·5 - 61)/99·5 = 39\%$) and this may be subdivided into a fault shortening component of 15% and a layer parallel ductile component of 24%. Although the extent of the ductile deformation was insufficient to generate a cleavage in these particular limestones, in thin section the individual calcite grains are moderately strongly twinned.

Other geometric features of these two duplexes concern the average shape of the horses: the thickness to length ratio is 1 to 4·5 and the average angle between the imbricate thrusts and the bounding floor and roof thrusts is 23°. Quite a lot more detail can be obtained from these various duplexes and it is recommend that the student reads the original publications which first described these geometric features (Cooper *et al.*, 1982, 1983).

It should be clear from the above discussion that area balancing can be used quite independently of line length balancing. It therefore gives a valuable cross check for correctness of any line length balance. Area balancing is particularly important where layers have been folded because in

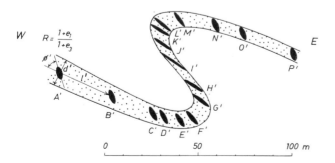

Figure 24.11. *Strained layer in a fold. See Question 24.5.*

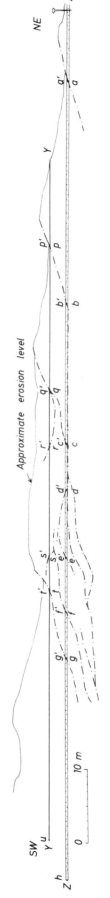

Figure 24.12. *Line length reconstruction of the Hénaux quarry outcrop of Figure 24.3.*

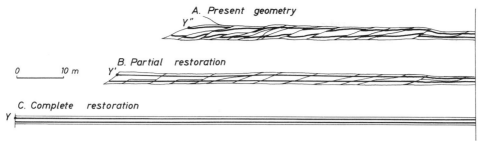

Figure 24.13. *Geometry of the lower duplex at Hénaux. After Cooper et al. (1983).*

folds bed lengths only remain unchanged in special classes of folds (concentric, Class 1B produced by layer parallel slip, or in certain types of kink folds). The principal limitations on area balancing methods are that the original layer thickness must be known and that the cross sectional area before and after deformation must be equal. Knowledge of the original thickness of different layers in a structure is not always available and in some depositional environments individual layers can change thickness and facies in a marked manner. In certain geological environments rocks can change volume by compaction processes proceeding during tectonic deformation and also by solution and redeposition of certain mineral species by the process of pressure solution. Layers situated deep in a sediment pile usually have least pore space because of diagenetic compaction as a result of the load of the overlying material. However, layers found in the upper parts of sediment accumulations often have a very high porosity. This is particularly true of sediments containing clay minerals and in clastic rocks where the sorting of the clastic grains is so good that intergranular pores remain unfilled with small grains. The geometric effects of tectonic pressure solution can be complex, and may or may not lead to volumetric changes. In some environments the material removed from highly stressed grain boundaries may be redeposited locally in pressure shadows (perhaps even on the parent grain) or in vein systems. If this type of redistribution of material is on a local "give and take" basis then the overall volume change will be zero. However in high tectonic levels evidence for pressure solution may be strong (e.g. stylolites cross cutting bedding surfaces, mutually dissolved grains or pebbles) but no volumetrically comparable depositional site can be found. In these circumstances it may well be that material has migrated away from the local system to be deposited in a quite separate tectonic environment or the dissolved material may even have been removed in the ground water system and lost in river flow. Significant volumetric and cross sectional area changes will then have taken place and area

balancing may give rise to reconstructional errors. These errors will be especially great under conditions of plane strain because the volumetric change Δ_v is directly equal to the area change Δ_A in the XZ profile plane (from Equations E.21 and 10.15 where $e_2 = 0$).

Now return to Question 24.2

Line length balancing: Example 1

Answer 24.2

The geometric construction of Buxtorf's section through the Jura mountains (Figure 24.4) is remarkable in many ways. Not only did it introduce the basic concept of décollement, it should also be clear that the fold forms have been reconstructed at depth in a way that clearly indicates that he was aware of the geometric features of line length balancing. The measurements of bed length in this section are as follows:

Contact U. Jura–M. Jura	48·6 km
Contact M. Jura–L. Jura	46·0 km
Contact L. Jura–Trias	46·1 km
Bed in M. Trias	45·0 km
Lowest Trias	36·6 km
Trias–Basement contact	36·5 km

The beds above the lowest Trias (Buntsandstein) all show a rather consistent total length, with the lower parts of this stratal group showing somewhat shorter bed lengths than the upper part. These slight differences show that the beds are rather well line length balanced. The systematic changes with depth might arise because of drafting errors. However, in the field one sees features that could also account for the discrepancies. In the lower Jurassic and upper and middle Triassic formations very short wavelength folds (5 m to 10 cm) exist on the outcrop scale. These undulations have to be averaged out on the scale of the profile and therefore true bed lengths are probably somewhat greater than those measured in this profile. In these rocks one sometimes finds cleavage developed at a high angle to the bedding. This structure is usually interpreted as being the result of a ductile deformation with shortening perpendicular to the cleavage planes (Session 10). It therefore implies that measured bed lengths have been reduced by layer parallel strain and bed thicknesses have been correspondingly increased.

It is clear that the lowest part of the Triassic succession (Buntsandstein) has a bed length much shorter than that of the overlying strata but in accord with the length of the unconformity between the sandstone and the underlying metamorphic complex. It follows that the décollement horizon lies above the Buntsandstein.

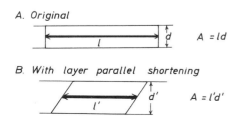

Figure 24.14 *Geometric features of a rock element showing the changes taking place in bed length and bed thickness as a result of layer parallel shortening keeping constant cross sectional area A.*

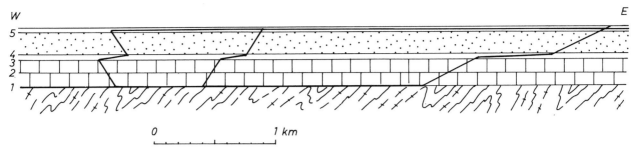

Figure 24.15. Reconstructed template of the section shown in Figure 24.5.

The total shortening of the cover strata above the décollement relative to the basement is probably best determined from the differences in bed lengths between the uppermost stratal levels and the Trias–Basement unconformity, 100 (36·5–48·1)/48·1 ∼ 25% shortening.

The probable reasons for the independent, foreland-directed sliding of the Mesozoic and Tertiary cover relative to the basement have been extensively discussed in the literature of Alpine geology. There is no doubt that the basement and cover were once attached in a coherent manner as part of the northern (European) plate of the Alps. The geometry of this section implies that there must be an excess of basement stripped of its cover and lying somewhere to the south of this profile, and it is generally thought that this basement has been subducted under the main Alpine mountain ranges as a result of crustal collision with the southern (Italian) plate of the Alps. Perhaps we should point out that there does seem to be some evidence that the sub-Jura basement might not have been so completely undeformable as this classic section of Buxtorf suggests and that locally some shortening has occurred by the formation of low angle thrust faults.

Line length balancing, Example 2

Answer 24.3★

Because the bed thicknesses in the section of Figure 24.5 are rather constant through the section and the folds show a kink-like parallel geometric style the bed length balancing

method should be appropriate for a restoration. However when the bed template is made from a pin line position in the east of the section the reconstructed positions of the cut off thrusts show a systematic increase in dip towards the west and eventually to dip reversals of the faults (Figure 24.15). There is no simple way to explain these changes in kinematic or dynamic terms and this section is therefore written off as being a false reconstruction. What we would like to emphasize here is that a section which at first sight looks to be reasonable and to obey the rules contains geometric features which are quite inconsistent with the geometry of thrust tectonics. The geometric feasibility of the uppermost thrust units cannot in any case be rigorously checked by constructional methods because not enough data exist to enable the higher and lower units to be connected. This is often one of the most critical features of a section and one which renders reconstructions impossible. It should also be clear that the validity of sections which have unseen parts because of depth of burial (and therefore no surface exposure) or which have been lost by erosion is extremely difficult to verify, and that calculations of shortening components made with such sections should be treated with great circumspection.

Calculation of the depth to detachment

Answer 24.4

Figure 24.16 shows the graphical solution to the problem. The uplifted cross section area of the Mont Terri anticline between points p and q is 0·645 km². The actual length of the

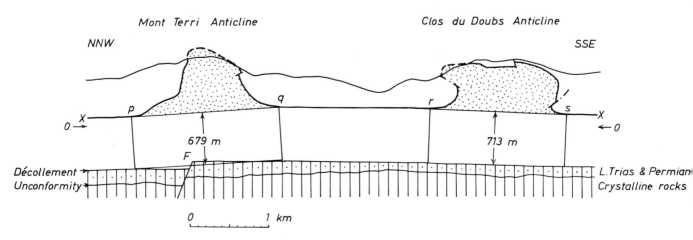

Figure 24.16. Answer 24.4. Construction of uplift areas (stippled) and calculations of depth to décollement below marker horizon XX.

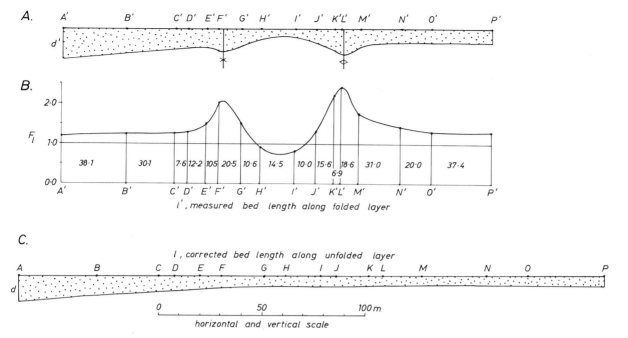

Figure 24.17. Answers to Question 24.5 and 24.6★. A: Variations of measured orthogonal thickness with position in the fold of Figure 24.11. B, shows a correction factor graph constructed from the data of Table 24.1 and the curves of Figure 24.9. C, shows a reconstruction of the original length and thickness of the folded layer.

folded and faulted marker layer is 27·65 km, whereas the direct length between point p and q is 18·15 km, giving a shortening of the layer of 9·50 km. Dividing the uplift area by the shortening gives the *average* depth to basement of 679 m. The cross section uplift of the Clos du Doubs anticline is 0·791 km². The bed length between r and s is 28·31 km, the present distance rs is 17·23 km, so the shortening is 11·08 km. Dividing the uplift area by the shortening gives a basement depth below the rs join of 713 m.

The average depth of the marker horizon in the syncline NNW of the Mont Terri anticline is distinctly low in its topographic level compared with its position at points q, r and s, and the pq join line slope makes an angle of about 4° to the horizontal. This might be accounted for by having a similar slope in the underlying décollement surface, but another interpretation more in accord with the regional geometry and local tectonics is illustrated in Figure 24.16, with the overall dip of the décollement keeping fairly constant and the irregularity being generated by an extension fault at F.

The accuracy of calculations such as we have made here depends upon several factors. It is vital that the constructional data are as well constrained as possible by surface outcrops so that the bed lengths before and after deformation and the volumes (or areas in section) of uplifted rock are based on the best possible observations or inferences. The presence of faults clearly poses problems because their development leads to significant change in bed lengths and in uplift areas.

The décollement horizon calculated in this problem lies at the top of the arkosic sandstone member of the Trias, the actual cover–basement unconformity between the Triassic and Permian sandstone and the crystalline basement lying well beneath this glide horizon.

Line length balancing in strained rocks

Answer 24.5

Figure 24.17A shows the variation of measured thickness d' of the layer in the fold recorded against position along the layer. It is clear from this graph that some thickness variations appear to be geometrically correlated with position in the fold (note the increased thickness at the hinges of the folds), whereas others might be independent of fold location and possibly original variations. Figure 24.17B shows a correction factor graph derived from the data of Table 24.1 and the values of the areas under the correction factor curve between any two adjacent localities. These data have been used to calculate a bed length correction for each successive location and the total bed length AP of 284 m is shown reconstructed in Figure 24.17C. The present direct join length $A'P'$ is 125 m, so the total shortening is 159 m which can be expressed as $1 + e = 0·44$ or 56% shortening. The power of this technique to make accurate constructions is that it does not make any assumptions about the original stratigraphic thickness of a layer and that the knowledge of area change can be used to modify the correction factor if necessary. Its main limitation arises from uncertainties of correction factor value where the principal extension of the strain ellipse in the section lies close to the bedding (ϕ' low) and where the strain ratio R is high ($R > 6·0$). In this region of the correction factor curves (Figure 24.9) the value of F_l is particularly sensitive to orientation values ϕ'.

An example of using this method to restore a cross section of the very variably strained group of rocks which form the Morcles nappe of the Helvetic zone of the Swiss Alps is illustrated in Figure 24.18 (Huggenberger, 1985). The profile section (Figure 24.18A) was established using the methods

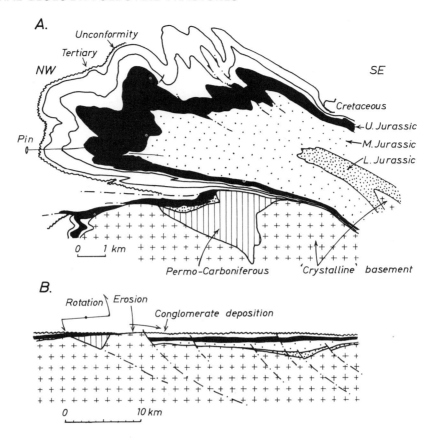

Figure 24.18. *Palinspastic reconstruction of the Morcles nappe (after Huggenberger, 1985). A illustrates the present distribution of the rocks in fold profile and B shows a palinspastic reconstruction of the Mesozoic and Tertiary rocks.*

of down-plunge projection (Session 18). The rocks are strained, sometimes very strongly so (Figure 11.10), and the layers are shortened or lengthened at different localities in the structure together with accompanying changes in layer thickness. The overturned limb of this structure shows especially intense strains ($X:Z$ ratios of 50:1 are common) and the original stratigraphic thicknesses here have been drastically reduced. Measurements of deformed objects and the forms of pressure shadows indicate that over most of this section the strains are close to plane ($e_2 = 0$). Using a combination of strain correction and area balancing techniques Peter Huggenberger has been able to reconstruct the original stratal thickness and the interrelationships of the various stratigraphic units. Because of strong differential shearing on the fold limbs the reconstruction pin line was chosen along the axial surface of the anticline shown. Original bed lengths and thicknesses were then computed and a palinspastic section constructed (Figure 24.18B). This section strongly suggests that, before the compressional tectonics developed the Morcles nappe, an earlier period of extensional tectonics with associated block rotations gave rise to rather complex series of basins where sedimentation took place. A particularly interesting outcome of this reconstruction is that it provides a method of block rotation leading to basement uplift and erosion and a possible source of the Tertiary conglomerates found now in the overturned limb and frontal sector of the Morcles nappe (Figure 7.24). This example clearly shows how careful geometric recon-

structions may be used to restore palinspastically regions of extremely complex tectonic structure. However, we would emphasize that large amounts of field data are necessary before such a reconstruction can be attempted.

In the example of the Morcles nappe the initial starting profile contained very few unknowns because every point on it was derived by projection techniques of surface data. In regions which do not have such strong topographic relief or such inclined fold axes strain correction techniques for layer lengths can also be applicable to build up a palinspastic picture, but the final section may be incomplete, consisting only of partial reassemblages. For example the folded layers shown in Figure 24.19A may be incompletely exposed. It may be possible, using strain data from a succession of localities in different stratigraphic levels, to interconnect the initial line lengths and original bed thicknesses so as to arrive at a partial reconstruction as shown in Figure 24.19B.

Correction of layer thickness in strained rocks

Answer 24.6★

Figure 24.17C shows the corrected thicknesses of the original folded layer using the inverse correction factor of Equation 24.3. It will be apparent that the layer did show initial thickness variations and that the region between localities A and G showed a marked increase in thickness from east to west.

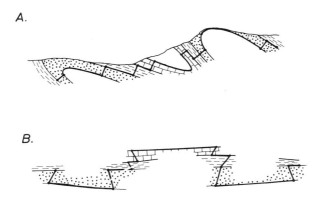

Figure 24.19. Method of reconstruction of an incompletely exposed section by layer side-step techniques. A are the data. The heavy points and lines represent localities and layer sections which can give strain correct layer lengths and thicknesses. B shows the incompletely restored section. Note how the originally orthogonally situated data localities in A are "back sheared" in the reconstruction B. The amount of back shear is computed from the available strain data.

CONCLUDING COMMENTS

The techniques of section balancing, especially those developed over the past decade, have undoubtedly led to great improvements in the objectivity and reliability of geological sections and, perhaps most important, have made us think much more critically than in the past about the problems of drawing a cross section. However in some circles there seems to have developed an uncritical acceptance of certain of the constructional rules which has led to a rather monotonous uniformity of geometric style in terrains often of differing lithologic and tectonic environments. For example, many recently published sections through the frontal zones of the Rocky Mountains, Appalachians, Caledonian chains of northwest Europe, Alps and Pyrenees are practically interchangeable. Although we agree that certain aspects of these deformation zones might have something in common, we are of the opinion that much of the apparent geometrical correspondence arises from identity in the methodology used to make these constructions and not because of any correspondence in rock rheology or mechanical behaviour.

The most extreme examples of this approach treat rock layers as if they were identical sheets of paper with their final fault and kink-like fold forms being controlled only by the orientation of some previously made set of scissor cuts.

No geologist likes to be accused of having drawn an unbalanced section, but we should be honest in admitting that practically all the sections we construct from field and seismic data have uncertainties of different types and magnitudes. Beware of the geologist who announces in his talk or in his paper that his sections are "perfectly balanced" and who gives an impression that these sections are above reproach from any source. If you meet such a character you can be pretty sure that he is inexperienced, naive, dishonest or combines these traits.

Remember that the choice of method used to construct geological sections which will approach reality or to undeform a given sector to its predeformational state depends critically on the type of geological environment of the region and that there are no universally correct techniques applicable to all types of terrain.

KEYWORDS AND DEFINITIONS

Balanced section or Balanced cross section
A section which has been constructed using geometric constraints which have been chosen as realistic for the particular tectonic environment of the section. Sections may be constructed by **line length balancing**: a technique which assumes that the original bed lengths are unchanged after deformation. **Area balancing** assumes that the original cross sectional area of any bed in the section is unchanged (Figure 24.14). **Strain balancing** uses the known finite strains in a structure to correct layer lengths and bed thicknesses (Figure 24.10).

Compatibility
The geometric constraints necessary to make a rock mass coherent after deformation. See Session 3, p. 53.

Décollement fault
See p. 518 and Figure 23.21.

Fault bend fold
The bending of strata in the hanging wall of a fault as the layers pass over a non planar fault surface. These folds are generally open structures and some are

characterised by axial surfaces which migrate through the beds (Figure 23.30).

Fault propagation fold   Folds developed at the tip of a propagating fault. They are generally tight and strongly asymmetric with steep or overturned beds in the forward facing limbs.

Palinspastic reconstruction   A restoration of the stratigraphic layers to their initial pretectonic disposition.

Pin line   A fixed reference line taken across a series of beds where there is no interbed slip between adjacent layers and which is used as a fixed reference line in a restored cross section. Pin lines are normally chosen in undeformed sections or sometimes along the axial traces of folds.

Restorable cross section   A cross section which may be undeformed according to some given geometric rules so that the beds attain their original pre deformational position in a coherent way to give a correct palinspastic restoration. In contrast **unrestorable sections** are those which, due to the choice of section line, cannot be simply undeformed.

Thin skinned tectonics   A tectonic style characteristic of certain parts of the uppermost crustal levels where a sedimentary cover succession has been detached from and shortened independently from an underlying basement.

KEY REFERENCES

Boyer, S. C. and Elliott, D. (1982). Thrust systems. *Am. Assoc. Petr. Geol.* **66**, 1196–1230.

We recommended this as a key reference in Session 23 for a description of many important geometric features of thrust faulting. It also gives an excellent review of section balancing and includes new cross sections through the Blue Ridge of the southern Appalachians and through the central European Alps. Our principal criticism of both sections is that they both propose sub-planar sole thrusts within the basement, with the basement underlying these thrusts apparently in an undisturbed condition. We suggest that it is incorrect to continue the deformation styles of thin skin tectonics into the middle and lower crust.

Cooper, M. A., Garton, M. R. and Hossack, J. R. (1982). Strain variation in the Hénaux Basse Normandie duplex, northern France. *Tectonophysics*, **88**, 321–323.
Cooper, M. A., Garton, M. R. and Hossack, J. R. (1983). The origin of the Basse Normandie duplex, Boulonnais, France. *J. Struct. Geol.* **5**, 139–152.

These two papers describe the geometric features of almost completely exposed duplex structures and describe in detail the methods used to determine the pre-deformational geometry by line length and area balancing. These papers would form excellent commentaries on Answer 24.1.

Dahlstrom, C. D. A. (1969). Balanced cross sections. *Can. J. Earth Sci.* **6**, 743–747.

The paper describes the constraints generally employed for section balancing and figures a number of real examples of the use of balanced sections in solving problems in the Rocky Mountain Foothills of Alberta.

Gibbs, A. D. (1983). Balanced cross-section constructions from seismic sections in areas of extensional tectonics. *J. Str. Geol.* **5**, 153–160.

The concept of balanced cross sections were mostly evolved in compressional terrains. This paper shows how the ideas behind the calculations of depth of décollement, line and area balancing can be applied to extensional fault terrains.

Hossack, J. R. (1979). The use of balanced cross-sections in the calculation or orogenic contraction, a review. *J. Geol. Soc. Lond.* **136**, 705–711.

As well as outlining the main techniques of section balancing this review gives useful discussions on the problems that arise when volume loss of rock material takes place as a result of diagenetic and tectonic compaction and pressure solution.

Laubscher, H. P. (1962). Die Zweiphasenhypothese der Jurafaltung. *Eclog. Geol. Hevl.* **55**, 1–22.

Laubscher, H. P. (1965). Ein kinematisches Modell der Jurafaltung. *Eclog. Geol. Helv.* **58**, 231–318.

For those who read German these two papers can be very highly recommended. They set out, in an extremely thorough way, methods of fold reconstruction for the determination of shortening and depth to basement décollement in the Jura Mountain region of north Switzerland. Many quantitative reconstructions are made to determine a likely kinematic model for this classic fold and fault zone and, of especial interest, are the developments of the implication of block rotations in the overall kinematic modelling.

Ramsay, J. G. (1969). The measurement of strain and displacement in orogenic belts. *In* "Time and Place in Orogeny" (P. E. Kent *et al.*, eds) 43–79. Special Publication 3. Geological Society, London.

This paper discusses several topics relevant to this session including descriptions of several methods of measuring shortening and some of the effects of convergence and divergence in the movements of thrust sheets. A discussion is given of the technique of line length balancing in strained strata and this describes the main effects of area changes in the cross section and how they are related to overall volume changes that would form extra background reading for Answer 24.5.

Woodward, N. B., Boyer, S. E. and Suppe, J. (1985). An outline of balanced cross-sections. *In* "Studies in Geology II," 2nd edn, 170 pp. University of Tennessee Department of Geological Science.

This extremely useful manual was especially prepared for a short course at an Annual Meeting of the Geological Society of America. It reviews techniques and includes an interesting historical perspective. Eight exercises are given. This publication is highly recommended.

SESSION 25

Mechanical Analysis of Fractures

After a brief introduction to the terms used to describe fractures the concepts of force, stress and elastic deformation of a rock are discussed. Laboratory methods of rock testing under controlled stress conditions are introduced and the empirical methods of defining critical failure stress with the aid of Mohr stress circles and the Mohr circle envelope are developed. Aspects of the mechanical principles governing rock fracture are discussed and pore fluid pressure is shown to play a very important role in controlling the conditions of rock failure. It is shown what practical methods are available for estimating the orientations of principal stresses acting in the Earth's crust using the techniques of P-wave first motions from earthquakes. Finally, the special types of fractures which result from impacts of solid bodies from space on the Earth's surface are described and it is shown how the directions of shock waves can be derived from their geometry.

INTRODUCTION

In Session 23 we discussed the characteristics of faults from a purely geometrical standpoint. In this session we wish to establish some of the fundamental mechanical ideas on brittle failure of rocks. **Fractures** are the structures developed by brittle failure, and are extremely widespread in the upper 10 km of the crust where temperatures and confining pressures are relatively low (0–300°C, 0–4 kb). Fractures show an incredible range of sizes from megalineaments with lengths of hundreds or even thousands of kilometres, to microcracks with lengths of fractions of a millimetre such as may be seen in thin sections of rock under the microscope. The understanding of the nature of these fractures and their relationship to the stresses causing them is not only important for the geologist interested in understanding the tectonic activity of the past and the present. Fractures play a key role in allowing fluids of many types to migrate through the crust and in the mining and petroleum industry an understanding of the geometry of fracture is extremely important. In the petroleum industry special methods of increasing the extent of underground fracture systems have been developed so as to allow for more effective transport of petroleum and gas through reservoir rocks. Rock fracture systems are also of great importance to the rock mechanics engineer as fractures can very greatly influence the feasibility of building a dam or a road in a particular terrain and the costing of quarrying or tunnelling through rocks containing fractures rests very heavily on the frequency, nature and stability of the fractures.

We will use the term **fracture** to cover all discrete breaks in a rock mass where cohesion was lost. This general term covers **faults**, where the two sides are displaced relative to each other, **joints** where the two sides show no differential displacement (relative to the naked eye), **healed** or **sealed joints** where fluids passing through the rock have partially or completely joined together the adjacent sides by the deposition of crystalline material, and **veins** where a considerable thickness (> 1 mm) of filling material occupies the region between the fracture walls.

Force and stress

Before we can discuss the mechanical aspects of rock fracture we must first investigate the concept of **stress**. Stress arises in the Earth's crust as a result of the action of different **forces**. The two most potent forces are those of **gravity** and a second, often termed **tectonic force**, which generally arises from the cell-like motions of the fluid and plastic material in the internal parts of the planet resulting from thermal convection. Gravitational force acts vertically whereas tectonic forces generally act in the crustal zone in a sub-horizontal manner. At any locality the total force vector is given by the vector sum of the force components. Stress is defined as force per unit area. In Appendix E we set out the mathematical background for those stresses acting in two and three dimensions in a small element of the crust. In such a small element (mathematically these elements are best considered so small that the stresses act at a point) we find that stress, like strain, is a tensor quantity. Stress requires six components for its definition. In any small cube-shaped element we need to have three **normal stress components** σ_x, σ_y, and σ_z acting perpendicular to the opposite pairs of cube faces, and three **shear stress components** τ_{xy}, τ_{yz}, τ_{zx} acting parallel to the pairs of faces. In a general state of stress it can be shown that there are three mutually

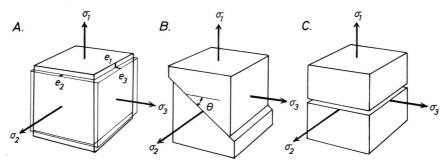

Figure 25.1 A *Elastic deformations* e_1, e_2 *and* e_3 *in an elemental cube under stresses* σ_1, σ_2 *and* σ_3. B *Failure by the production of a shear fracture with angle* $\theta°$ *from the principal compression direction.* C *Failure by the production of extension fracture.*

perpendicular planes on which no shear stress acts, and these are known as the **principal planes of stress**. The normal stresses acting perpendicular to these principal planes have the values of the maximum, intermediate and minimum that are possible. They are termed the **principal stresses** and their values are designated σ_1, σ_2 and σ_3 in order of tensile magnitude. The total stress tensor can then be resolved into six features; the three values of, and the three terms defining the orientation of the principal axes of stress. Sign conventions are often a problem and, in geological literature, one often finds the order of magnitude of these stresses defined in the opposite way to the one we adopt. We have adopted the standard convention of engineering mechanics because we think it more logical to relate the maximum *positive* tensile stress σ_1 with the maximum *positive* elongation e_1. There are many mathematical correspondences between stress and strain because they are both second order tensor quantities, but the geologist should on no account confuse them or use the terms interchangeably. First, they have quite different physical dimensions: strain is a scalar quantity measured only by geometric lengths, whereas stress is measured in units of Newtons per square metre. Second, we have seen that finite strain requires nine terms in its three-dimensional formulation whereas stress requires only six. In some situations the orientations of the principal axes of stress coincide with those of the principal strain axes but, apart from small elastic deformations, this coincidence is uncommon. It should be apparent that, over periods of geological time, the principal stresses will change attitudes and values: we can never refer to *the* stress in geology, but only the particular stress state acting at a given time. In contrast, we have seen that the finite strain state is unique at one point no matter how complex are the deformation increments that built this state. If you want to refer to the stress state, please first pause and consider if you are really discussing the instantaneous *dynamics* of a situation (involving stress), or the overall geometry or *kinematics* (involving displacements and strain).

Because there are certain mathematical similarities between the equations describing stress and strain we find that it is possible to represent the variations of shear stress τ and normal stress σ in a similar way to that used for variations in the parameters γ' and λ' for finite strain. The variations in τ and σ on any plane making a specific angle with the principal stress directions can, for a given stress state, be represented in a **Mohr diagram** or **Mohr circle construction**. We recommend that the student works

through Appendix E at some stage during the course of this session to obtain a sound basis of stress analysis.

Stress–strain relationships in a homogeneous, isotropic, elastic body

For our comprehension of mechanical fracture we need to have some idea of the relations of stress and strain in a solid, non-flowable material. If we consider a small solid elemental cube and subject it to a stress system it will change shape (Figure 25.1). It will suffer a stretch parallel to the axis of maximum tensile stress σ_1 and a shortening parallel to the axis of least tensile stress σ_3. The changes of length parallel to the intermediate stress axis σ_2 can be positive or negative. In a truly elastic material, usually termed a **linearly elastic material**, these strains will be small in value (less than 1% deformation) and will be linearly related to the principal stress values. In order to describe the relationships we need to have two parameters describing the linearity. These are the **longitudinal modulus E** or **Young's modulus of elasticity** defined like the extension e of an elastic spring under a stress σ

$$\sigma = Ee \qquad (25.1)$$

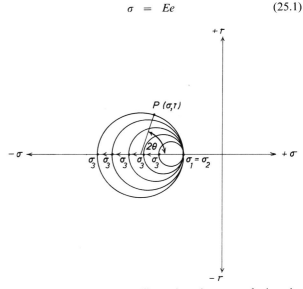

Figure 25.2. *Mohr circles illustrating the type of changing stress states produced during a rock test on a cylindrical specimen. The confining stress* $\sigma_1 = \sigma_2$ *is kept constant and an increasing axial compression eventually leads to failure on a shear fracture (point P) at an angle of* θ *to the maximum compression direction.*

and a second modulus describing how the material contracts sideways perpendicular to the applied normal stress and termed **Poisson's ratio** v, where

$$v = \frac{\text{extension normal to the applied stress}}{\text{extension parallel to the applied stress}} \quad (25.2)$$

In the unit cube of Figure 25.1 the changes in length perpendicular to the upper and lower faces (extension e_1 in the σ_1 direction) will be influenced not only by the principal stress value of σ_1, but also by the values of σ_2 and σ_3 which work in the opposite sense to that of σ_1. We can write three *equations of linear elasticity* relating the strains and stresses in the cube:

$$e_1 = \frac{\sigma_1}{E} - \frac{v\sigma_2}{E} - \frac{v\sigma_3}{E}$$

or

$$e_1 = \frac{1}{E}[\sigma_1 - v(\sigma_2 + \sigma_3)]$$

and

$$e_2 = \frac{1}{E}[\sigma_2 - v(\sigma_3 + \sigma_1)] \quad (25.3)$$

and

$$e_3 = \frac{1}{E}[\sigma_3 - v(\sigma_1 + \sigma_2)]$$

One important feature of these results is that a condition of **plane stress** ($\sigma_2 = 0$) does not necessarily induce a state of elastic **plane strain** ($e_2 = 0$).

Adding together the three Equations 25.3 we find that

$$e_1 + e_2 + e_3 = \Delta_v = \frac{(1 - 2v)}{E}(\sigma_1 + \sigma_2 + \sigma_3)$$

where Δ_v is the elastic volume dilation or, using the concept of **mean stress** $\bar{\sigma} = (\sigma_1 + \sigma_2 + \sigma_3)/3$,

$$\Delta_v = \frac{3\bar{\sigma}(1 - 2v)}{E} \quad (25.4)$$

if there is a volume change we can express this as a **bulk modulus of elasticity** K.

$$\bar{\sigma} = K\Delta_v \quad (25.5)$$

where $K = E/3(1 - 2v)$. If there is no volume change, Poisson's ratio $v = 0.5$.

As we change the state of stress the shape of the elemental rock cube changes (Figure 25.1). However, there comes a critical stress state where the elastic properties of the material can no longer accommodate the shape change. At some particular stress level the **elastic limit** of the material is reached and the cube deforms in a non-elastic way, either by undergoing irreversible deformation by **plastic flow** or by **brittle failure** and the development of **fractures**. High temperature and slow strain rate (or rate of changing the stress conditions) favour the former, whereas low temperature and fast strain rate favour brittle failure. In this session we will concentrate our attention on brittle behaviour. In laboratory experiments it is found that when the stresses are tensile, failure often takes place by the development of an **extensile fracture** (Figure 25.1C) forming normal to the maximum tensile stress direction σ_1.

Under compressive stress failure occurs on a **shear fracture** (Figure 25.1B), the surface of which contains the intermediate stress axis σ_2 and makes an angle of θ with the principal compression direction σ_3 which is always less than 45°, and typically in the range $2\theta = 40\text{–}80°$. It is important to note that this shear fracture does not form along the surface of maximum shear stress in the material, which is always found in conjugate positions, intersecting in the σ_2 direction but making angles of $\pm 45°$ with σ_1 and σ_2 (see Appendix E, p. 671, and Figure E.7). At the time of fracture initiation there is usually a sudden drop in the stress levels and in laboratory experiments it is found that the rock sample can never again withstand those stress states that led to the initial fracture formation: the fracture has induced a fundamental weakness into the material.

The stress levels for the initiation of fracture change with rock type and with the external physical factors under which the experiment takes place, and it should not be surprising either that different rocks have differing strengths or that the strength varies under different conditions. The problem facing the experimentalist is that the number of rock types and physical variables is too large to allow a thorough examination of all the geological potentials.

From these initial comments it should be clear that the fracture surface is developed along a plane characterized by a particular shear stress τ, which is less than the maximum possible under the given stress conditions, and that this plane has a normal stress σ acting across it. Our first question relates to experimental methods which enable us to define the ranges in values of these critical factors.

QUESTIONS

Laboratory tests for rock failure

Question 25.1

Experiments are carried out in the rock testing laboratory on small cylinders of rock cored from a piece of intact rock material (with no pre-existing fractures). These cylinders are carefully prepared so that there are no surface irregularities which might induce stress variations, and they are placed inside a ductile metal jacket, generally of copper. This cylinder is then placed into a *triaxial testing rig*, a machine which enables us to produce a uniform confining stress around the cylinder and then to produce an axial load from which we are able to calculate the stress along the cylinder axis which is in excess of the confining stress. We do not want to go into the somewhat complex engineering technology needed to do this, but we should mention that the increase of load can be applied at a constant rate by using a down-screwing platten device, and that methods are available for controlling the temperature and sometimes the fluid pressure inside the specimen. The stress system on the sample is triaxial, but the radial stresses are equal to the confining stress. The stress systems $\sigma_1 = \sigma_2 > \sigma_3$ can be represented on a Mohr diagram by a single circle (see Appendix E, p. 671, and Figure E.5) which gives the stress conditions on every longitudinal section containing the cylinder axis. We keep the confining stress constant (the σ_1 point on the Mohr circle remains fixed) and as we increase the compressive (axial) stress the Mohr circle expands its radius by the movement of the σ_3 point left-wards along the σ-axis of the diagram (Figure 25.2). At some particular axial stress the specimen fractures and loses coherence. We can measure the orientation of the failure surface to the cylinder axis (θ) and plot the stress conditions that were acting in the specimen at the time of failure (see Figure 25.2: the normal to the plane at point P makes an angle of 2θ with σ_1). By conducting a series of experiments with differing confining stress levels we can make a number of limiting fracture-inducing Mohr circles and separate which combinations of σ and τ are stable from those which lead to fracture.

Table 25.1.
Deformation of Quintnerkalk: data for Question 25.1

Experiment	Confining stress $\sigma_1 = \sigma_2$ (MPa)	Axial stress at time of failure σ_3 (MPa)	Angle θ (deg)
1	−0·10	−450·0	14
2	−20·0	−550·0	20
3	−40·0	−640·0	19
4	−100·0	−750·0	20
5	−200·0	−900·0	25

NB. The negative sign arises from our notation of compressive stress.

Table 25.1 sets out a series of results made on a fine grained Jurassic limestone (Quintnerkalk) in our ETH rock mechanics laboratory by our colleague Ueli Briegel. This rock was deformed under dry conditions at room temperature (20°C). The axial load was provided by a geared rig producing an axial strain rate of $\dot{e} = 10^{-5}\,\text{s}^{-1}$. The stresses are given in megapascals (1 MPa = 10^6 Pa = 10 bars).

Plot these data as Mohr circles, and determine the geometric envelope of all the stress failure circles, known as the **Mohr envelope**. Answer the following questions:

1. Do rocks require a greater or lesser stress difference to bring fracture failure as the mean compressive stress $\bar{\sigma}$ is increased?
2. Could an intact specimen of Quintnerkalk be expected to be stable and without fracture development with sresses given by $\sigma_1 = -750\,\text{MPa}$, $\sigma_2 = -400\,\text{MPa}$, $\sigma_3 = -130\,\text{MPa}$?
3. What would occur, in terms of rock stability, if we added a state of positive hydrostatic stress on the stresses of Question 2 (i.e. changed the stresses to $\sigma_1 + \sigma$, $\sigma_2 + \sigma$, $\sigma_3 + \sigma$)? What might be the implications of increasing the *fluid pressure* contained in the rock pores on the development of fractures?
4. It is calculated that a joint surface passing through the rock in a tunnel wall is likely to be subjected to a normal stress of $-530\,\text{MPa}$ and a shear stress of $+300\,\text{MPa}$. Would this situation be stable or would the joint be likely to undergo shearing movement?

Now proceed to the Answers and Comments section and then continue with the text below.

The relationships between faulting and earthquakes

The study of the Earth's seismicity has played a key role in understanding the overall structure of the planet and also in understanding and modelling the tectonic activity that is going on today. Earthquakes provide a major means by which tectonically accumulating elastic stresses are released. Their study gives much information on **neotectonic activity**. From today's pattern of earthquake distribution it is clear that earthquakes are concentrated in certain geographic zones of the Earth's crust and upper mantle. We will assume that the student is aware of this pattern of location and depth distribution of earthquake **foci** and is already acquainted with the general theory of **plate tectonics**. We know that three major types of tectonic regime predominate, and that these are connected with compressive plate collisions, plate separations and sideways or transform motions between plates. It should be emphasized, however, that in continental regions the surface distributions of epicentres and the depth patterns of earthquake foci are often very complex. In continents earthquakes often occur in rather broad diffuse zones, and there is usually a higher frequency of seismic events in the upper 15 km of the crust than in the lower crust, with a secondary increase of frequency in the upper mantle. This uneven vertical distribution is probably best explained in terms of rock composition and rheology. The silicate minerals that dominate the crust are generally brittle in the upper crust but become ductile in the higher temperature environments of the middle and lower crust. In contrast, laboratory experiments on olivine rocks (peridotites, etc.), which are known to dominate mantle rheology, show that these materials can be brittle at temperatures of up to 800°C. It is this brittle behaviour which is probably effective in the upper mantle.

Earthquake analysis provides many fascinating ideas and facts, and we strongly hold the opinion that seismologists and structural geologists should work together much more than is often the case today. These research fields have much common ground: the study of current deformational activity is almost certainly a key to understanding many of the deformations of the past, while the observations made by structural geologists have implications for rock rheology and the physical constraints that might play a role in present earthquakes. The mapping of recent surface ruptures along earthquake faults before these features have been removed by erosion (e.g. Tchalenko and Ambraseys, 1970) has much in common with those observations made by structural geologists determining the movement history along ancient faults.

In this session we would like to spend some time to show how we might obtain more information on the elastic deformations associated with earthquakes and how we might use these data to make deductions about the stress field existing today.

Earthquake motions are recorded on a **seismogram**. It is well known that a number of distinct types of elastic waves are developed during a seismic event: each type has distinctive physical characteristics and travels at different speeds of propagation through any given rock. The seismogram records the earthquake as successive arrivals of **P-waves** (primary, compressional and extensional particle motions parallel to the direction of propagation), **S-waves** (secondary, with sideways particle motion perpendicular to the direction of propagation) and the largest amplitude **surface-waves** (waves travelling along or near the surface and responsible for the maximum damage during an earthquake). When a number of seismic records are available

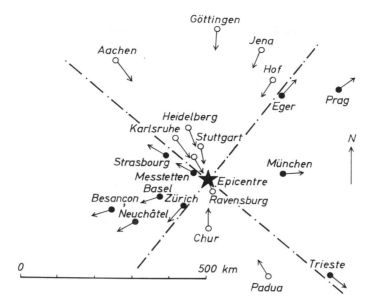

Figure 25.3. *Map showing the first motion P-waves which were compressional (filled circles) or dilatational (open circles). The dash–dot lines show the traces of the nodal planes. After Hiller (1936).*

from recording stations arranged around a particular epicentre it is found that the first arrivals of the P-waves at some stations indicate that the particles initially suffered a motion towards the station (a push or compression), whereas other stations have records indicating an initial motion away from the station (a pull or dilation). When the distribution of compressions and dilations is put on a map, it is found that the two types of first motion are grouped very systematically (Figure 25.3) and can be separated by two lines. When a three-dimensional analysis of the data is carried out it is seen that the map lines represent the traces of two perpendicular planes, the **nodal planes**, which separate the first motions

into four separate volumetric zones (Figure 25.4A). If we have arrival characteristics from a number of recording stations *A* to *J*, it might be possible to determine, using projection techniques, the orientations of these nodal planes (Figure 25.4B), a technique known as **P-wave first motion analysis**. Because the earthquake focus lies at the intersection of the nodal planes, it is generally considered that one or other represents the fault plane along which the slip that generated the earthquake occurred. From the seismic data alone it is not possible to select one as against the other, the reason being that right-hand slip motion on one plane gives an identical quadrant solution to left-hand

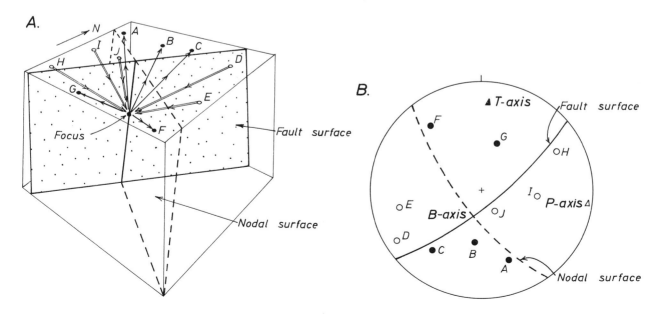

Figure 25.4. *A Three-dimensional representation of the two perpendicular nodal planes which divide the body into compressional and dilatational sectors. One of the nodal surfaces represents the fault plane, the other has only mathematical reality. B is an equal area projection of the compressional (filled circles) and dilatational (open circles) data from the various surface stations of A showing the fault plane and perpendicular nodal plane as great circles.*

A.

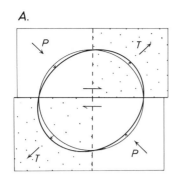

B.

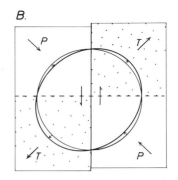

Figure 25.5. Schematic diagram to show that movements on either of the nodal planes sets up similar first motion displacement sectors.

slip motion on the other (Figure 25.5). The selection is generally made (if at all) on geological knowledge of the existing fault pattern, from the traces (if present) of surface breaks accompanying the seismic activity, or perhaps from a planar alignment of aftershock earthquake foci thought to represent repeated slip at different localities on the same fault.

The bisector of the compressional initial motion sector is generally known as the T-axis, that of the dilational sector is the P-axis, and the intersection of the nodal planes as the B-axis. The geometric significance of the T- and P-axes is shown in Figure 25.5 with reference to the first motion ellipsoid. Probably T-, B- and P-axes correspond very crudely to the principal stress orientations σ_1, σ_2 and σ_3 respectively. However, you should immediately see problems in a one-to-one correlation. We know that fault surfaces are not initiated at angles of 45° to σ_1 and σ_3 and, because of static friction, movement along a pre-existing fault might not imply that the maximum and minimum principal stresses were oriented at 45° to the fault surface.

P-wave first motion analysis

Question 25.2

The data of first motions illustrated in Figure 25.6A, B and C come from the Helvetic zone of the Valais, one of the most seismically active regions of the Swiss Alps. They were provided by one of our colleagues at the ETH Zürich, Dr Nazario Pavoni. The scale of the data area is

considerably smaller than that illustrated in Figure 25.3, but the seismic pattern and its significance is the same. The data were gathered using portable seismic recording stations in a region where **aftershocks** were taking place after a major earthquake (1946, magnitude 5·7) had occurred. The focal depths of the microearthquakes investigated ranged from 1 to 10 km (located in the carbonates of the Helvetic nappes or in the upper part of the underlying autochthon—see Figure 11.10), and their magnitudes were in the range 0·2 to 2·5. Construct the nodal planes from the projections, and determine the T-, P-, and B-axes.

Proceed to the Answers and Comments section and return to the discussion below.

Fractures produced by bombardments of the Earth's surface by objects from space

It is well known that extraterrestrial objects have, in the geologically recent past, struck the surface of the Earth with considerable momentum producing an **impact crater** or **meteor crater**. Probably the best known is the Barringer Crater in Arizona, about a kilometre in diameter, some 200 metres deep and formed about 25 000 years ago. Many such geologically recent craters have now been recognized. Some contain fragments of nickel–iron but, due to vaporization of the meteor on impact and the rapid weathering of such metallic material in certain climatic regions, not all of these craters contain such obvious clues to their origin. It seems highly likely that, during the Earth's past history, many such impact craters or **Astroblemes** (Dietz) must have formed,

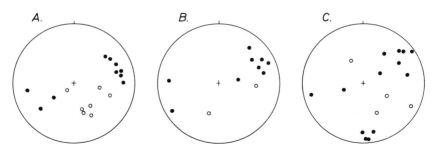

• *Compressional (away from source) first motion*

○ *Dilatational (towards source) first motion*

Figure 25.6. First motion data from three stations in the Helvetic region of the Valais, Switzerland. See Question 25.2. (Data from Pavoni, 1980).

Figure 25.7. *Distribution of the known continental impact craters. Open circles are craters with meteoritic fragments, filled circles are structures with shock metamorphic features. After Grieve, 1982.*

and many circular structures have geological relationships which suggest that they were formed hundreds of millions of years ago. Figure 25.7 shows the distribution of the known continental impact craters. The distribution is uneven, but looking at this map it should be clear that the absence of craters in regions covered by tropical forest, tundra or ice is clearly explainable. Perhaps what is of more interest is the general absence of such structures in the more recently produced Alpine fold belts (e.g. Western USA and western Canada), suggesting that high concentrations elsewhere might be generally related to the stability of continental shield areas over large time periods. The boundary between the Cretaceous and Tertiary Systems is a plane where rather spectacular biological extinctions took place (Alvarez *et al.*, 1980) and this horizon is also characterized worldwide by a particularly high iridium anomaly, locally showing increases of 16 000%. It has been suggested that these features were the result of the disintegration of an extremely large asteroid object about 10 km in diameter impacting the Earth at this time. However, no suitable large crater has yet been found as a candidate for this event. It is possible that it landed in the ocean and has perhaps been subducted or covered with lava flows from the upwelling asthenosphere.

It should be expected that such craters on the continents were ephemeral, being rapidly eroded or filled with debris during the normal course of exogenic processes. Their better preservation on the Moon's surface relates to the deficient atmosphere and the absence of the types of tectonic remoulding processes which are characteristic of the Earth's surface.

Structures with circular forms can be developed in many ways: by volcanic activity and the intrusion of steeply outward dipping sub-cylindrical **ring dykes** or steeply inward inclined **cone sheets**; by **diapiric rise of low density**

magmatic plutons or rock salt; and by single phase or superposed phases of folding. We need to consider what special features might be expected in or around impact craters that act as finger prints for this particular process. Much depends on the size and momentum of the extraterrestrial object, but certain special features might be expected:

1. **Meteoritic fragments**.
2. **Upturned crater flaps** (contrast with volcanic caldera).
3. Fallback **breccias**.
4. Local rock melting by friction on fractures, and the formation of **pseudotachylite**.
5. Highly **fractured minerals** (e.g. quartz with shock deformation lamellae).
6. **High density mineral polymorphs**. Particularly common are high pressure polymorphs of quartz: **Coisite**, a form only stable at pressures of 30–70 kb and with a density of 2·9 (cf. quartz 2·6), and **Stichovite** formed at pressures of 80 kb and with a density of 4·3. It is interesting to note that such polymorphs have been found in nuclear explosion centres at the Nevada test site. It is clear that the stresses required to form these minerals are quite out of the range of crustal and upper mantle conditions where the confining pressures come about by overburden load.
7. **Shatter cones**. These are fracture surfaces of conical shape cutting the rock. The surfaces do not generally form complete cones, but show cone sectors from about 10° to 40° of a complete cone (Figure 25.8). The partial cone sectors form interesting, intersecting and interlocking arrays, the sizes of individual surfaces varying from a few centimetres to about 2 m. The cone sectors are characterized by the presence of radiating

Figure 25.8. Shatter cones developed in a fine grained siltstone, NW Vredefort, Transvaal, South Africa.

bundles of striae with a horsetail effect. These frond-like striae never cross, and adjacent striae always converge towards the apex of the cone on which they lie. The spacing of the striae varies somewhat with rock type, the most closely spaced being generally found in rocks with a fine grain size. Although at first sight the differently oriented conic sectors and converging striae give an impression of a somewhat chaotic geometric organization, when these features are systematically measured it is found that the overall angular relations are consistent with the surface of a simple cylindrical cone. Perhaps the most remarkable feature of the geometry of shatter cones is their *sense of polarity*: the apex of the cone always points in a consistent direction at any one locality. The circular cone geometry of these structures must be related to some system of axial stress $\sigma_1 = \sigma_2 > \sigma_3$ or $\sigma_1 > \sigma_2 = \sigma_3$), but the polarity of the fracture surface is not so easily explained by the types of stress distribution that occur in geological environments or normal laboratory tests. For example when a rock cylinder is subjected to an axial compression, conical fractures are sometimes developed, but these cones are related to the non-directional symmetry of the stress distribution and can point *in either direction* along the maximum compression axis. The clue to the significance of their polarity cones from the observations that when large masses of dynamite are exploded or an underground nuclear test takes place such polar structures are, in fact, developed. They appear to be characteristic of very high deformations with wave velocities of about $5000\,\mathrm{m\,s^{-1}}$. Their polarity is related to the sense of movement of

the wave front through the rock, the opening of the conic form being consistent with this movement sense. Such shock waves would be expected to arise from the impact of large meteoritic bodies on the Earth's surface. Although explosion activity is not infrequently associated with certain types of volcanic activity, the energy release in these environments does not appear to generate shock waves of the magnitude required for the formation of shatter cones.

Shatter cone geometry

Question 25.3

The Vredefort dome is situated in the Transvaal region of South Africa about 120 km south of Johannesburg (Figure 25.7). It is a circular structure of about 50 km diameter formed in Precambrian rocks consisting of about 13 000 m of more or less conformable sediments and lavas (the Transvaal and Witwatersrand Groups of about 1800 to 2000 million years in age) lying unconformably on a series of Archaean basement rocks, mostly of granitic gneiss aspect. The Archaean gneisses now form a central core of a sub-cylindrical dome and the overlying sediments are steeply upturned (and locally overturned) to form a collar-like structure around the gneiss. The southeastern part of the structure is mostly covered by Permo–Carboniferous sediments of the Karroo Formation, but extensive drilling in the region shows quite convincingly that the dome structure continues beneath this formation, and that it has a slightly elliptical form the long axis of which trends NW–SE.

Table 25.2 gives the orientation of the striae found on small shatter cone sectors in a quartzite horizon located in

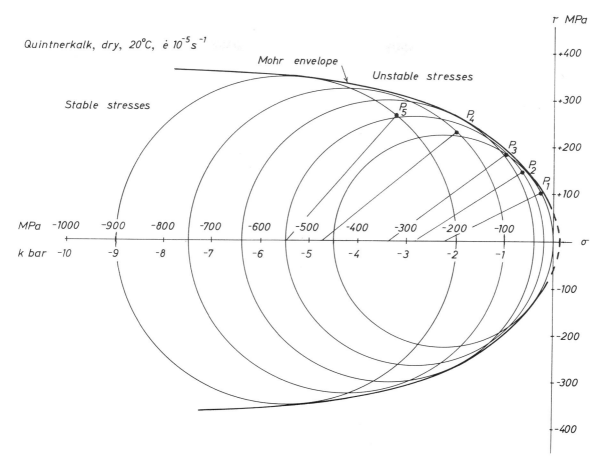

Figure 25.9. *Answer 25.1. Mohr circle and Mohr envelope.*

the northeast of the dome. The polarity sense of these striae has been recorded with respect to the sense of convergence of immediately adjacent striae, those plunging downwards are indicated *d*, those upwards as *u*. Plot these data on a projection. Determine the small circle representing the cone surface, and plot the cone axis. What is the apical angle of the cone? Does the cone apex point upwards or downwards?

The bedding plane at this locality has a strike of 146° and dips 33° to the southwest. Cross bedding structures in the quartzite show that the beds are inverted. Rotate the shatter cone to a position where the bedding is horizontal and normal. Assume that the beds achieved their inverted position by a body rotation about a rotation axis parallel to the

strike. Does the cone apex point upward or downward in its restored position?

ANSWERS AND COMMENTS

Answer 25.1

Part 1. Figure 25.9 shows the **Mohr circles** for the experimental data, and the Mohr envelope constructed from them. Because the data are real the Mohr circles cannot be made to touch the envelope at all points, but the misfits are very small. The **Mohr envelope** forms a curved locus

Table 25.2.
Orientation of striae on shatter cone surfaces, NE Parys.

Locality	Azimuth	Plunge	Polarity	Locality	Azimuth	Plunge	Polarity	Locality	Azimuth	Plunge	Polarity
1	229	57	u	11	202	7	u	21	278	65	u
2	57	21	d	12	287	50	u	22	72	31	d
3	84	17	d	13	197	25	u	23	107	4	d
4	269	29	u	14	44	25	d	24	292	39	u
5	288	21	u	15	194	39	u	25	298	34	u
6	200	51	u	16	98	20	d	26	73	16	d
7	30	5	d	17	89	20	d	27	286	3	u
8	46	25	d	18	296	38	u	28	294	8	u
9	70	28	d	19	305	45	u	29	294	42	u
10	262	47	u	20	75	27	d	30	36	19	d

Table 25.3.
Rock strength as defined by failure stress at low and high confining pressures and room temperatures (after Paterson, 1978).

Rock type	σ_3 when $\sigma_1 = \sigma_2 = 0$	σ_3 when $\sigma_1 = \sigma_2 = 1$ k bar
1. Low porosity dolomite and quartzite	3 k bar	5–10 k bar
2. Igneous and high grade metamorphic rocks	1–2 k bar	5–8 k bar
3. Low porosity sediments, low and medium grade metamorphic rocks	0.5–1 k bar	2–3 k bar
4. High porosity sediments	0.1–0.5 k bar	variable c. 1 k bar

closing towards the right-hand side of the figure, and from the data available it is rather uncertain how it continues into the positive tensile stress part of the field. The Mohr envelope is the separation line between fractured rock (unstable stresses) and intact rock (stable stresses). Once the envelope surface has been constructed it is possible to make predictions to see if a given stress system is likely to lead to fracture formation or not, always bearing in mind that the envelope is characteristic for a particular set of physical constraints. Because the envelope opens towards the left this implies that rocks at high mean compressive stresses are stronger than at low compressive mean stresses. **Strength** is here defined as the stress difference $\sigma_1 - \sigma_3$ that the rock can withstand before breaking. Different rock types have difference strengths depending upon their mineralogical structure, porosity and nature of grain to grain contacts. Typical values of resistance to shear failure of a number of rock categories is set out in Table 25.3.

Unfortunately it is very difficult to obtain good information from the tensile side of the Mohr envelope and much discussion is still going on as to the form of, and mechanical implications of, this part of the graph. In principle we should deduce that the angle of any shear failure to the maximum compression direction should be higher at high compressive confining stresses than at low compressions (Figure 25.10). At the rapidly curving part of the Mohr envelope we should expect a rapid decrease in θ (perhaps with less definitive angles as occurs under high confinement) to the intersection of the envelope on the σ axis where extension fractures should be expected. The actual measured data in the compression field do show an increase in θ angles with confining compressive pressure, but the data points P_1 to P_5 are all smaller than those we should predict from the touching points of the Mohr circles and the Mohr envelopes. (e.g. P_1 predicted $\theta = 20°$, measured $\theta = 14°$; P_2, $25°$ and $20°$; P_3, $32°$ and $19°$; P_4, $39°$ and $20°$; P_5, $38°$ and $25°$.) The reason for this is almost certainly that the angles of the actual measured angles are controlled by the geometric configuration of the sample. When a photoelastic test is made on a cylinder of elastic, optically active material, it is always found that the maximum stress differences are not uniform, but are concentrated at the periphery of the circular ends of the rock cylinder. It is at these positions where fractures are likely to initiate, and the angle they make with the cylinder axis during propagation across the sample is, to a large extent, influenced by the proximity of the specimen walls as well as the differential stresses existing in the sample.

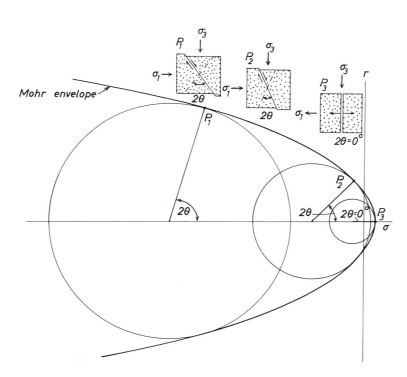

Figure 25.10. *Relationships between the angle θ and differing stress states defining a curved Mohr envelope.*

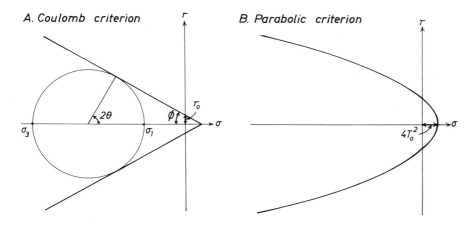

Figure 25.11. A *The classical Coulomb criterion for failure, and* B *the parabolic Mohr envelope predicted from Griffith crack theory.*

The Mohr circle **failure criterion** we have established is a completely empirical one which expresses the interrelationships of the numerical parameters during our experiments. There has been much discussion into the form of the envelope as expressed mathematically in a relationship between normal and shear stresses at the time of failure:

$$\tau = f(\sigma) \qquad (25.6)$$

The simplest "classical" envelope is a pair of straight lines (Figure 25.11A): the **Coulomb criterion** (sometimes also called the Coulomb–Navier or Coulomb–Mohr criterion) which can be expressed in a linear relationship:

$$|\tau| = \tau_0 + \sigma \tan \phi \qquad (25.7)$$

The significance of the terms τ_0 and $\tan \phi$ in the graph is clear, τ_0 is the intersect of the lines on the shear stress axis and the two lines have slopes of $\pm \tan \phi$. The term τ_0 is known as the **cohesion**, and $\tan \phi = \mu_i$ is the **coefficient of internal friction**. Although the results of our particular experiments do not fit at all well with this criterion, many materials do fit this quite well (soils, granular aggregates, granular rocks). The relationship of the angle between a developing fracture and the principal axis of compressive stress is clear from the Mohr diagram, $\theta = 45° - \phi/2$ and this remains constant for all differential stresses. The coefficient of internal friction is not a real frictional parameter in any physical sense because, before fracture occurs, no surfaces exist on which friction may be defined. Generally μ_i has values of 0·5 to 1·0 (giving ϕ angles of 30° to 17°) and is usually slightly higher in value than the **coefficient of static friction** μ_s along actual surfaces in the material. Other mathematical expressions have been suggested for non-linear Mohr envelopes (see discussion in Paterson, 1978):

$$\tau^2 = C_0 + C_1 \sigma \qquad (25.8)$$

$$\tau = C_0 + C_1 \sigma^n \qquad (25.9)$$

$$\sigma_1 = C_0 + C_1 \sigma_3^n \qquad (25.10)$$

$$\sigma_1 - \sigma_3 = C_0 + C_1 \sigma_3^n \qquad (25.11)$$

$$\sigma_1 - \sigma_3 = C_0 (\sigma_1 + \sigma_3)^n \qquad (25.12)$$

where C_0, C_1, and n are material constants (not identical in the various equations). Some of these expressions can be derived from physical considerations of the likely stability of microcrack systems in the rocks and, although we have insufficient space to enter into any detailed discussions of the theories of rock failure, we should mention the important theory of failure developed by Griffith (1920). He suggested that the strength of any brittle material was controlled by the presence of small cracks, which have been termed **Griffith cracks**. Anyone looking at a rock under the microscope will be well aware of such flaws in natural rocks: boundaries between the constituent minerals, weaknesses controlled by crystallographic planes (mineral cleavage) or rather randomly oriented cracks passing through individual crystals and perhaps crossing from one grain to another. Griffith pointed out that when such an array of cracks was stressed, very high local stresses can be built up, particularly at the tips of planar flaws. These high stress levels can lead to propagation of the microcracks, to interconnections between cracks, and finally to the development of through-going discontinuous shear fractures. When the mathematical analysis of this situation is made a quadratic relationship of the form of Equation 25.8 results and the Mohr envelope shows a parabolic form.

$$\tau^2 = 4T_0^2 + 4T_0 \sigma \qquad (25.13)$$

where T_0 is the uniaxial tensile strength of the material (Figure 25.11B)

There are several important outcomes of these empirical and mechanical considerations in relation to the geometrical features of natural shear fracture systems. From the form of the Mohr envelope and the concept of an internal coefficient of friction it is clear that *conjugate fault sets should always form with their acute bisector subparallel to the maximum compressive stress*. There is an overwhelming amount of observation supporting this conclusion (Figure 25.12). E. M. Anderson, a geophysicist based in Edinburgh took these relationships further (Anderson, 1951). He pointed out that, because the surface of the Earth has zero shear stress (the shear stresses produced by wind or water motion being totally insignificant in terms of the shear stresses which exist in rocks) only three stress regimes are dominant. Either the $\sigma_1\sigma_2$, $\sigma_2\sigma_3$ or $\sigma_3\sigma_1$ principal stress planes must lie parallel to the Earth's surface, situations which give rise to **normal**

Figure 25.12. *Profile through a conjugate fault system developed by late stage Alpine deformation in banded gneisses, Ascona, S Switzerland. Note the sequential development of the conjugate faults (cf. Figure 23.17) and the characteristic acute 2θ angle between the fault planes.*

faults, overthrust faults or **strike-slip faults** respectively (Figure 25.13). We now have a mechanical reason for the purely geometric subdivision of faults made in Session 23. Of course, as we go deeper into the crust the orientations of the principal stresses will no longer be perpendicular and parallel to the land surface. The orientations and values of the principal stresses will change in ways governed by consideration of stress equilibrium (Equations E.20) and compatibility. We will leave a deeper analysis of the potentials of these variations to Volume 3 of these series.

Part 2. The Mohr diagram for a state of stress where the three principal stresses are not the same is shown in Figure

25.14. Three circles which touch each other at points on the σ-axis represent the states of stress on the three principal planes of stress. It can be proved mathematically that all possible variations of shear stress and normal stress lie in the stippled region of this diagram (an analogous proof for this, relative to the Mohr circles for finite strain, can be found in Ramsay, 1967, pp. 149–153). From this diagram the circle that touches the Mohr envelope first will always be that related to the stress state on the $\sigma_1\sigma_3$ plane, and it is generally agreed that it is the maximum differential stress $\sigma_1 - \sigma_3$ that is all important in controlling rock fracture. Theoretically the value taken by σ_2 should not influence the failure conditions; this is found to be almost, but not quite true in

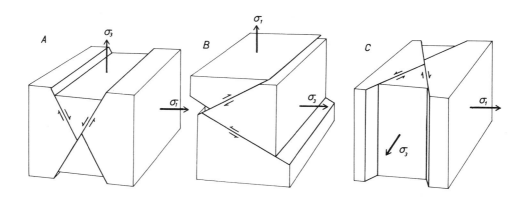

Figure 25.13. *The three main types of conjugate faults expected to develop close to the surface of the Earth. A, normal faults; B, thrust faults; and C, strike-slip faults.*

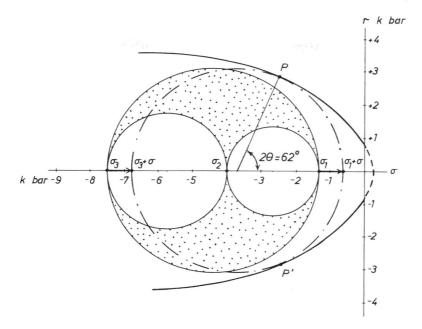

Figure 25.14. *Mohr circles for a stress state $\sigma_1 > \sigma_2 > \sigma_3$: all possible values of σ and τ for this state lie in the stippled area between the three circles. By increasing all stresses by a factor σ the stress circles move towards the right-hand side of the diagram so that the $\sigma_1 \sigma_3$ circle will eventually touch the Mohr envelope at P.*

experiments. In our uniaxial loading test machine it is not possible to make such experiments, but it is possible to modify the experimental set-up to vary all three principal stresses by subjecting a rock cylinder (usually hollow) to torsion. However, there is no general agreement on the effects of varying σ_2 except it appears that the angle of shear failure θ increases as the value of σ_2 decreases (becomes more compressive).

With the stress values of this question, the Mohr circle does not touch the failure envelope, so intact rock will not fracture.

Part 3. If we progressively increase the values of the three principal stresses equally, the Mohr circles of the second part of this question move towards the right of the diagram (Figure 25.14). Where the value of the uniform addition to the stress field is $+0.7\,\mathrm{kb}$ the $\sigma_1 + \sigma$, $\sigma_3 + \sigma$ stress circle touches the Mohr envelope at point P and the rock will fail on shear surfaces oriented at $\theta = 31°$ to the principal compressive stress. The geological implications are that if we are able to change the pore pressure in a rock we may be able to go from a situation of stability to one where fractures will be generated.

Practically all rocks have a porosity controlled by the gaps between particles (especially in clastic sediments), misfitting grain boundaries, or the presence of microcracks. These pores are usually filled with fluids; generally water, petroleum or gas. In igneous and some metamorphic rocks the infill may be rock melt. The pores generally interconnect in some way. The fluid material is in communication through the rock and gives the rock a characteristic **pore fluid pressure**, p. In laboratory experiments with rocks and soils this pressure exerts a very great influence on the mechanical properties of the material. The fluid pressure p usually interacts with the general or local and particular stress states to produce a so-called **effective stress**. In isotropic mate-

rials the principal stress values are changed to $\sigma_1 + \alpha p$, $\sigma_3 + \alpha p$, where α is a constant depending upon the material ($\alpha \approx 1.0$ in soils and ≈ 4.0 in some sandstones). This increase in pore pressure can lead to rock failure, as we have seen above.

Pore pressure can increase in rocks in many ways, and we list some of the most important below.

1. Rapid **sediment burial** in regions of high sedimentation rate is a particularly important mechanism for increasing pore pressure in sedimentary basins. High pore pressures are very well known from drilling (e.g. Gulf basin, Great Valley of California). They are in fact characteristic of practically all thick sediment sequences, especially where thick shale sequences or evaporites act as impermeable barriers to outward flow of fluids.

2. **Tectonic deformation** can close the pore spaces in one site and this fluid can migrate along some tectonic gradient to induce high pore pressure at some other site. The speed of the pore closing process has to be sufficiently fast for the fluid not to be ineffectively bled away, or the system needs to have some impermeable cappings at certain levels to retain the pressure build up. One important example of this type occurs in thrust belts, where one mass is overthrust over another to build up a high fluid pressure in the foot wall. This high pressure can lead to failure at some lower level in the foot wall and the development of a new, active thrust. It seems very likely that this is the mechanism that can best account for the progressively outward (towards the foreland) migration of active thrusting in the frontal parts of orogenic belts.

3. **Rapid agitation** of any loosely packed uncemented sediment enables the individual grains to take up a more closely packed order and to the build-up or

expulsion of intergranular fluid. This is the classic method for developing quicksands. It is a mechanism particularly important in neotectonic regions undergoing deformation by sudden seismic events, and it is quite common to find the developments of mud flows and mud volcanoes in such regions.

4. When a fluid containing dissolved material is heated a high pressure is built up by a process known as **aquathermal pressuring**. The increase of pore pressure by aquathermal pressuring is, in sediments, often closely linked with increases of pore pressure by burial, because burial generally takes the sediment into higher temperature environments. This mechanism can also have local importance where sediments are in contact with hot rising masses of salt at the contacts of salt domes, and it often plays an important role in the contact aureoles of igneous intrusions.

5. **Mineral phase changes**, as a result of diagenesis and regional or contact metamorphism may be very effective in increasing fluid pressure. Particularly important are the transformations of montmorillonite to illite, gypsum to anhydrite, serpentine dehydration and in the various mineral transformations resulting from dehydration reactions in the classic increase of pressure and temperature in the facies sequence from anchimetamorphism to greenschist, amphibolite and granulite facies. Calculations suggest that large quantities of fluid can be evolved in subduction zones where rocks are progressively metamorphosed. Probably most of this fluid finds its way to higher levels of the subduction zone where, providing suitable impermeable barriers exist, it can lead to a pore pressure increase.

6. The **crystallization of magmatic rocks** always leads to the expulsion of large quantities of fluid. The abundant fracture and vein systems found around plutonic bodies almost certainly owe their origin to fracture instabilities driven by such high pore pressure.

Fluid movement during earthquake activity: its implication on fracture development

Before, during and after an earthquake it is observed that severe changes take place in the fluids of the rocks around the earthquake focus and on the overlying surface. Careful observation of changes in ground water levels have a considerable potential for predicting the imminence of an earthquake event. Before an earthquake occurs there is generally a decrease in water well levels and in spring activity, whereas dramatic increases in surface water flow from deeper levels takes place during and after an earthquake. This sudden influx of water (often warm and containing brines) can lead to the liquefaction of unconsolidated sediments, the formation of sand dykes and mud volcanoes and the development of water pressure induced landsliding.

It is well known that earthquake motion is episodic. Many of the largest earthquake fault zones are characterized by periods of violent activity with high magnitude earthquakes and surface displacements of 1–5 m followed by periods of relative quiescence with time scales of the periods measured in tens or even hundreds of years. The San Andreas fault activity has been extensively studied by trenching and by investigating the buried scarps and structures such as sand

blows and fluidizations structures in the Holocene sediments in the immediate vicinity of the fault plane (Sieh, 1978). Using radiocarbon dating it has been shown that major earthquake events took place at 545 ± 45, 665 ± 80, 860 ± 35, 965 ± 50, 1190 ± 45, 1245 ± 45, 1470 ± 40, 1745 ± 25 and 1897 (all these dates are AD). These data give an average major seismic event cycle of about 160 years and give an average movement of 30–60 mm per year. It is also known that during the periods of quiescence there is a variable amount of microseismic activity which is not usually recorded in the form of land surface ruptures but which is detected with seismic instruments. One model which has been proposed for the periodicity of major fault **stick–slip motion** and for the near surface observations is a **dilatency-fluid diffusion mechanism** (Sibson et al., 1975). When a pre-existing fault is subjected to an increasing tectonic shear stress the rocks adjacent to the fault plane will show a dilation as a result of an increase in pore space. In unconsolidated sediments or weakly cemented rocks dilatency can be produced by the overriding of individual granular components, whereas in lithified rocks dilatency can be produced by inducing openings by grain boundary sliding, or by microcracking and the production of extension veins. Such dilatant phenomena are well known in laboratory experiments in soils and rocks and lead to non-linear elastic effects with progressive changes in the measured elastic parameters (see summary in Paterson, 1978). In the laboratory, the dilatency effect often gives rise to acoustic emmisions, and it is quite likely that the microseismic activity of currently active fault zones arises from microcracking accompanying dilation. As a result of the increase in pore space in the fault zone, the fluid pressure is decreased. Because this increases the compressive normal stress across the fault, there is an increased resistance to movement on the fault plane. Gradually fluid is drawn from the surrounding rocks into the dilatant spaces and the fluid pressure in the vicinity of the fault will rise. Seismic failure

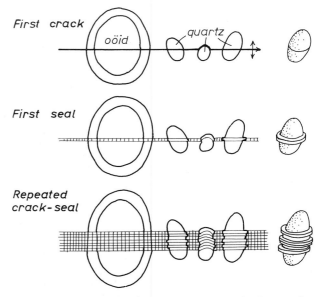

Figure 25.15. *Sequence of crack-seal events leading to the formation of composite veins as seen in a deformed pisolitic ironstone. The right side of the figure shows the predicted three-dimensional forms of the sedimentary quartz grains. After Ramsay (1980).*

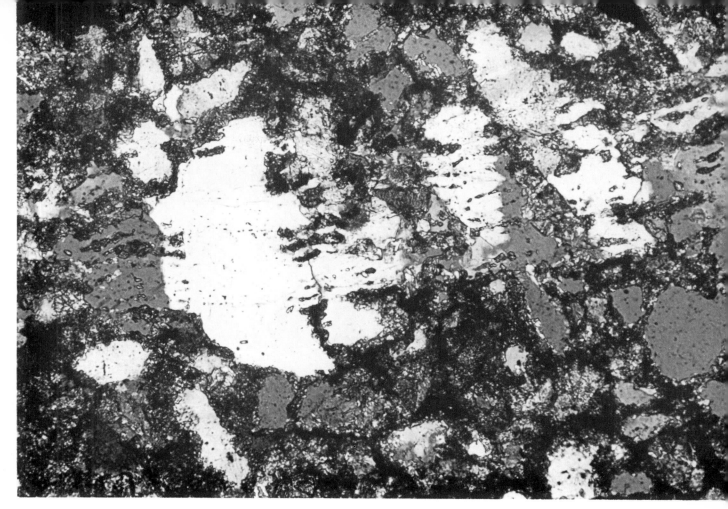

Figure 25.16. *Thin section (crossed polars) of composite crack-seal veins passing through originally sub-spherical sand grains in a greywacke. Note the optical continuity of the seals with the quartz grains through which they cut. Hope Cove, south Devon, England. Enlargement ×35.*

Figure 25.17. *Thin section (crossed polars) of solid inclusion trails and wall parallel inclusion bands in a quartz–calcite crack-seal vein. Note the similarity of shapes and the optical parallelism of the small calcite inclusions in each trail. Jurassic calc-phyllite, Ilanz, Graubünden, Switzerland. Enlargement ×100.*

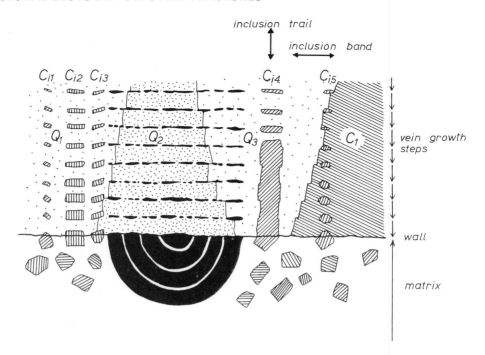

Figure 25.18. *Diagram to illustrate the development of inclusion trails and inclusion bands in crack seal veins from the Windgällen region, northern Swiss Alps. Q_1, Q_2, Q_3, represent the main quartz fibres and C_1 a major calcite fibre. C_{i1} to C_{i5} are calcite fragments arranged in trail form; their orientation is the same as that of crystals found in the vein wall. The black object in the matrix is a chamosite ooid, and the black bands associated with it are composed of chlorite. The spacing of the inclusion bands is about 50 μm.*

will again occur when the static frictional resistance cannot withold the increasing tectonic shear stress and decreasing effective normal stress across the fault. The sudden seismic event and the relief of the shear stress is likely to bring the rock particles into closer contact, partially closing the dilation spaces. The excess fluid will be expelled rapidly in the direction of lowest confining pressure, generally upwards. Providing there is a continued tectonic stress other cycles can be generated in the same way. This mechanism provides a method for periodically developing earthquake failure on a fault and for sequentially injecting quite large volumes of pore water upward into the higher levels of the fault zone. Such **seismic pumping** (Sibson *et al.*, 1975) could also provide a mechanism for forming **hydraulic fractures** in the fault zone, and deposition of ionic materials in solution from the rising (and cooling) fluid injections into open fissures can lead to the formation of **vein fillings** in these fissures.

Vein formation and the crack-seal mechanism

In Session 13 we briefly introduced the concept of the crack-seal mechanism of vein growth. This is a way of building veins of considerable width by repetitions of a sequence of events as follows. First a crack forms in the rock, as a result of the build up of elastic strains in the material and the release of this deformation by a fracture process. The walls of the fissure separate slightly (typical separation distances are 10–100 μm almost irrespective of rock type) as the elastic deformation of the walls is released. Fluids are introduced into the crack and new crystalline material is deposited on the walls of the opening eventually to produce a more or less complete solid binding together with the walls. This crystalline material may be of the same species as those crystals

found in the walls, it may differ, or it may be of mixed type. Once the walls of the crack are sealed the tectonic stresses have continuity across the veinlet and can continue to build up again until some critical failure stress is reached. This stress failure level might be the same as or different from that necessary to break the originally intact rock. Because the initial vein has introduced some planar anisotropy into the rock it is particularly common for the first vein to control the location of the subsequent veins. The new vein often forms inside the earlier one, splitting it into two parts, or forms along one or other of the vein–wall contacts. The process of sealing now continues until there is wall to wall cohesion between the separations induced by the new fissures. The cycles are repeated and can lead to the formation of wide veins made up of composite microveins (Figures 25.15, 25.16).

The step increment method of building veins is often very clearly appreciated by the preservation of various types of screens of solid inclusions inside the veins. Two main types of inclusion geometry are common, they may form separately or together. **Inclusion bands** (Figures 25.16, 15.18) are inclusions grouped in narrow zones parallel to the walls of the microveins. They can usually be compositionally related to the crystals found in the wall rock, either because they represent thin detached screens of wall rock, or because they are parts of an earlier vein, the crystal components of which were controlled by the wall composition (Figure 25.18). **Inclusion trails** are solid inclusions running sub-perpendicular to the walls, each trail being formed of isolated crystals generally of the same species and in optical continuity (Figures 25.17, 25.18). As a result of this process fibrous crystal vein infillings are generally developed, the contacts between the crystals being characterized by a saw tooth form, the teeth widths being the same as the

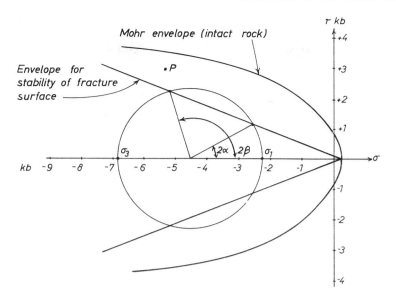

Figure 25.24. *Relationship between the Mohr envelope and the curves defining the stress states necessary for regenerating movement on a fracture surface. For a state of strain given by the $\sigma_1\sigma_3$ Mohr circle here orientations with the range between α and β would be unstable, and undergo shear displacement.*

their kinematics are close to those of strike-slip faulting with right-hand shear on the E–W striking nodal plane, and left hand shear on the N–S plane. The P-axis and T-axis orientations trend NW–SE and NE–SW respectively and are sub-horizontal. Although there is no clear evidence of active faults at the surface today associated with these earthquakes there are many steeply inclined ENE–WSW striking steeply inclined geological faults. The surface morphology of some of these suggests that they might be neotectonically active and some show sub-horizontal striae. It should be emphasized that this is a region with a very complex pattern of geological faults, including many with dip-slip motion, so the correlation of the E–W nodal plane with the similarly trending geological faults is not absolutely conclusive.

It is especially interesting that some of the last tectonic deformations imprinted in these rocks are conjugate shear zones with en-echelon vein arrays which also indicate an overall NE–SW extension and a NW–SE shortening, so the directions of the seismic stress field today have geometrical affinities with the stresses acting 5–15 million years ago.

It is usual to present the nodal plane solutions in simplified projection form with the compressional sectors stippled or made black, the corresponding dilational sectors left unornamented. The three main types of fault geometry correspond to three main nodal plane patterns (Figure 25.26A, B and C). More complex patterns arise when com-

binations of these various types occur together (D). Such tectonic maps are quite easy to read and an example from the Switzerland produced by Pavoni (1980) is shown in Figure 25.27. In this example most of the solutions show steeply inclined B-axes and gently inclined NW–SE P-axes, features indicative of a predominant strike-slip mode. The azimuth trends are also given for the P-axes. These are also fairly systematically oriented, suggesting that the present day stress field is fairly systemically oriented across the main Alpine trends. We should emphasize here that these P-axes should not be directly correlated with the axes of maximum compression. The σ_3-axes could be located anywhere in the white sectors of the projections.

Shatter cone geometry

Answer 25.3

An equal area plot of the shatter cone striae is shown in Figure 25.28A, filled circles representing those directions with downward polarity, open circles with upward polarity. These data can be fitted on to a small circle, the cone axis of which is shown as an open star symbol, and plunges 20° towards 246°. The cone polarity is upwards, pointing radially away from the dome centre (Figure 25.28B). This is a characteristic feature of all the shatter cones developed in

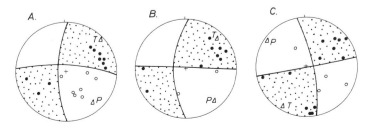

Figure 25.25. *The nodal planes and P- and T-axes for the data of Question 25.2. After Pavoni (1980).*

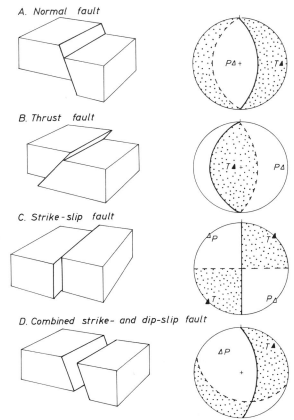

A. Normal fault

B. Thrust fault

C. Strike-slip fault

D. Combined strike- and dip-slip fault

Figure 25.26. *Fault plane solutions and their relationship to main types of faults.*

angle is 145°. The cone axis must also be rotated through the same angle, but this pole moves along a small circle about the 146° striking horizontal rotation axis. The restored position now shows a plunge of 10° towards 50°, but the cone axis still has an upward polarity (Figure 25.28C). In making this rotation we have assumed that the rocks suffered only a body rotation without undergoing strain. Because of the general radial disposition of the shatter cone axes about the dome centre this technique will not severely modify the azimuth. However, if internal deformation of the rocks did take place (as seems likely from compatibility constraints on the geometry), the restored angles of plunge might not be correct. Manson (1965) has shown that all the restored shatter cones point upward at angles generally ranging from 3–30°, and that there is a tendency to focus on a central location situated at 14 ± 3 km above the present land surface (Figure 25.29B).

A suggested evolutionary sequence for the development of the Vredefort dome structure is shown in Figure 25.30. The first impact (A and B) produced a high energy shock wave front moving downwards and outwards from the impact site. This shock wave gave rise to the downward opening (upward pointing) conical fractures, the axes of these shatter cones probably forming normal to the wave front. As the shock wave moved downwards (C) its energy became dissipated and at some particular frontal stage the energy was insufficient to produce more shatter cone fractures. At the same time as the wave front was passing through the deeper crustal levels an outwards explosion and expulsion of the upper parts was taking place, the fragmental debris probably building a crater rim as well as a wide regional accumulation of fall-out material. The rebound of the crater walls led to an outward dipping orientation of the originally horizontal sediments and a rotation of the already formed shatter cones. Because of the isostatic imbalance caused by the removal of over-burden material a slow upward viscous flow of the lower crust and upper mantle took place (D), leading to the uplift

this region (Figure 25.29A). The apical angle of the cone is about 90°. Because the bedding planes are overturned it is necessary to rotate the bedding plane pole π along an equatorial great circle through the horizontal, to the net centre, to restore it to its original position π_0. The rotation

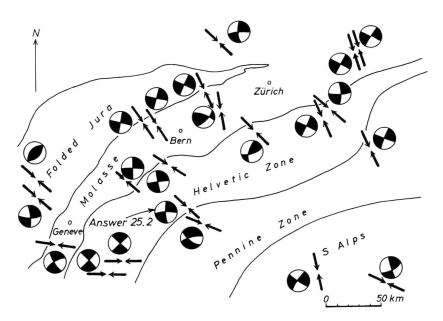

Figure 25.27. *Regional pattern of first motion solutions giving the neotectonic activity in Switzerland. The arrows indicate the trends of the P-axes of the solutions. After Pavoni (1980).*

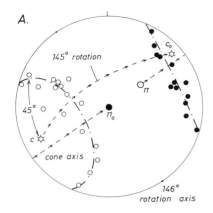

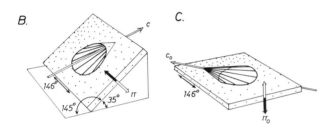

Figure 25.28. A *Equal area projection of shattercone data of Question 25.3. The bedding pole π is restored to a horizontal position π_0 by rotation about the bedding strike line, and the cone axis rotated accordingly from c to c_0. B and C show the geometrical features before and after retsoration.*

of the crater and further steepening of the sedimentary rocks in the collar. The presence of mantle material is recognized today by a strong positive gravity anomaly in the centre of the dome. The final stages of the evolution of the region (E) were the erosion of all the crater debris and the development of an unconformable cover of Karroo sediments.

The Vredefort dome shows many features of an impact structure formed over 250 million years ago. Its unique form, presence of shatter cones, coesite and pseudotachylite melt together form good "fingerprint" evidence for an impact. It has been calculated that the energy release was at least a million times greater than accompanied the catastrophic Krakatoa volcanic explosion, and many thousand times larger than the greatest possible earthquake.

Although we tend to think that such meteoritic impacts are of rare occurrence we believe that, although sporadic and often unpredictable, they are important features of crustal rocks, and that future research is likely to find evidence that many more exist than are shown in the map of Figure 25.7.

Products of rock fracturing

To complete this session we will comment on the various different types of rock produced by localised and intense fracturing, and make a link into the next session by considering some of the transitional products of shear zones in both brittle and ductile regions.

The names given to the rock types associated with brittle and ductile shear zones have often intermixed descriptive and genetic ideas. We will pick out and recommend those terms which are, in our opinion, most relevant to field and

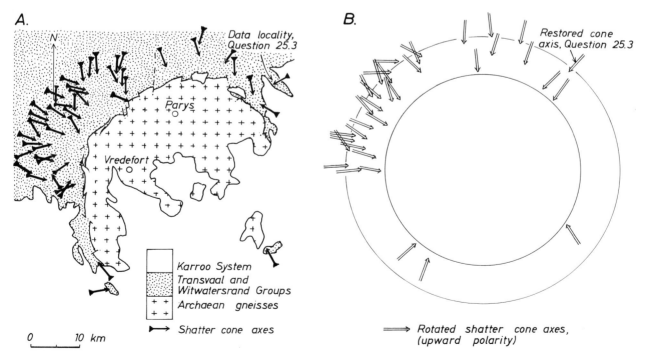

Figure 25.29. A *Geological map of the Vredefort dome showing the present orientations of the cone axes and their polarity. B illustrates the restored position of the cone axes. All now point upward to a central point above the ring structure. After Manton (1965).*

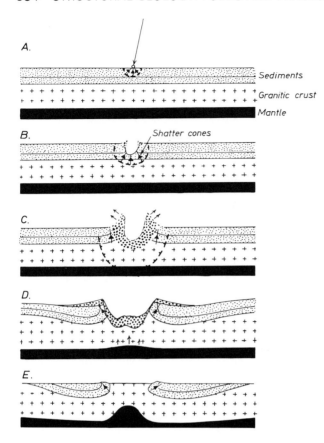

Figure 25.30. Suggested sequence of events which formed the Vredefort structure. A First impact, shock wave front advances, shatter cones formed (see polarity symbol and axis orientation). B Shock wave advance. C Expulsion of near surface material, shock wave with reduced energy. D Fall-back of material into crater, isostatic readjustment of mantle leading to collar steepening. E Situation today.

microscopic observations and which are in most common use. For a fuller discussion we recommend that the student wishing to involve himself deeper in this topic should read the excellent review articles by Higgins (1971), Sibson (1977) and Wise *et al.* (1984).

Movement along a fault surface a high levels in the crust (< 5 km depth) generally results in the fracturing and breaking of wall rock fragments as a result of the process of **cataclasis**. Angular blocks of material, with interblock spaces, or with the blocks packed together with more finely comminuted material, produces an originally cohesionless mass known as **fault breccia** (Figures 25.31, 25.32) or **micro breccia** (largest fragments < 1·0 mm). Where the comminuted material is granulated to a fine flour (0·1–100 μm) the product is known as **gouge**, or when consisting to a large extent of clay-like phyllosilicate as **clay gouge**. Although both breccia and gouge are cohesionless materials, they can become impregnated and sealed by crystal growth in the voids to produce **cemented breccia** or **cemented gouge** (Figure 25.31).

Melange is a special type of breccia containing local or exotic competent blocks (some of which may be very large indeed—often of kilometric dimensions) embedded in a less competent matrix in regions where high level incompletely consolidated and unlithified sediments have been disturbed

by imbricate faulting or gravitational gliding (Greenly, 1919; Hsü, 1968). **Olistostromes** are chaotic sub-marine mud debris flow deposits formed on the ocean floor where recently deposited sedimentary materials have moved in a semi-fluid condition.

Cataclasite (Grubenmann and Niggli, 1924) is the general name given to fine grained products of cataclasis which possess internal cohesion (Figure 25.33). The dominant mechanical processes involved are microfracturing and microbrecciation of crystals with intergranular sliding and rotation of grains. Because of the irregular behaviour of the fragmental components these crystals show no optically preferred orientation. Because of their general fine grain size and the moderate temperature and pressure conditions under which they form the individual components cohere together to form a rock. Banding is absent (although see Chester *et al.*, 1985). The proportion of recognizable parent material to totally crushed rock has been used for further sub-division: **protocataclasite** (> 50%) **cataclasite** (10–50%), **ultracataclasite** (< 10%). In the older literature, particularly that coming from British geologists, cataclasites were often termed flinty crush rock.

Where rapid movement occurs along a fault surface located in the upper crust, the frictional heat generated can lead to the production of a rock melt (McKenzie and Brune, 1972; Sibson, 1975). The rock produced from this melt is known as **pseudotachylite** (Shand, 1916). It is generally a black or dark coloured glass and, although the name tachylite refers to the composition of a basic volcanic glass, no such petrographic or chemical limitations are placed on pseudotachylite. In fact most pseudotachylites are chemically more allied to acid pitchstones, being derived from the melting of granitic parents. Pseudotachylite may be localised along the fault surface on which it was generated, but it is often intruded into the surrounding walls taking on dyke-like forms (Figure 25.34). These dykes frequently contain sub-angular or rounded fragments of wall rock material carried as xenoliths (Figure 25.35) sometimes of quasi-conglomerate appearance. In thin section pseudotachylite is seen to be a truly isotropic glass (Figure 25.36) often containing broken wall fragments as microxenoliths. However, like volcanic glass, pseudotachylite can undergo divitrification to produce microlite crystals, spherulites and acicular overgrowths on the contained fragments (Figure 25.37). Many fine-grained dark coloured ultracataclasites resemble pseudotachylite in the field, but in thin section they will not show the characteristic features of a true glass. Calculations have been carried out by McKenzie and Brune to assess the environmental conditions likely for melting and they conclude that pseudotachylites could be developed at pressures of 1 kb for displacements as small as one millimetre whereas at 10 kb melting is impossible for any displacement value. Sibson suggested that at depths of around 5 km, "single jerk" faulting could lead to the production of a certain thickness, *d*, of pseudotachylite by a shear movement of distance *s* governed by the equation

$$s = 436d^2 \qquad (25.15)$$

To develop a 1 cm thickness of melt requires a slip of about 5 m.

The term **mylonite** now has a rather different definition from that suggested by Lapworth (1885). The name comes from the greek *mylon*: to mill, and Lapworth was under the

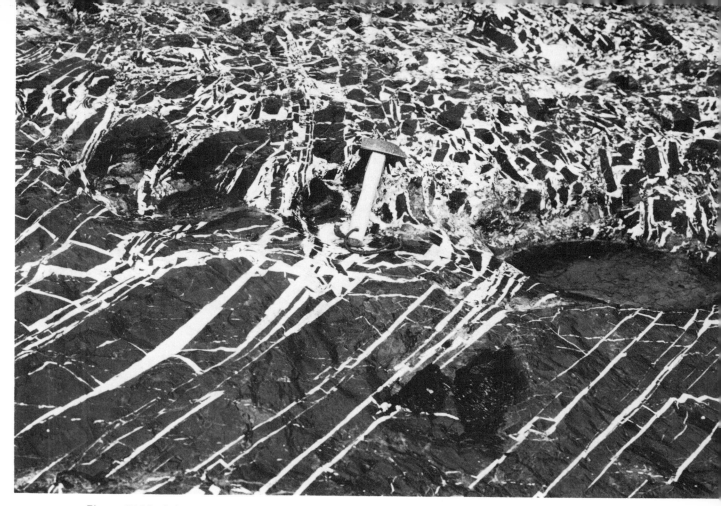

Figure 25.31. Calcite cemented fault breccia in limestone. Grosse Windgällen, central Switzerland.

Figure 25.32. Thin section of microbreccia and fine grained gouge. Tarskavaig thrust, Skye, Scotland. Enlargement ×50.

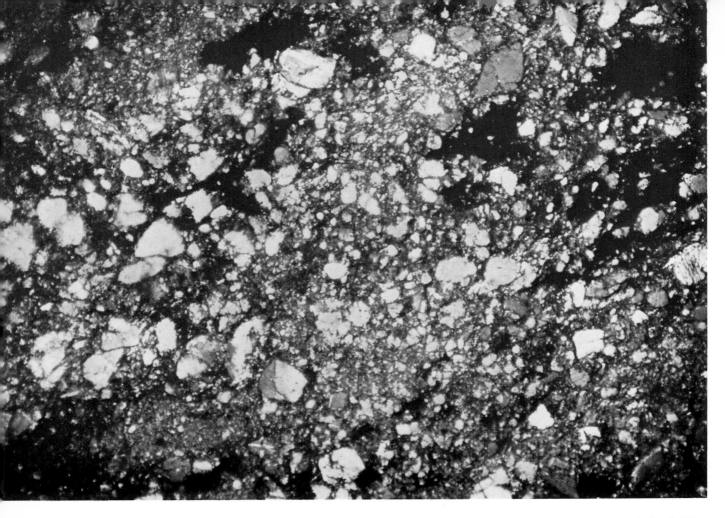

Figure 25.33. *Thin section of cataclasite with broken fragments of quartz and feldspar. Nelspruit, east Transvaal, South Africa. Enlargement ×35, crossed polars.*

Figure 25.34. *Injection veins of pseudotachylite into granite gneiss. Parys, Vredefort dome, South Africa.*

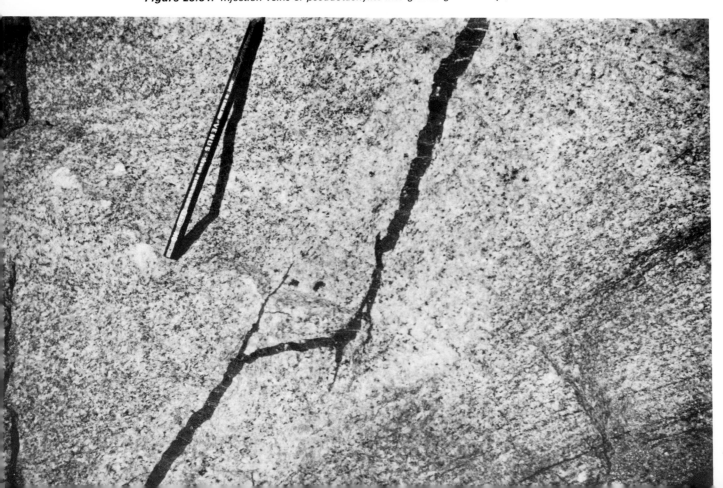

Figure 25.35. Rounded xenolithic blocks of country rock gneiss in a black pseudotachylite glass. West Greenland.

Figure 25.36. Thin section of pseudotachylite glass containing flow oriented particles of broken country rock granite gneiss. Outer Hebrides, Scotland. Enlargement ×50, crossed polars.

Figure 25.37. Thin section of a faulted plagioclase feldspar with crystallite overgrowths in a devitrifying pseudotachylite glass. Outer Hebrides, Scotland. Enlargement ×50, crossed polars.

Figure 25.38. Banded mylonite derived from granite gneiss, with prophyroclasts of feldspar. North of Tamanrasset, Hoggar, Algeria.

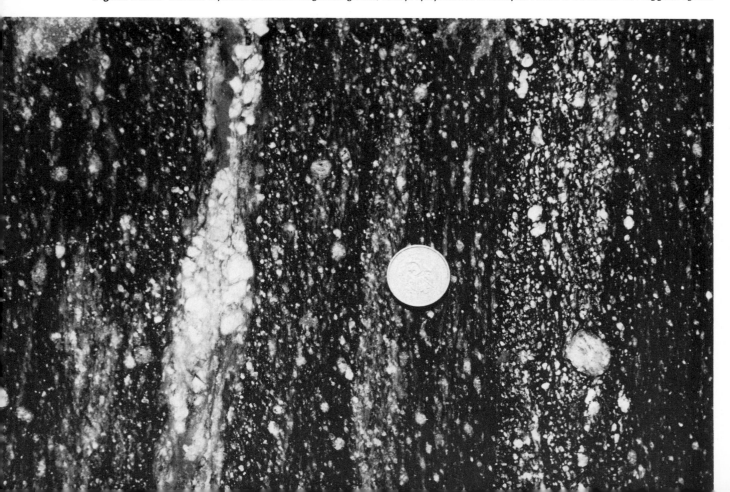

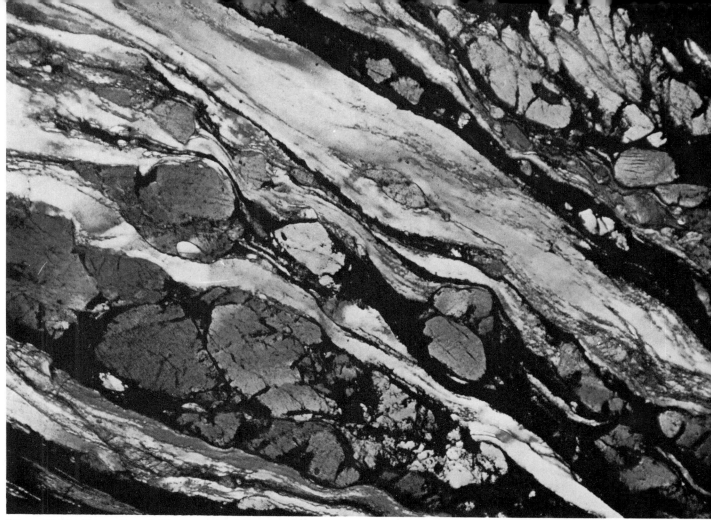

Figure 25.39. Thin section of ribbon quartz and broken feldspar grains in a mylonite derived from granitic gneiss. Outer Hebrides, Scotland. Enlargement ×80, crossed polars.

Figure 25.40. Ductile shear zone developed in metagabbro, showing the formation of a banded hornblende schist with characteristic sigmoidal schistosity. N Uist, Outer Hebrides, Scotland.

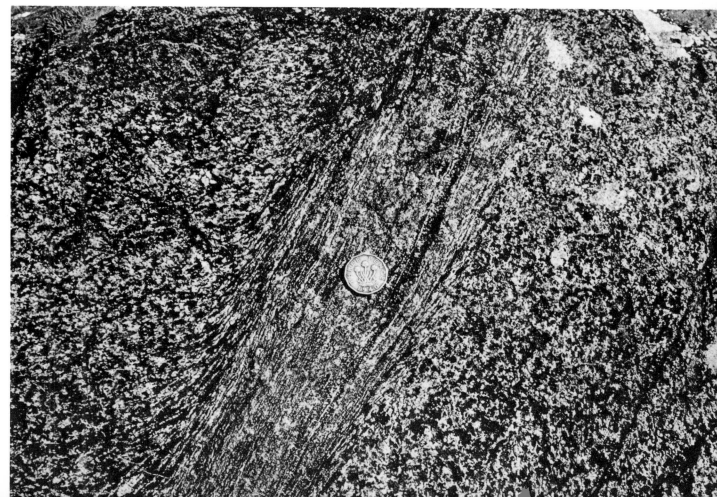

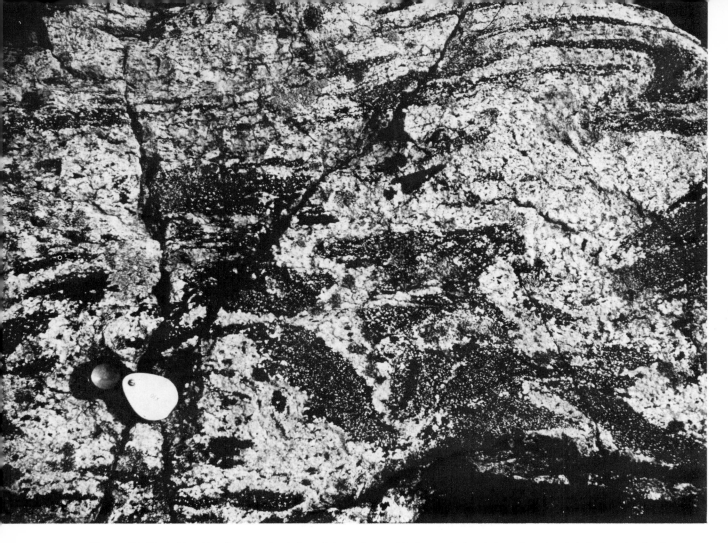

Figure 25.41. A *Hornblendic agmatite before deformation and* B *intensely banded gneiss after Caledonian ductile shear zone deformation. Glenelg, northwest Highlands of Scotland.*

impression that the finely laminated schists localized in the Moine Thrust zone of northern Scotland were the result of crushing and grinding of the larger size original crystals. The grain size of mylonites is certainly smaller than that of the original grains (generally $< 50 \mu m$), but it is now realized that these small grains are derived from dynamic recrystallization of plastically deforming grains, and not by simple cataclasis. Because of this crystal plastic deformation mechanism the grains usually show characteristically preferred optic orientations which are best observed under the microscope with the aid of a quartz plate. Mylonites are banded rocks (Figure 25.38) even when the parent materials are unbanded: the banding developed by a separation of different mineral species depending upon their individual ductility. The *banding* is often termed **foliation**. This may be parallel to the alignment of the long axes of individual crystals, defining the **schistosity**, but it is not uncommon to find the two planar structures with cross cutting relationships. Banding and schistosity are often accompanied by intense linear orientations of the mineral constituents, a feature generally interpreted as the result of a **stretching fabric** parallel to the X-axis of the finite strain ellipsoid. In many mylonites derived from quartz rocks the quartz is drawn out into long narrow **quartz ribbons** (Figure 25.39) separated by zones comprised of silicate minerals. Mylonite often contains large residual crystals, **porphyroclasts** which represent relatively unmodified individual, or groups of, parent crystals. Although cataclasis is not the primary mechanism producing mylonitic rocks, the porphyroclasts often show microfracturing and separation of the individual fragments (Figure 25.39). Mylonites like cataclasites, are sometimes subdivided according to the proportions of recognizable parent material and thoroughly dynamically recrystallized matrix: **protomylonite** ($> 50\%$), **mylonite** (10–50%) **ultramylonite** ($< 10\%$). Mylonites are localized in ductile shear zones and in these zones it is not uncommon to find extensive recrystallization after deformation. This recovery can lead to the production of fairly large,

usually equidimensional grains with **polygonal intercrystalline relationships** and occasionally to the formation of particularly large crystals or **porphyroblasts** easily visible to the naked eye. When such recovery recrystallization takes place the strong preferred orientations that characterized the dynamic recrystallization process are weakened or lost. A special type of mylonites, termed **superplastic mylonites** (Boullier and Gueguen, 1975) show large porphyroclasts embedded in a fine grained matrix without optic alignments. These rocks are interpreted as being characteristic of high temperature environments and where aggregate deformation has proceeded by grain boundary sliding.

In regions where ductile shear zone development has taken place under conditions of amphibolite grade metamorphism the dominant rock types are not mylonites but medium grain sized **schists** and **gneisses**. They are usually schistose and often banded even when the parent material shows no planar anisotropy (Figure 25.40). This banding appears to be best interpreted as the result of plastic deformation with the shearing, smearing and interlinking of the more ductile crystal components between the adjacent less ductile crystals (Jordan, 1987). In ductile shear zones, a fine banding can also be produced as the result of the deformation of previous coarse banding or heterogeneous mineral disposition (Figure 25.41A and B). In ductile shear zones the deformation increments are preferentially developed in previously deformed rock rather than in the relatively undeformed walls. This phenomenon is known as **strain softening** (Ramsay and Graham, 1970), in contrast to the well known effect of cold working of metals termed **strain hardening**. Several reasons for strain softening have been proposed: grain size reduction during deformation, changes in mineral phases during deformation from deformation resistant to deformation acceptant mineral species, the development of schistosity or banding anisotropy, and the localization of fluid migration along the shear zone.

KEYWORDS AND DEFINITIONS

Cataclasis	The process of fracture and breaking apart of rock particles and crystal components, together with the rotation and mechanical mixing of the particles. The products may be incoherent clastic materials: **breccia** or **gouge**, or form cohesive rocks: **cataclasites**.
Failure criterion	An empirical or theoretical expression which relates normal and shear stresses at the time of initiation of rock fracture. It can be graphically expressed by the **Mohr envelope**, a line which forms the limiting boundary to the **Mohr stress circles** at the time of failure.
Mylonite	A fine grained, banded, cohesive rock formed by localized processes of plastic flow and dynamic recrystallization and generally characterized by very high finite strains.
Pore pressure	The hydrostatic stress exerted by fluid material contained in the rock pores. Generally this modifies the stresses existing in the solid rock mass to a new state known as the **effective stresses**.
Pseudotachylite	A rock melt produced by frictional heating along a fault plane.

Shatter cone	A conical shaped fracture surface, the conical form of which has a definite polarity sense, believed to be formed when a high energy shock wave passes through a rock.
Stick-slip	A jerky or unsteady displacement motion along a shear surface. The slipping motion is believed to go together with earthquake phenomena along crustal faults.

KEY REFERENCES

Papers which give good background reading directly related to the Questions of this session

Question 25.1

Anderson, E. M. (1951). "The Dynamics of Faulting and Dyke Formation." Oliver and Boyd, Edinburgh, 206 pp.

This is the second edition of a book that has had a very great influence on geologists seeking mechanical accounts of different types of fracture development.

Cloos, E. (1955). Experimental analysis of fracture patterns. *Geol. Soc. Am. Bull.* **66**, 241–256.

This is one of the classic papers describing laboratory experiments on model materials to assist our understanding of fracture geometry.

Gretener, P. E. (1981). "Pore Pressure: Fundamentals, General Ramifications, and Implications for Structural Geology." American Association of Petroleum Geologists Education course note Series 4, revised edn, 131 pp.

This is a perfect book for the student, conveying the basic information on pore pressure in a very clear way, and providing an extensive list of references. Of particular interest are the excellent practical examples taken from known oilfields and the sound discussion on the application of the principles to general tectonic problems.

Hubbert, M. K. (1951). Mechanical basis for certain familiar geologic structures. *Geol. Soc. Am. Bull.* **62**, 355–372.

Most structural geologists regard this as a classic work, being one of the first attempts to make geologists aware of the theory of stress and its implications for the analysis of fracture patterns.

Paterson, M. S. (1978). "Experimental Rock Deformation— the Brittle Field." Springer Verlag, Berlin, Heidelberg and New York. 254 pp.

We are very conscious that lack of space has not enabled us to attempt any detailed discussion on the geological significance of laboratory experiments with rock materials. This provides a superb review of the results of many years of experimental work and includes a very comprehensive list of references.

Price, N. J. (1966). "Fault and Joint Development in Brittle and Semi-brittle Rock." Pergamon Press, Oxford. 176 pp.

This book can be highly recommended. It covers a lot of material from theoretical, experimental and field sides and the written style is very clear.

Sibson, R. H. (1981). Fluid flow accompanying faulting: field evidence and models. *Am. Geophys. Union, Maurice Ewing Series* **4**, 593–603.

A discussion of the fracture pattern and vein systems that develop as a result of changing fluid pressure in the vicinity of major faults.

Question 25.2

Molnar, P. and Chen, W. P. (1982). Seismicity and mountain building. *In* "Mountain Building Processes" (K. J. Hsü, ed.) 41–57. Academic Press, London.

This modern review concentrates particularly on neotectonic problems and how they may be partially resolved using first motion studies.

Pavoni, N. (1977). Erdbeben im Gebiet der Schweiz. *Eclog. Geol. Helv.* **70**, 351–370.
Pavoni, N. (1979). Investigation of recent crustal movements in Switzerland. *S.M.P.M.* **59**, 117–126.
Pavoni, N. (1980). Comparison of focal mechanism of earthquakes and faulting in the Helvetic zone of the Central Valais, Swiss Alps. *Eclog. Geol. Hevl.* **73**, 551–558.

These papers will provide more interpretative details to the answer we made for Question 25.2.

Sieh, K. E. (1978). Prehistoric large earthquakes produced by slip on the San Andreas Fault at Pallett Creek, California. *J. Geophys. Res.* **83**, 3907–3939.

This work provided a research breakthrough in setting out data to establish the movement history of the San Andreas Fault since the sixth centry AD. The data was derived from trenching the sediments deposited near to the fault to discover the indications of past major seismic events.

Tschalenko, J. S. and Ambraseys, N. N. (1970). Structural analysis of the Dasht-e Bayaz (Iran) earthquake fractures. *Geol. Soc. Am. Bull.* **81**, 41–60.

This is one of the classic pieces of work on mapping surface breaks along an earthquake fault. It is a model of how to observe and interpret a very complex set of surface rupture features.

Question 25.3

Bucher, W. H. (1963). Cryptoexplosion structures caused from without or within the Earth ("Astroblemes" or "geoblemes"). *Am. J. Sci.* **261**, 597–649.

Dietz, R. S. (1961). Vredefort ring structure: Meteorite impact scar? *J. Geol.* **69**, 499–516.

Hargraves, R. B. 1961. Shatter cones in the rocks of the Vredefort Ring. *Geol. Soc. S. Africa Trans.* **64**, 147–153.

All three of these publications give pertinent discussions to the problems posed in Question 25.3.

Manson, W. I. (1965). The orientation and origin of shatter cones in the Vredefort ring. *Ann. N.Y. Acad. Sci.* **123**, 1017–1048.

This paper sets out in detail the results of a study of shatter cones, the techniques for measuring these features, and a discussion of their significance.

Papers for widening the scope of the discussions in this session, some with particular reference to wide implications of fracture mechanics

Davies, D., Suppe, J. and Dahlen, F. A. (1983). Mechanics of fold- and thrust-belts and accretionary wedges. *J. Geophys Res.* **88**, 1153–1172.

Dahlen, F. A., Suppe, J. and Davies, D. (1984). Mechanics of fold and thrust belts and accretionary wedges: cohesive Coulomb theory. *J. Geophys. Res.* **89**, 10 087–10 101.

These two publications discuss the large scale mechanical features of moving wedges of sedimentary material over subducting plates with particular reference to the tectonics of Taiwan. They would provide material for widening the discussion on the application of mechanical principles to the geometric problems of overthrust tectonics.

Hast, N. (1969). The state of stress in the upper part of the Earth's crust. *Tectonophysics* **8**, 169–211.

Most stress determinations in near surface rocks have been carried out by rock mechanics engineers interested particularly in problems arising during mining and engineering constructions. This paper, however, interprets stress measurements in a way which has more interest to geological sciences and the interpretation of stress distributions arising from tectonic activity.

Lisle, R. J. (1979). The representation and calculation of the deviatoric component of the geological stress tensor. *J. Str. Geol.* **1**, 317–321.

Although there have been many diagrams proposed for representing finite strain variations there have been few proposed for stress. Richard Lisle here proposes two types of stress diagrams, counterparts of strain diagrams, for representing the values of the principal stresses and which enable the properties of a particular stress state to be quickly visualized.

Reches, Z. and Dieterich, J. H. (1983). Faulting of rocks in three-dimensional strain fields. I. Failure of rocks in poly-axial, servo-control experiments; II. Theoretical analysis. *Tectonophysics* **95**, III–132; 133–156.

We concentrated our analysis of fracture development using two-dimensional constraints. These papers are recommended for widening the view to consider the mechanics of non-plane-strain situations, and it is shown that four sets of faults are required to accomplish the necessary shape changes.

Siddans, A. W. B. (1984). Thrust tectonics, a mechanistic view from the west and central Alps. *Tectonophysics* **104**, 257–281.

This is a stimulating analysis of the mechanics of thrusting of plastic nappe sheets where thrust surfaces are localized along weak décollement levels.

Woodcock, N. H. (1986). The role of strike-slip fault systems at plate boundaries. *Phil. Trans. R. Soc. Lond.* **A317,** 13–29.

This paper points out that most plate boundaries have movement vectors which are oblique to the boundary trace, and suggests that some of the kinematic approaches to plate tectonics problems need to be adjusted to allow for this obliquity.

Papers describing and defining the products of shear zone deformation

Higgins, M. W. (1971). Cataclastic rocks. *Prof. Paper U.S. Geol. Surv.* **687**

Sibson, R. H. (1977). Fault rocks and fault mechanisms. *J. Geol. Soc. Lond.* **133,** 191–213.

Wise, D. U., Dunn, D. E., Engelder, J. T., Geiser, P. A., Hatcher, R. D., Kish, S. A., Odom, A. L. and Schamel, S. (1984). Fault related rocks: suggestions for terminology. *Geology* **12,** 391–394.

SESSION 26

Ductile and Brittle Shear Zones

Various types of shear zones are defined and their geometry described. The concept of strain factorization is evaluated and its usefulness discussed. It is shown how the shear strain and dilational factors of ductile shear zones may be evaluated using the orientations of line elements both outside and inside the zone, combined with data related to the principal orientations and values of finite strain. The geometric features of wedge-shaped shear zones are discussed and related to examples in the Swiss Alps. A discussion of compatibility constraints controlling the geometric features of conjugate shear zones lead to several types of possible models where the shear zones result in the development of large bulk strain. Shear zones characterized by en-echelon extension vein arrays are described and models are suggested to account for the range of observed small scale structural features and angular relationships of crossing conjugate zones. The session concludes with a description of the practical methods for determining the directions of the axes of principal stress from the orientations of conjugate and single sets of brittle shear zones.

INTRODUCTION

Shear zones are defined as planar or curviplanar zones of high deformation which are long relative to their width (length to width ratio greater than 5:1) and which are surrounded by rocks showing a lower state of finite strain. They may be subdivided into **ductile shear zones**, where the deformation state varies continuously from wall to wall through the zone, **brittle shear zones** or **faults** where the walls are separated by a discontinuity or fracture surface, and various intermediate types known as **brittle–ductile shear zones** combining these geometric features in different proportions (Figure 26.1). In brittle–ductile shear zones the continuous and discontinuous deformational features may or may not have developed at the same time. Individual shear zones often acquire their geometric features as a result of several distinct movement phases which might be separated over long time periods. Figure 26.2 shows an example of shear zones developed on a crustal scale and illustrates how an old shear zone can be reactivated. In the southern part of the region steeply inclined mostly ductile shear zones cut through various Precambrian gneisses and plutonic rocks and divide the region into compartments showing differing geometrical and geological characteristics. These shear zones probably penetrate to the base of the crust. They are quite wide (1–10 km), show very large displacements (tens to hundreds of km) and are frequently characterized by the formation of intensely banded **mylonite** derived from the deformation of the surrounding country rocks. The main phase of movement leading to their formation was of Precambrian age because they are unconformably overlain by Lower Palaeozoic strata. However, in these cover sediments the trace of the underlying major shear zone continues into the overlying sediments as a shear zone with a much lower displacement between the walls and, locally, sets of en-echelon fold forms are found in this cover. It is clear that the deep seated basement shears were rejuvenated during a period of much later tectonic activity.

Ductile shear zones

The deformation geometry of ductile shear zones takes several forms. In the area under study the zone may be parallel sided and the strain profiles determined from strain markers or from the orientation and intensity of tectonically induced fabrics such as cleavage or schistosity may remain almost constant in differing transects through the zone (Figures 26.3A; see also Figures 3.3, 3.18). In contrast, shear zones can show walls which converge and diverge and which show deformation profiles at differing localities along the zone which are not the same (Figure 26.3B). Generally most shear zones show this type of non-parallelism and heterogeneous strain near their ends unless their movements are taken up by some transform type of shear zone.

We will first examine some of the geometrical implications of the parallel sided constant strain profile types. In our previous studies of the geometry of ductile shear zones (Session 3) we introduced the concept of strain compatibility to show that there are a limited number of deformation models which are consistent with this geometry. In these zones it can be shown mathematically that combinations of three main types of strain field are possible (Ramsay

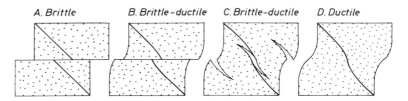

Figure 26.1. Different types of shear zones.

and Graham, 1970). Figure 26.4 illustrated these types schematically. They are:

A. Heterogeneous simple shear with shear plane orientations parallel to the walls.
B. Heterogeneous volume change with volumetric change developed by displacements perpendicular to the shear zone walls.
C. Uniform homogeneous strain of any type affecting the shear zone and its walls.

These three components can be combined *in any propor-*

tions irrespective of the individual strain values taken up by the three models without infringing the basic geometric features of the zone. Figure 26.5 shows the combinations of A and B, A and C, B and C, and finally all three acting together. It follows that we should anticipate any of these three components in a naturally formed parallel sided shear zone and the question we want to pose is whether it is possible to discover the proportions of the various components and, in doing so, be in a position to evaluate the total displacement geometry. This concept is known as **strain factorization**. Mathematically we are seeking

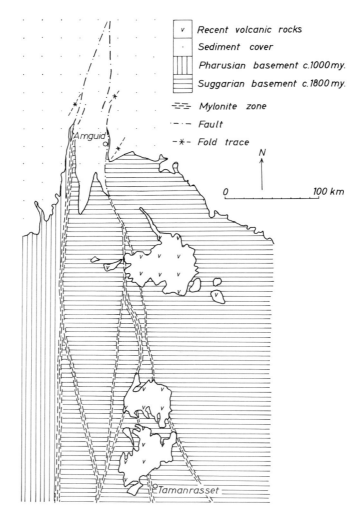

Figure 26.2 *Map of the Hoggar region of Algeria showing mylonites developed in deep ductile shear zones and the faults and folds in the sedimentary cover produced by rejuvenation of these zones. The emplacement of Tertiary and Recent intrusions and lavas was probably also controlled by the location of the shear zones.*

A. *Planar shear zone geometry*

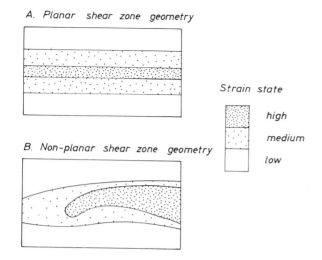

B. *Non-planar shear zone geometry*

Strain state

high

medium

low

Figure 26.3. A: *The constant geometric features of planar shear zones. B: Irregular shear zone with non-parallel walls and showing variable strain profiles.*

componential deformation matrices which, when multiplied together, give the total strain matrix. To make such a scheme we have to *specify the order of multiplication* of the three factors because matrix multiplication is a non-commutative process (see Session 4, p. 65). We can change the order of the multiplication scheme at our convenience but the actual numerical terms appearing in the factors will

then differ. For the simplest description we choose the total strain to be represented by a homogeneous strain followed by a simple shear followed by a volumetric dilation and an area dilation in the shear zone profile. If we choose a two-dimensional coordinate scheme with the x-axis parallel to the length of the shear zone the total strain matrix has the following factors

$$\begin{bmatrix} 1 & 0 \\ 0 & 1 + \Delta_A \end{bmatrix} \times \begin{bmatrix} 1 & \gamma \\ 0 & 1 \end{bmatrix} \times \begin{bmatrix} a & b \\ c & d \end{bmatrix} = \begin{bmatrix} A & B \\ C & D \end{bmatrix}$$

| area change (third) | simple shear (second) | homogenous strain (first) | total strain |

$$(26.1)$$

The dilation term Δ_A and the simple shear term γ will both be different functions of variable y (the distance across the shear zone), and the four displacement components a, b, c and d will be constants independent of x and y values. Because we will not be able to specify the rotation component ω of the overall homogeneous strain the first matrix can be simplified so that $b = c$ (see Equations B.16, C.15). The total strain matrix can now be specified using the rules of matrix multiplication (Equation C.22) as

$$\begin{bmatrix} a + b\gamma & b + d\gamma \\ b(1 + \Delta_A) & d(1 + \Delta_A) \end{bmatrix} \qquad (26.2)$$

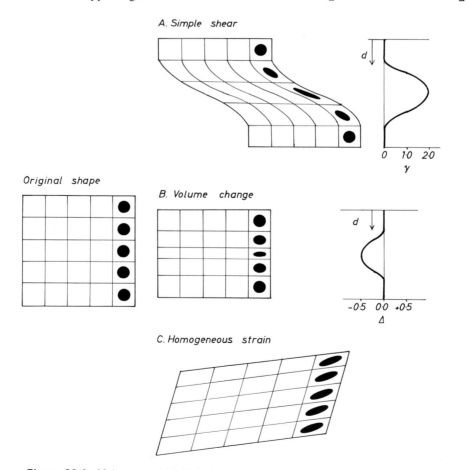

A. *Simple shear*

Original shape

B. *Volume change*

C. *Homogeneous strain*

Figure 26.4. *Main types of strain factors that can exist in planar ductile shear zones.*

A x B   Simple shear & volume   change

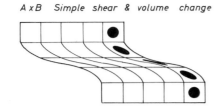

A x C   Simple   shear   &   homogeneous   strain

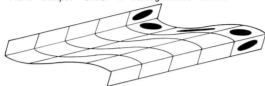

B x C   Volume   change   &   homogeneous   strain

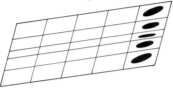

A x B x C   Simple   shear   &   volume   change   &   homogeneous   strain

Figure 26.5. *Various combinations of the three strain factors that can be found in planar ductile shear zones.*

If the walls of the shear zone are unstrained this simplifies to

$$\begin{bmatrix} 1 & \gamma \\ 0 & 1 + \Delta_A \end{bmatrix} \qquad (26.3)$$

QUESTIONS

Separation of shear and dilational components using changes in line directions

Question 26.1

Figure 26.6 shows the geometric changes taking place in an initial element of material as a result of applying a shear strain γ and a volumetric dilation Δ_v recorded on the shear zone profile as an area dilation Δ_A ($\Delta_A = \Delta_v$). Two lines initially making angles of α and β with the shear zone walls take up new positions to make angles of α' and β' after deformation. Application of simple trigonometry (or using Equation B.13a) gives

$$\cot \alpha' = \frac{\cot \alpha - \gamma}{1 + \Delta_A} \qquad (26.4)$$

$$\cot \beta' = \frac{\cot \beta - \gamma}{1 + \Delta_A} \qquad (26.5)$$

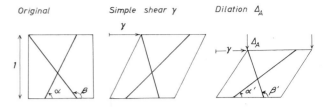

Figure 26.6. *Geometric relations of two lines with initial orientation α and β which are subjected to a shear (γ) and a dilation ($1 + \Delta_A$).*

Remember the sign convention for shear strains (left-hand positive and right-hand negative). Combining these equations, eliminating either $1 + \Delta_A$ or γ, gives

$$\gamma = \frac{\cot \alpha' \cot \beta - \cot \beta' \cot \alpha}{\cot \alpha' - \cot \beta'} \qquad (26.6)$$

$$1 + \Delta_A = \frac{\cot \alpha - \cot \beta}{\cot \alpha' - \cot \beta'} \qquad (26.7)$$

If we know the orientations of two differently oriented lines (traces of bedding planes, dykes, axial surfaces of pre-shear zone folds, etc.) both inside and outside the shear zone, it is possible to compute the values of the two factors at different points at a specific distance from the shear zone walls. These equations are correct where the walls are unstrained, but they are also correct where an overall homogeneous strain (first factor in the general factorization scheme) is present because although such a strain will lead to a difference of direction of the initial orientation of the lines, it will not cause the lines to become curved.

Figure 26.7 shows a sketch of two aplite dykes of differing orientations cutting through a granite mass which, when traced into a shear zone, change their orientations. In the granite surrounding the shear zone the dykes make angles of $\alpha = 52°$ and $\beta = 100°$ with the direction of the shear zone. The granite inside the zone is schistose and shows a linear stretching fabric in the plane of this section. Table 26.1 gives the angles α' and β' of the dyke traces inside the zone at different localities which are paired to be equidistant (d) from the shear zone walls (A and A', etc.). Determine the shear and dilational components from these data using Equations 26.6 and 26.7. A programmable desk computer will be very helpful. Plot the data in graphical form. Determine the total shear and dilation across the whole shear zone.

Now proceed to the Answers and Comments section and return to the starred Questions below or to Question 26.5.

Table 26.1.

Locality	d(cm)	α' (deg)	β' (deg)	θ' (deg)
Wall	0·0	$\alpha = $ 52	$\beta = $ 100	—
AA'	18·0	59	111	142
BB'	38·0	91	136	148
CC'	58·0	113	150	158
DD'	74·0	120	152	159
EE'	93·0	111	146	154
FF'	118·0	73	126	146
Wall	140·0	52	100	—

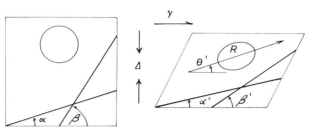

Figure 26.8. Diagram showing how an initial undeformed element is modified as a result of shear and dilation: the incorporation of strain data.

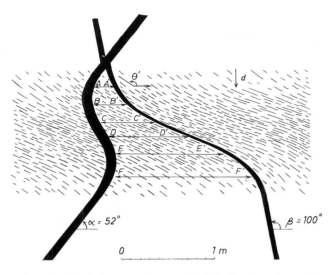

Figure 26.7. A shear zone in a granite deflecting two differently oriented aplite dykes. See Questions 26.1 and 26.2★.

Separation of shear and dilational components using strain data: 1. unstrained wall rocks

Question 26.2★

Where a dilation component exists in a shear zone the orientations, values and ratios of the principal strains will depart from those of simple shear alone. Using the general equations for these features (Equations B.14, B.19, B.20) with the particular values of Equation 26.2 of the three factor matrix we obtain

$$\tan 2\theta'' = \frac{2(1 + \Delta_A)[b(a + d) + \gamma(b^2 + d^2)]}{(a + b\gamma)^2 + (b + d\gamma)^2 - (b^2 + d^2)(1 + \Delta_A)^2}$$

(26.8)

$$(1 + e_1)^2 = \tfrac{1}{2}(a^2 + b^2 + 2\gamma b(a + d)$$
$$+ (d^2 + b^2)[\gamma^2 + (1 + \Delta_A)^2]$$
$$+ \{a^2 + b^2 + 2\gamma b(a + d) + (d^2 + b^2)$$
$$\times [\gamma^2 + (1 + \Delta_A)^2]^2$$
$$- 4(1 + \Delta_A)^2(ad - b^2)^2\}^{1/2})$$

(26.9)

$$(1 + e_2)^2 = \tfrac{1}{2}(a^2 + b^2 + 2\gamma b(a + d)$$
$$+ (d^2 + b^2)[\gamma^2 + (1 + \Delta_A)^2]$$
$$- \{a^2 + b^2 + 2\gamma b(a + d) + (d^2 + b^2)$$
$$\times [\gamma^2 + (1 + \Delta_A)^2]\}^2$$
$$- 4(1 + \Delta_A)^2(ad - b^2)^{1/2})$$

(26.10)

where θ'' refers to the angle between the strain ellipse long axis and the zone walls *inside* the shear zone. This notation is recommended because a different state of strain (with orientation θ') exists in the walls of the shear zone.

If the walls are unstrained ($a = 1$, $b = 0$, $d = 1$) these simplify to

$$\tan 2\theta'' = \frac{2\gamma(1 + \Delta_A)}{1 + \gamma^2 - (1 + \Delta_A)^2}$$

(26.11)

$$(1 + e_1)^2 = \tfrac{1}{2}(1 + \gamma^2 + (1 + \Delta_A)^2$$
$$+ \{[1 + \gamma^2 + (1 + \Delta_A)^2]^2$$
$$- 4(1 + \Delta_A)^2\}^{1/2})$$

(26.12)

$$(1 + e_2)^2 = \tfrac{1}{2}(1 + \gamma^2 + (1 + \Delta_A)^2$$
$$- \{[1 + \gamma^2 + (1 + \Delta_A)^2]^2$$
$$- 4(1 + \Delta_A)^2\}^{1/2})$$

(26.13)

Figure 26.8 shows the various features of angular change and strain features that can be used to factorize the total deformation into γ and Δ_A components: a knowledge of *any two* of these enables the factorization to be made. This is quite an easy matter in principle, but somewhat time consuming in practice because of the long winded nature of the solutions to the combined equation pairs. A programmable desk computer is essential, but once the programs have been compiled the required solutions may be obtained rapidly. For example, if the walls of the shear zone are unstrained we can combine Equations 26.4 and 26.11, eliminating $1 + \Delta_A$.

$$\gamma^2 \tan^2\alpha \,[\tan 2\theta''(1 - \tan^2\alpha') - 2 \tan \alpha']$$
$$+ 2\gamma[\tan 2\theta'' \tan \alpha \tan^2\alpha' + \tan \alpha \tan \alpha']$$
$$+ \tan 2\theta'' (\tan^2\alpha - \tan^2\alpha') = 0$$

(26.14)

In Table 26.1 we give the orientation θ'' which can be used in the computation. This was derived directly from the orientation of the schistosity in the granite. Using data from one of the aplite dykes (α and α' values) and θ'' find the shear and dilation components in the shear zone. Check that they are consistent with those derived in Question 26.1. The main problem here is that for each input of α, α' and θ'' we make in Equation 26.14 we obtain two values of γ, and when these γ values are put in Equation 26.4 two values of $1 + \Delta_A$ are found. The two mathematically possible solutions arise from the structure of the equation giving $2\theta''$. You will remember than the orientation equation gives two directions which are mutually perpendicular. When the θ'' values are put into Equation 26.14 one answer pair arises from θ'' and another pair from $\theta'' + 90°$. Generally the selection of which answer pair is correct is quite simple when the geo-

Figure 26.9. Aplite and granite deformed in a shear zone, Laghetti, Swiss Lepontine Alps. See Question 26.3★ and Table 26.2.

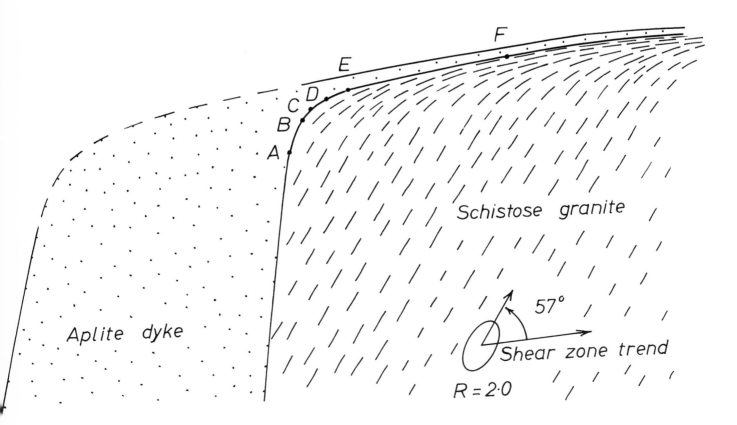

metric forms of the problem data are checked (e.g. consistency of sign of γ and the shear deflection of the planar structure or with the schistosity orientation, or the fact that $1 + \Delta_A$ must always be positive).

Separation of shear and dilational components using strain data: 2. strained walls

Question 26.3★

Figure 26.9 shows an aplite dyke passing from a region of low strained granite into a highly strained situation at the margin of a shear zone. Analysis of deformed granite in the walls of this shear zone show that the strain ellipse long axis makes an angle of $\theta' = 57°$ with the trend of the shear zone and the aspect ratio of this ellipse is $R = 2\cdot0$. The data (Table 26.2) obtained along the dyke contact refers to the angle α' made by the dyke and the shear zone and the local orientation of the schistosity θ'' in the shear zone. The section of this exposure appears to be a true profile of the shear zone because an intense lineation in the shear zone lies parallel to the outcrop surface. We need to find the shear and dilation components. To do this we have to put the equation for the general shear zone matrix (Equation 26.2) in a form suited to computation. First we must express the values a, b and d of the wall strain matrix in terms of the orientation and ellipticity of the strain ellipse. These are found from Equation C.17

$$\left.\begin{aligned} a &= R^{1/2}\cos^2\theta' + R^{-1/2}\sin^2\theta' \\ b &= (R^{1/2} - R^{-1/2})\sin\theta'\cos\theta' \\ d &= R^{1/2}\sin^2\theta' + R^{-1/2}\cos^2\theta' \end{aligned}\right\} \quad (26.15)$$

Solving Equations 26.4 and 26.8 for γ by eliminating $1 + \Delta_A$ we obtain the quadratic equation

$$\gamma^2(b^2 + d^2)\tan^2\alpha\,[\tan 2\theta''(1 - \tan^2\alpha') - 2\tan\alpha']$$

$$- 2\gamma\,\{\tan 2\theta''\,[b(a + d)\tan^2\alpha - (b^2 + d^2)\tan^2\alpha'\tan\alpha]$$

$$- \tan\alpha'\tan\alpha(b(a + d)\tan\alpha + b^2 + d^2)\}$$

$$+ \tan 2\theta''[(a^2 + b^2)\tan^2\alpha - (b^2 + d^2)\tan^2\alpha']$$

$$- 2b(a + d)\tan\alpha\tan\alpha' = 0 \quad (26.16)$$

Using the data of Table 26.2 together with Equations 26.4, 26.15 and 26.16 determine the shear and dilational component in the shear zone which are in addition to those strains found in the shear zone wall. A programmable desk computer is essential to solve these equations effectively.

Table 26.2.

Locality	Dyke contact α' (deg)	θ'' (deg)
Wall	$\alpha = 77$	$\theta' = 57$
A	70	51·5
B	47·5	36·5
C	34·5	28·5
D	24·5	21·5
E	13	12
F	4·5	4·5

Calculation of shear zone components in sections other than profiles

Question 26.4★

Where a natural section of a shear zone structure, either on an outcrop or on a map, does not contain the shear direction, the equations set up for the solution of the previous three questions cannot be used.

Instead we must have some other features from which we can construct the shear vector direction and amount and compute the dilation across the zone. In calculating the displacement direction and amount across a simple fault surface we have seen in Session 23 that we need to seek two differently oriented planar features of known position on either side of the fault (Figures 23.11 and 23.12; Question 23.2). If we need to consider the possibility of determining the volume dilation across the zone two planar features are insufficient: we also need a third plane differently oriented from the other two. Such situations are extremely rare in naturally deformed rocks; however they do occasionally exist and one such is illustrated in Figure 26.10. This outcrop consists of a slightly deformed and schistose xenolithic granite cut by three non-parallel aplite dykes. These dykes pass from one wall through the shear zone and can be identified in the other wall. First, look at the complex relationship of the dykes as they pass through the shear zone. The dykes show a clear cut displacement discontinuity but also a bending in the zone adjacent to the fault truncation. The bending sense and size of offsets of each dyke is different. This geometry should indicate to the unwary geologist the dangers of determining the relative shear sense of the zone walls without first determining the direction of the shear vector. It should be apparent in this outcrop that the displacement vector must pass through the outcrop surface. The different apparent shear senses of dykes A and B arise because the dykes make different angles with the shear vector and the axes of the **drag folds** formed in the ductile part of the shear zone have differing orientations controlled by the intersection of the shear zone and each dyke. The amplitude of the drag fold formed in dyke C is much less than those folds in dykes A and B. This probably implies that the shear movement vector lies closer to the intersection of the shear zone and dyke C than with the dyke intersections A and B with the zones (see discussion of the amplitude of folds formed in superposed fold systems, pp. 501–502 and Figure 22.25).

Because of the intricate orientation relationships of the three dykes with the shear zone it is possible to obtain a triangle of intersection of the dykes on one side of the zone and compare this with the triangle of intersection on the other side. If the triangles match in size there is no volumetric dilation and the shear vector can be obtained by matching any of the three apices. If there is a mismatch there must be removal or addition of material across the zone and the dilation can be found by moving one side perpendicular to the shear zone until its intersection triangle does match that on the other side.

The outcrop surface has been mapped very accurately using a 10 cm grid: it is not horizontal but has a strike of 12° and a dip of 35°E. The orientation of the three dykes were computed from their apparent dips on this surface and other joint faces and are given by: A, strike 50°, dip 25° NW; B, strike 166°, dip 55° W; and C, strike 42°, dip 61° NW. The

Figure 26.10. *Three differently oriented dykes passing through a brittle–ductile shear zone. Laghetti, Swiss Lepontine Alps. See Question 26.4★.*

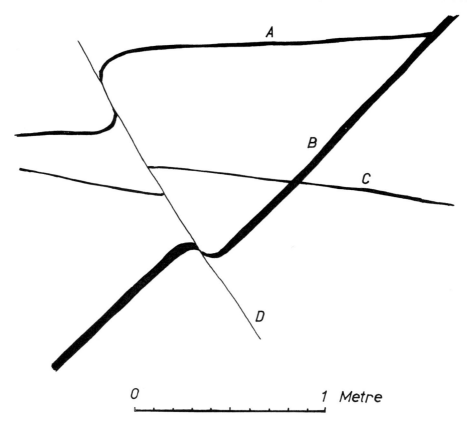

shear zone orientation is: strike 92°, dip 79° S. The total offset of each dyke across the shear zone in the plane of the surface was determined by projecting the wall orientations on to the shear zone trace (removing the local effects of the drag fold bending). These are *AA'* 51 cm, *BB'* 21 cm and *CC'* 21 cm. Using equal area or stereogram projection methods together with graphical construction compute the shear vector direction and amount and the dilation component across the whole shear zone. Be warned that the necessary computations to find a solution are complicated and time consuming, and require very careful three-dimensional thinking. This is perhaps a double star quality question!

Brittle–ductile shear zones: the development of en-echelon vein arrays

Question 26.5

In Sessions 2 and 3 we discussed the geometric principles governing the formation of en-echelon vein systems as a result of the opening of extension fractures. We saw how the activity of the shear component and a possible dilation component led to local orientations of the principal axes of the strain ellipse which were oblique to the shear zone walls (Figure 3.21). We also saw how continued deformation in the shear zone could lead to a change in orientation of the veins (line rotation) to form sigmoidally shaped veins and how later formed en-echelon veins might cross cut the curved portions of the sigmoidal veins. The development of successive cross cutting veins in single shear zone systems often leads to the break up and separation of the country rock screens between the initial veins and the production of fragments of country rock with lozenge-shaped cross sections isolated on all sides by vein material (Figure 26.11). The end result of this process is to form a type of breccia where the fragments are not separated or mixed in a chaotic manner but where they retain some continuity of geometrical features inherited from the order of the initial country rock. However, when the shear strain is very high, the rotations of individual country rock fragments can be quite large and the breccia so formed can be rather unordered at first sight (compare Figure 26.11 at the start of this process with the disorder of the breccia of Figure 25.31).

Brittle and brittle–ductile shear zones, like ductile shear zones, are often found in conjugate sets, one group with left-hand shear sense, the other with right-hand sense (Figure 26.12). The orientations of the tips of the veins in each system are often sub-parallel, whereas the central parts of the veins in each shear system are rotated in opposite senses away from the tip direction (Figures 26.12, 26.13). The vein tips in each zone sometimes link up with a regional set of extension veins which bisects the acute angle between the crossing shear zones (Figure 26.13). Where conjugate sets meet, they may cross each other, generally one (earlier) being displaced by the other set, or one set may make a bridge-like link between two adjacent and parallel zones (Figure 26.12).

Where the shear zones are more or less of equal intensity, as measured by their frequency, comparable intensity of veining and sigmoidal vein forms, it seems geometrically correct that the bisectors of the angles between the conjugate zones will give the directions of maximum and minimum bulk strain of the rock mass, and the zone intersection will

Table 26.3.
Vein orientations in Figure 26.5. Data for Question 26.5.

Locality	Strike	Dip
A	139°	75° SW
B	104°	79° S
C	130°	75° SW
D	155°	80° SW
E	145°	77° SW
F	43°	60° NW
G	80°	71° NW
H	92°	84° N
I	44°	62° NW
J	27°	65° NW
K	56°	62° NW
L	62°	60° NN

lie close to the intermediate strain axis. Although it might be valid to correlate these axes with the principal stress directions ($1 + e_1$, $1 + e_2$ and $1 + e_3$ coaxial with σ_1, σ_2 and σ_3 respectively) we should always be wary of making this step too soon. For example, if the intensity of the conjugate vein systems was uneven (Figure 26.14) would such a correlation be justified? What might be the directions of bulk strain in this figure and what might the shear zone dihedral angle bisectors mean?

Figure 26.15 illustrates a conjugate shear system of en-echelon calcite veins (black in the diagram) cutting through limestone containing pre-existing calcite veins (stippled in the diagram). The early formed veins are of two generations and consist of blocky crystals of white calcite, whereas the en-echelon vein sets in the shear zones consist of finely crystalline calcite with a fibrous habit. Although the age relationships of the various calcite veins are generally clear it is interesting to note that where the fibrous veins cut earlier veins their crystals connect (the fibres have the same optial orientation as the crystals of the walls) in a somewhat similar manner to the geometric features of Figure 13.8. We would therefore like to emphasize that, because of this crystal linkage, it is not always a simple matter in the field to say which vein is early and which is late. This is in contrast, for example, to crossing igenous dykes where the later intrusion is chilled against the earlier dyke.

The orientations of the two shear zones in Figure 26.15 are *P*, strike 162°, dip 76° SW, and *Q*, strike 120°, dip 78° SW. The orientations of the veins at various localities *A–L* are given in Table 26.3. Plot these data on an equal area projection. Discuss the significance of the geometric features shown by the various veins and the conjugate shear zones and determine the principal axes of bulk strain.

Now go to the Answers and Comments section and then return to Question 26.6 below.

Brittle shear zones: calculation of the principal stresses

Question 26.6

In Session 25 we found that brittle failure of a rock generally sets up conjugate shear fractures which, because of internal friction effects, generally have their acute dihedral angle bisected by the maximum compressive stress direction σ_3 and their obtuse dihedral angle bisected by the minimum

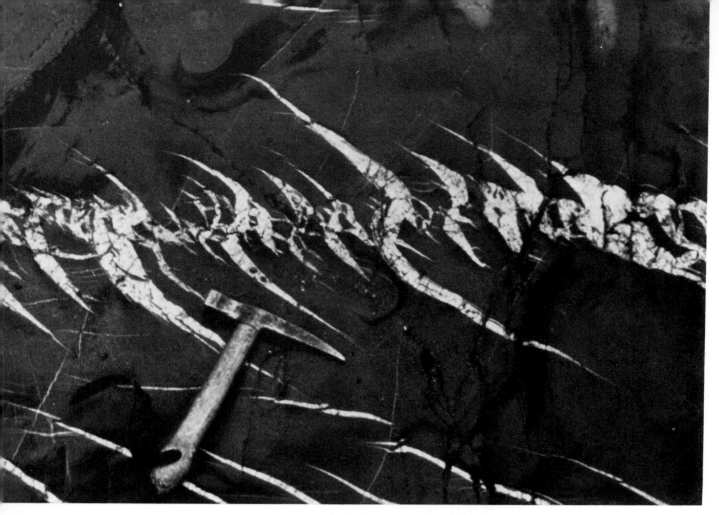

Figure 26.11. *Brittle–ductile shear zone with sigmoidal en-echelon quartz veins in a greywacke. Millook Haven, N Cornwall, UK.*

Figure 26.12. *Conjugate sets of en-echelon calcite filled veins in Upper Jurassic limestone, Wildhorn nappe, Melchtal, Switzerland. The height of the rock face is about 20 m.*

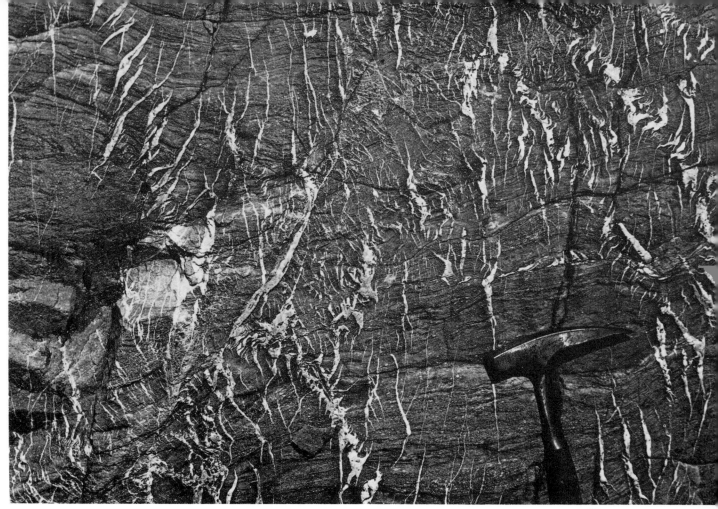

Figure 26.13. Conjugate zones of en-echelon quartz veins linked with non-conjugate regional extension veins. Middle Jurassic sandstones. Lötschen pass, Central Switzerland.

Figure 26.14. Uneven development of en-echelon vein systems in Upper Jurassic limestone. Urbachtal, central Switzerland. See Question 26.5.

Figure 26.15. Conjugate shear zones cutting Cretaceous limestone. Grand Cor, Morcles nappe, Valais, Switzerland. See Question 26.5 and Table 26.3.

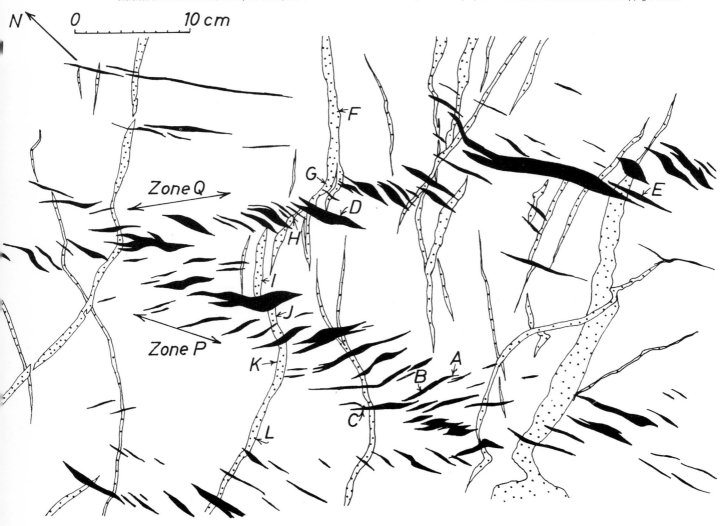

Table 26.4
Orientations of shear fibres. Data for Question 26.6.

Locality	Shear plane strike and dip	Lineation azimuth and plunge	Relative motion of upper side of fault
1	65°/20° NW	281°/14°	Up and to SE
2	165°/27° NE	65°/32°	Up and to SW
3	145°/18° NE	105°/5°	Down and to SE
4	175°/53° E	95°/53°	Up and to W
5	180°/28° E	90°/23	Down and to E

compressive stress direction σ_1. In Session 9 we showed how these geometric features could be used to construct the directions of principal stress producing the conjugate faults.

The problem we would like to consider here is, knowing only the orientation of one of the fault plane surfaces and having the movement vector defined by striae on the slickenside surface, or the presence of shear fibre veins on the surface, how we can determine the principal stresses. Basically the problem consists of resolving the total geometry of Figure 26.16A into the geometric features of the incomplete geometry of Figure 26.16B. To construct the principal stresses we proceed as follows:

1. σ_2 lies on the fault plane oriented at 90° to the movement direction.
2. σ_1 is situated in a plane (great circle with pole σ_2) at an angle of θ (where 2θ is the acute bisector of a conjugate fault set) from the slip direction on the fault. The choice of sense of the angle θ is given by the differential movement sense across the fault plane. To make such

a construction we must have some previous knowledge of the angle 2θ and the relative sense as well as the orientation of the slip vector.

3. σ_3 is best constructed by finding a point on the great circle with pole σ_2 at an angle of $90 - \theta$ from the slip direction on the opposite side of the fault as that made for construction 2.
4. Check that σ_1, σ_2 and σ_3 are mutually perpendicular.

Table 26.4 sets out data derived from a field study of deformed sandstones at Muzaffarabad, NE Pakistan. The stratified rocks have been folded and are cut by narrow brittle shear zones which contain fibrous crystals of calcite. These fibres overlap each other and enable the sense of motion on the shear plane to be evaluated (see Figures 13.32 and 13.33). At the five independent localities only single shear zones were available. However, other work in this terrain showed occasional localities where conjugate shear systems could be found, and where an angle of $2\theta = 38°$ was characteristic of the fault geometry. Determine the principal stress directions at the five localities.

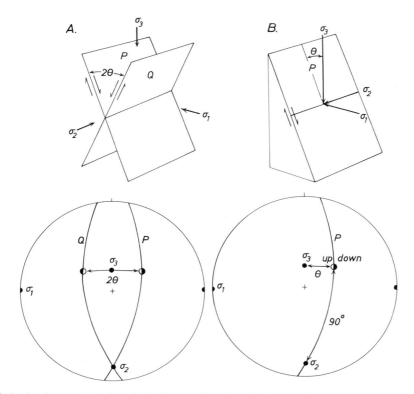

Figure 26.16. *Methods of computing the principal stress directions A, from conjugate or B, from single fracture surfaces.*

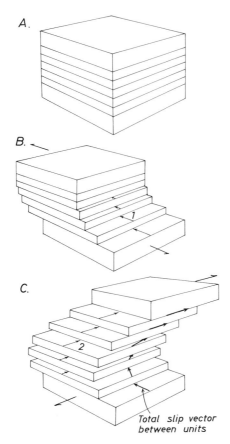

Figure 26.17. *A card sliding model to indicate how the total shear vector in a shear zone might be compounded from two separate simple shear motions. The original A is sheared first B by parallel movements in direction 1 and with a deformation gradient across the model. In C it is sheared in direction 2 again with a gradient across the block. The total slip vectors between the elements are those of simple shear, but the amount and direction of shear varies through the block.*

ANSWERS AND COMMENTS

Separation of shear and dilational components using changes in line directions

Answer 26.1

In this and in the next two questions the mathematical solutions employed require that the measurements be made

Table 26.5.
Answer 26.1.

Locality	Shear factor γ	Dilation factor $1 + \Delta_A$
A	0·20	0·97
B	0·80	0·95
C	1·09	0·73
D	1·21	0·73
E	1·12	0·87
F	0·50	0·93

in a true profile section of the shear zone, that is a section that contains the slip direction of the shear component of the deformation. Providing this constraint is met it does not matter what orientation the two line elements have with the shear zone walls, the only constraint being that they must have been planar before the formation of the shear zone. Such profiles are generally recognized by the presence of a stretching lineation produced by the shear components that lie in the plane of the section or, if the walls of the shear zone are unstrained, the curvature of the internal schistosity is cylindrical about a line perpendicular to the profile plane. We have already met the possibility of a shear zone undergoing rejuvenation at a time after its origination. If the rejuvenated shear components are oriented differently from those of the original motion it is possible to arrive at very complex geometric situations where no one profile across the shear zone can contain all the total displacement shear vectors. An analogy is a card deck which is sheared differentially in one direction and then superposed by a differential shear in a new direction (Figure 26.17). Depending upon the proportion of the component shears, the slip vectors between each pair of cards will have differing orientations. In a ductile shear zone of this type the maximum extension direction and accompanying stretching lineation cannot be aligned in a single plane. In such geometric situations we cannot simply use the techniques of angular changes and strain orientations to determine the shear and dilation components of the deformation.

The numerical answers to the question are set out in Table 26.5 and plotted in Figure 26.18. The positive (left-hand) shear components show systematically increasing values towards the centre of the shear zone, and are in accordance with the increase in intensity and change of orientation of the schistosity from the boundary of the shear zone into its centre. It is possible to find the total displacement across the zone using the integration methods discussed in Session 3 (p. 42). The dilation also shows a systematic change towards the shear

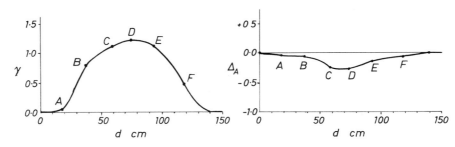

Figure 26.18. *Graphical distribution of the variation in shear strain γ and area dilation Δ_A across the shear zone of Figure 26.7.*

zone centre reaching a minimum value of 27% decrease at the position of greatest shear strain. It is possible to compute the total shortening normal to the shear zone from the graph of Figure 26.18 using a similar method to that used for the total shear calculation. The total shortening is given by

$$\int_0^{x=d} 1 + \Delta_A \, dx \qquad (26.17)$$

which is equivalent to the area *above* the curve of Figure 26.18 (the dilation being negative). Volumetric changes are not uncommon in shear zones. They are usually related to mineralogical transformations in the zone taking place as a result of changing environmental conditions or to removal of material by fluid phases passing through the zone during deformation. Good examples may be found in the Precambrian Lewisian gneisses of the Western Highlands of Scotland (Beach, 1976). Here an old granulite facies metamorphic complex (Scourian) is later deformed by shear zones in the Laxfordian orogenic event. In these shear zones the "dry", potassium-poor rocks are modified to an amphibolite facies mineralogy: hornblende replaces the original pyroxene and biotite and potassium feldspar appear (Figure 26.19). Here the shear zone develop-

ment appears to have been accompanied by the presence of aqueous fluids containing potassium ions.

Although we solved this question mathematically, the geometric features of Figure 26.7 offer solutions by graphical constructions. For example, by comparing the lengths AA', BB', etc. with the equivalent planar projection of the two dykes before deformation it is possible to evaluate the shear and dilation across the zone. However, with other types of problem this graphical method may not be applicable. When dyke deflections are combined with a pre-existing discontinuous rock layering (e.g. such as found in gneissic layering), or when different dykes of a swarm are used, it is generally only possible to use their angular changes to compute the strain factors.

The factorization scheme employed here is a special one in that the individual shear and dilation factors are compatible in any combination. Such a relationship could be termed an **independently compatible strain factorization** scheme. Not all strain factorization schemes have this property. For example, it is well known that any two-dimensional strain may be considered as the product of an irrotational strain component and a body rotation

Figure 26.19. *Example of an amphibolite grade shear zone cutting through granulite facies gneisses. The lower part of the photograph consists of large dark aggregates of garnet in a matrix of orthopyroxene, clinopyroxene, plagioclase and quartz. In the shear zone the original banding persists, but the mineralogical composition now consists of fine grained hornblende, biotite, plagioclase, orthoclase and quartz, and the grain size is much smaller than that seen in the zone walls. South Harris, Outer Hebrides, Scotland.*

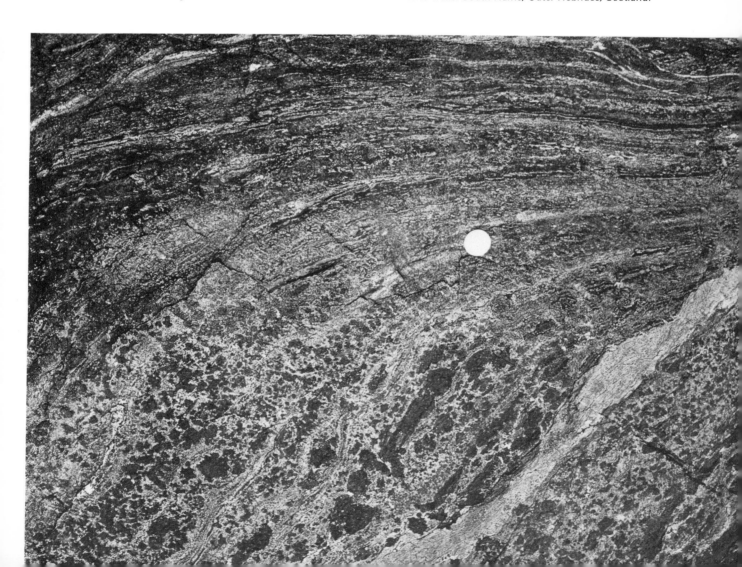

(Equation C.20 and Figure 26.21A). Heterogeneous simple shear could be represented by such a factorial scheme:

$$\begin{bmatrix} 2(\gamma^2 + 4)^{-1/2} & -\gamma(\gamma^2 + 4)^{-1/2} \\ \gamma(\gamma^2 + 4)^{-1/2} & 2(\gamma^2 + 4)^{-1/2} \end{bmatrix}$$

rotation

$$\times \begin{bmatrix} 2(\gamma^2 + 4)^{-1/2} & -\gamma(\gamma^2 + 4)^{-1/2} \\ -\gamma(\gamma^2 + 4)^{-1/2} & (\gamma^2 + 2)(\gamma^2 + 4)^{-1/2} \end{bmatrix} = \begin{bmatrix} 1 & -\gamma \\ 0 & 1 \end{bmatrix}$$

pure strain

(26.18)

In Figure 26.20 we have illustrated how these two factors vary independently through the body. The overlaps and gaps between the various composite elements here show their independent existence is not possible. This factorial scheme, although it can be mathematically justified, produces an **independently incompatible strain factorization** scheme.

There are an infinite number of ways that we can factorize the finite strain state at a particular locality. Several possibilities are set out in Figure 26.21. To set up a valid factor scheme the total number of variable terms in the factor

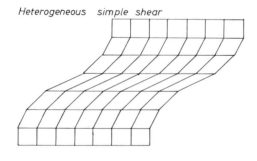

Heterogeneous simple shear

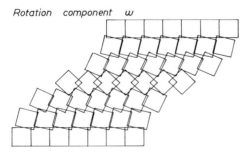

Rotation component ω

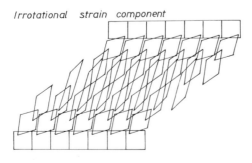

Irrotational strain component

Figure 26.20. *The factorization of heterogeneous simple shear into independently incompatible factors of body rotation and irrotational (pure) strain.*

Table 26.6.
Factorizing schemes for a general strain

Total strain	$abcd$ or $e_1 e_2 \theta' \omega$
Scheme A	1. $a'b'd'$
	2. ω
Scheme B	1. $a''b''d''$
	2. γ_x
Scheme C	1. $a'''b'''c'''$
	2. $e_x = e_1,\ e_y = e_2 = -e_1/(1 + e_1)$
Scheme D	1. $a''''b''''c''''$
	2. $e_x = e_2,\ e_y = e_1 = -e_2/(1 + e_2)$
Scheme E	1. $\alpha,\ \gamma_\alpha$
	2. γ_x
	3. Δ

scheme for a general strain must be four, because any strain requires four displacement or four strain components. Table 26.6 shows that each of the factor schemes does contain four displacement terms. Before we become involved in strain factorization techniques we should always consider the geological reason (if any) for separating factors; does it better help us to understand the overall geometric features of the system? If the answer to this question is affirmative, then we will probably have effective guidelines for choosing one particular factorization scheme. We are convinced, for example, that the separation of three independently compatible factors to describe the overall geometry of regular, parallel sided shear zones, as we have done in Session 3 and in the first part of Session 26, does greatly assist our understanding of the geometric complexity. The only danger in applying such a scheme is that there may be a tendency to think that the choice of order of matrix multiplication does relate to the time ordering of the separate strain factors in nature. That would clearly be quite false. One type of factorization scheme applied to shear zones that has become rather popular since the 1970s uses a wall parallel simple shear component together with a pure shear deformation with principal strain axes parallel and perpendicular to the shear zone walls according to the scheme illustrated in Figure 26.22, and described by the factorization scheme:

$$\begin{bmatrix} 1 + e_1 & 0 \\ 0 & 1 + e_2 \end{bmatrix} \times \begin{bmatrix} 1 & \gamma \\ 0 & 1 \end{bmatrix} = \begin{bmatrix} 1 + e_1 & \gamma(1 + e_1) \\ 0 & 1 + e_2 \end{bmatrix}$$

pure shear   simple shear

(26.19)

Several major problems arise with this concept, and most authors using this method appear to be unaware of them because they have focused attention on the geometry of one small element, neglecting the effects on interlinked elements. It should be clear that such a scheme can never compatibly coexist with undeformed walls (unless $e_1 = 0$, when $1 + e_2$ becomes $1 + \Delta_A$ and the transformation become identical with the concept of simple shear and wall perpendicular dilation examined earlier). Figure 26.23 illustrates the detachments along the shear zone walls that are necessary in adjacent small elements across such a shear zone (Sanderson, 1982). The second geometric problem is illustrated in Figure 26.24. Variable pure shear components perpendicular to the shear zone walls produce an overall

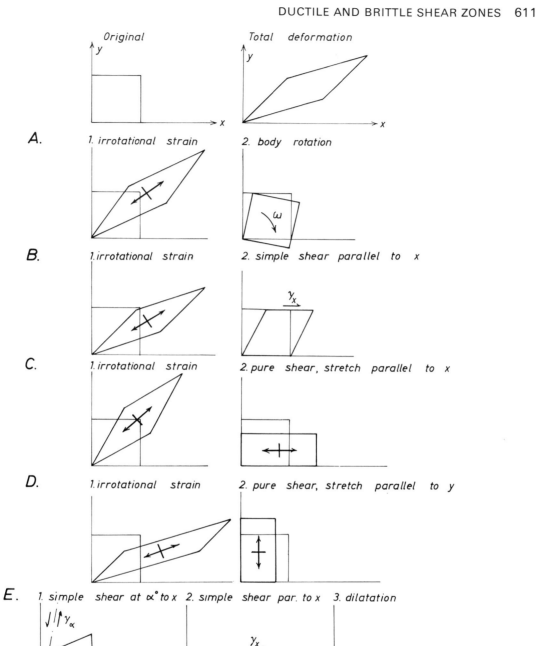

Figure 26.21. *Different types of possible factor schemes to describe a general deformation. Each scheme is valid because it contains four independent terms necessary to define a general translation and strain (see Table 26.2).*

displacement scheme which must either be accommodated by the formation of discontinuities between adjacent elements (if the strain in the elements remains irrotational as in Figure 26.24B), or produce a shear strain component with an amplifying value as one moves away from some fixed "pin line" marker (Figure 26.24C). In culinary terms this might be termed the "cream cake effect"—the extrusion of a central layer of cream when the top and bottom of the cake are pressed together. This geometry shows that a heterogeneous

pure strain component with axes parallel to the shear zone walls cannot exist by itself. The heterogeneity induces shear strains which vary as a function of distance from the pin line and as a function of the gradient of the "pure strain" component across the zone. Figure 26.25 shows the type of shear strain variation that occurs when the contractions normal to the walls show x values of displacement which are parabolic and sinusoidal. We are of the opinion that the usefulness of a factorial scheme of pure shear components is extremely

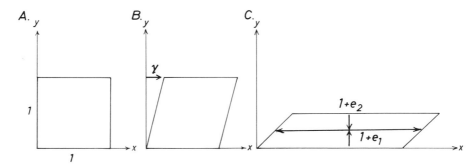

Figure 26.22. *A two component factorization of simple shear followed by an irrotational strain with axes parallel to the coordinate directions x and y.*

limited because variation in the first produces complex variations in the second. We term this type of variation an **interlinked factorization scheme** in which the factors are **independently incompatible**. Geologically we are of the opinion that this scheme sets up strain variations which do not appear realistic in our experience, and nowhere in the published examples of use of this model have the geometric and strain changes expected from the models set up here been described. The main type of geological situations where certain aspects of this model might be applicable are in various types of nappe and thrust tectonics where the two sides of a shear zone have differing strain states and form independent strain compatible units separated by a discontinuity. The geometry of the discontinuity can be accommodated in a number of ways (Figure 26.26). It is possible (Figure 26.26A) to have a shear zone with homogeneous or heterogeneous strain (analysed according to the geometry of Equation 26.19) with a discontinuity which progressively amplifies. If no discontinuity develops the shear zone sets up a heterogenous shear along its length which might be associated with a volumetric dilation. This model (B) might be one to account for the extensive pressure solution effects sometimes seen in certain regions of thrust tectonics. These two models might be combined in certain ways (model C) to produce a constant displacement along a discontinuity with uplift of the overlying block taking place if there is no dilation.

Conjugate ductile shear zones

Ductile shear zones generally occur in sub-planar conjugate sets. The zones intersect in a line which is generally perpen-

dicular to the shear displacement vector of both sets, and which is perpendicular to the directions of maximum stretching (X) in each zone. In conjugate sets it is generally found that one shear zone displaces another which is conjugate to it (Figure 26.27), but that the time relationships between the two conjugate sets is such that they appear broadly synchronous (Figure 26.28). This geometry arises because of compatibility problems; it is not possible to shear the two zones simultaneously across one another without setting up openings or regions of overlap at the intersection positions. The problem is geometrically identical with that discussed on p. 513 (Figure 23.16) arising during the development of conjugate fracture systems. With ductile shear zones the *acute angle sector between the zones lies in the direction of bulk stretching and the obtuse angle sector faces the bulk shortening direction.* This relationship is the opposite of that seen in brittle faults. By developing a conjugate set of shears it is possible to build up an overall bulk strain in a rock mass by forming localized planar shear zones with abnormally high finite strains separating regions which are unstrained or have a lower finite strain than that of the bulk deformation. The geological pattern that emerges is that of lozenge-shaped blocks of relatively unstrained country rock bounded by ductile shear zones. The scale of this pattern varies enormously from that of outcrops (Figure 26.27), to mountain sides (Ramsay and Allison, 1979) to sub-continents (Coward, 1980).

There are several possible geometric models which are internally compatible and which might account for the lozenge geometry. The main problem we have to tackle is whether the angles between the conjugate shear zones we observe today reflect the initial angular relations of the

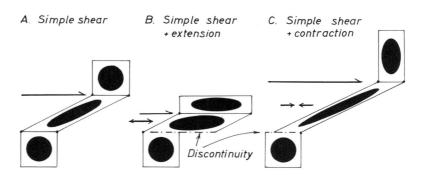

Figure 26.23. *Effects of combining simple shear and shear zone wall parallel pure shear. Where the walls are unstrained the shear zone strains are incompatible with the wall rock strains. After Sanderson (1982).*

A. Original grid

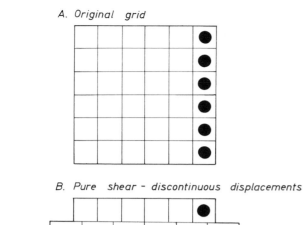

B. Pure shear – discontinuous displacements

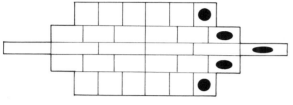

C. Pure shear – continuous displacements

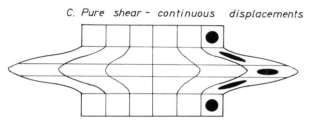

A. Parabolic field

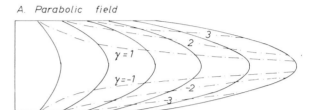

B. Sinusoidal field

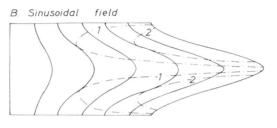

Figure 26.24. The "cream cake" strain effect as a result of trying to make a "pure shear" with principal strains parallel and perpendicular to the shear zone walls. In B, and C, a fixed vertical pin line for shear zone parallel displacement has been chosen in the centre of the model.

Figure 26.25. Different types of displacement field produced by differential wall perpendicular contraction in a shear zone, with the left-hand side of the diagram acting as a pin line for shear zone parallel displacements. The γ-numbers refer to the shear strains that are induced. They vary with distance from the pin line and with type of contraction field variation.

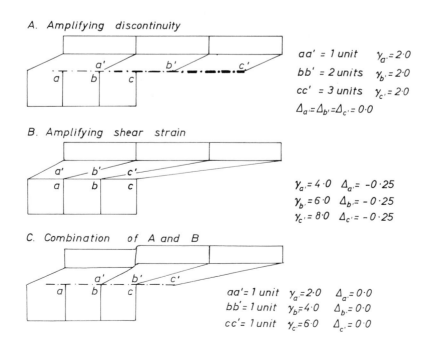

Figure 26.26. Geometric solutions of displacements, shear strain and dilations resulting from developing shear zones with differing wall strains.

Figure 26.27. *Granite containing aplite dykes being cut by conjugate shear zones. The zone with left-hand sense running from top left to the centre is deflected by a shear zone with right-hand sense running along the base of the photograph. Laghetti, Central Pennine Alps, Switzerland.*

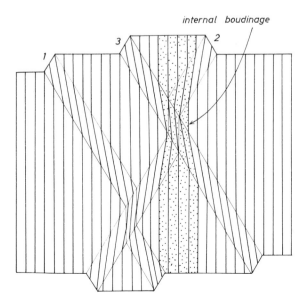

Figure 26.28. *Model of sequentially intersecting shear zones formed in order 1, then 2, then 3. Note how the stippled layer is thinned at the cross over point and develops an internal boudinage structure.*

zones, or are the initial angles modified during later stages of the overall deformation?

Figure 26.29 illustrates various possibilities. Models A and B show, on the left-hand side, the development of crossing ductile shear zones cutting through an initially circular marker, these zones having intersection angles of 90° and 120° respectively. We now continue to develop alternate shear zones with right- and left-hand shear displacement senses and the resulting overall geometry is shown on the right-hand side of the diagrams. With this scheme the early shear zones are locally deflected by the later zone, but *their original trends are practically unmodified.* Such a sequential shearing could proceed by activating new shear zones or reactivating old "dead" sectors of earlier shear zones. The originally circular boundary to the model takes on a discontinuous elliptical form recording the bulk strain of the mass. *The material inside each of the variously shaped lozenge-shaped blocks bounded by the shear zones remains undeformed* irrespective of the original angle between the shear zones, and the final model is characterized by *high strain contrasts between the zones and the intervening lozenges.* The shear zone geometry is constrained by two main factors: simple shear and dilation (Figures 26.4, models A and B or 26.5, model A × B).

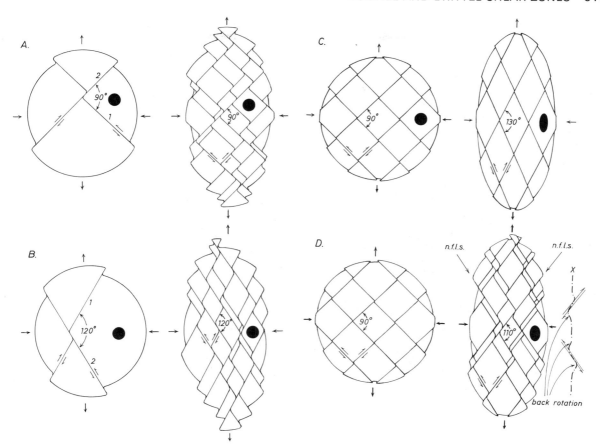

Figure 26.29. *Models illustrating how conjugate shear zones can develop large bulk strains. In A and B the initial and final angles between the shear zones remain unaltered, a feature which goes together with zero strain in the inter-shear zone lozenges. In C the shear zones undergo line rotation, and the finite strain in the lozenges becomes close to that of the bulk strain. In D the shear zones rotate and coincide with planes of no finite longitudinal strain of the bulk strain. The lozenge areas are strained but less so than that of the bulk strain. The line X shows how the local principal directions of maximum elongation have to be adjusted to make the shear zone strain and lozenge strain compatible, with the development of back rotations.*

The model shown in Figure 26.29C shows the opposite end of the deformation spectrum. Small displacement shear zones are initiated at some specific angle (here we chose 90° so that we can compare the final result with model A), and the whole body (shear zone and lozenge-shaped areas) undergoes a homogenous deformation. Both sets of *shear zones undergo line rotation* so that *the angle which faces the bulk shortening direction Z increases*, and the *intervening lozenges become strained*. The shear zones no longer coincide with the directions of no finite longitudinal strain, and generally will undergo a positive elongation along their length. Their walls are therefore deformed and their geometry is that shown in Figure 26.5 models A × C, B × C or A × B × C. In Figure 26.29C we have allowed the late deformation to proceed with little or no extra shear on the initial zones. This model is therefore characterized by *low strain contrasts between the shear zones and the intervening lozenges*.

There is an infinite range of geometric possibilities between these extreme models. However, one particular model (Figure 26.29D) does seem to have special geological significance. In model C there is an increase in length along the shear zones. It seems likely that once a shear zone is formed its initial weakness, together with the properties of simple shear, will continue to develop this zone as one of

predominantly simple shear. This implies that initial shear zones will tend to coincide with the lines of no finite longitudinal strain in the overall deformation. Model D shows the implications of this geometry: the initial shear zones (c. 90°) will change their interzonal geometry to coincide with the directions of no finite longitudinal strain in the total deformed body. The angular changes of the shear zones will therefore define the overall shape changes and finite strain in the inter shear zone lozenges. One especially important feature of this model is that *the average finite strain in any lozenge is no longer compatible with the condition of no finite longitudinal strain of the shear zone walls.* To arrange for a compatible solution it is necessary to rotate the strain ellipses in the lozenge area adjacent to the shear zones in a sense opposite to that of the rotations of the strain ellipses in the shear zones. In laboratory experiments with model materials such a back rotation phenomenon can be seen.

One important question arises from these various models. This is: what particular mechanical features influence the model preferences? It seems likely that the degree to which the rock **strain softens** will control which of the various models develops. In rocks with strong strain softening characteristics the shear zones will tend to operate to maximum effect, and models A, B or D are most likely. In rocks which do not show such well developed strain softening the

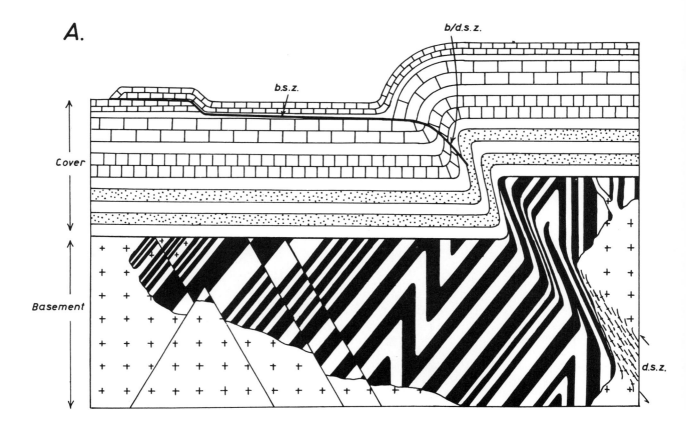

A.

b/d.s.z.

b.s.z.

Cover

Basement

d.s.z.

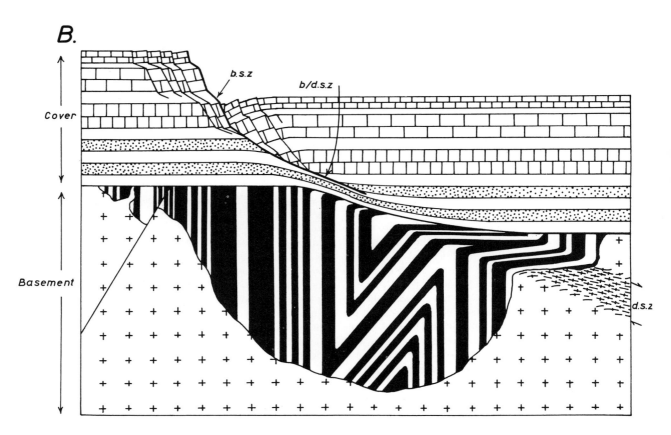

B.

b.s.z

b/d.s.z

Cover

Basement

d.s.z

Figure 26.30. *The interrelations between high level and low level shear zones in the crust in* A *regions of crustal compression, and* B *regions of crustal extension. After Ramsay (1980).*

shear zones and country rock will both be more or less equally susceptible to deformation and model C seems the most probable. Any factor which enhances or reduces the strain softening characteristics during the deformation is likely to have a great influence on the final deformation pattern.

Conjugate ductile shear zones have been recognized as an important deformation mode in many orogenic zones, particularly in those regions which have been deformed in upper greenschist or higher metamorphic conditions and where fluid phases are actively migrating through the crust. Such zones are generally characteristic of rocks of rather massive aspect which do not have a well developed planar lithology. They are especially abundant in continental crustal rocks, those generally termed basement or "crystallines", in massive gneisses where the banding does not provide a strong mechanical weakness and in plutonic igneous rocks. They do not develop in rocks which show a well developed mechanical anistropy: in such materials the instabilities produced during deformation are generally controlled by the layered structure and the competence contrasts, and folding and boudinage are more common. Clearly, in a layered medium, mechanical instabilities may be produced locally in competent layers involved in shear zone deformation (see below). However, one special feature of certain layer orientations involved in shear zones is a geometric thinning at certain points where conjugate shear zones cross one another. Figure 26.28 illustrates such structure, often termed **internal boudinage**. The special feature of this thinning is that it is entirely a result of geometric interaction of two zones and its localization is not governed by competence contrasts as in true boudinage.

In many continent–continent collision zones the overall geometry of the crust often shows a thickened aspect by the development of ductile shear zones at depth, these shear zones passing upwards into brittle overthrusts at higher levels (Figure 26.30A). Crustal thinning is also characterized by ductile shear zone formation at depth, and associated extension faults (often of domino-style) at higher levels (Figure 26.30B). In this extension model it seems possible that the relative rotation of the domino fault blocks is controlled by the stretching gradient in the underlying zones where ductile flow is taking place. In both the compression and extension models the change of angle of the shear zone as it passes from ductile, at depth, into brittle, at higher levels, seems to be connected with the characteristic angular relations of the shear zones with 2θ angles obtuse in the ductile field, and acute in the brittle field. Large sub-vertical shear zones are also known in conditions of relative sideways transport of crustal plates. The San Andreas fault is a typical example of such a structure in the upper crust. Although we cannot directly observe what happens at depth along this fault, it seems likely that this shear zone penetrates to the base of the crust as a ductile shear zone, perhaps showing the characteristics of those steeply inclined ductile mylonite zones previously described from the Hoggar region of the central Sahara (cf. Figure 23.1 and Figure 26.2).

Shear zones often play a major role in the later intrusion stages of large magmatic plutons and batholiths. Figure 26.31 shows a large composite pluton, the Chindamora batholith of central Zimbabwe, where the outermost early consolidated intrusions of tonalitic composition were stretched as a result of inflating the central part of the magma balloon with granitic magma. At the erosion level now exposed at the surface the stretching effect was mostly a radial expansion with tangential stretching setting up steeply inclined ductile shear zones. At some localities these are conjugate with interzonal angles of 2θ about $120°$. Using this angular relationship it is possible to calculate the directions of the principal bulk strains from single shear zones as well as from conjugate zones. An analysis of the maximum elongation direction (Figure 26.31B) shows that the outermost part of the pluton was extended parallel to the intrusion walls. Earlier we noted that shear zones often become the sites of fluid channel ways. Chemical analysis of the shear zones in this batholith indicate that those found in the central granite are depleted in potassium, whereas those at the periphery show potassium contents which are many times the values found in the country rock tonalite.

Shear zone terminations

Shear zones cannot continue indefinitely; they are terminated either in another shear zone or in undeformed material. With the second possibility the strain patterns we have described as being characteristic of planar zones must be modified where the shear zone tip comes to a stop in undeformed or homogeneously deformed rocks. The compatibility constraints that link the various strain states are somewhat complex, and vary depending upon the nature of the third dimensional effect, that is whether the overall strain is plane or not. Figure 26.32 illustrates two extreme models. At the centre of each is a simple shear zone with undeformed walls. The strains at the terminations can be evaluated from the displacements recorded in the originally square grid elements, and from the principal maximum elongation strain trajectories in the section plane. The plane strain model (A) shows terminations with one side of the shear zone passing into a shear direction normal shortening zone, and the other side passing into a shear direction normal compression zone. Between these regions one finds negative finite neutral points (at X and X'). If the strains are non-planar, so that material can be displaced normal to the shear direction, the geometry is constrained by having constrictional strains on one side of the shear zone tip, and flattening strains on the other.

Effects of deformation on competent layers in a shear zone

Until now we have only considered the geometrical effects of shearing on passive marker lines entering a shear zone. If layers of differing competence from that of their surroundings pass into a shear zone, the competence contrasts are likely to set up instabilities as a result of the enforced changes of length. Boudinage (Figure 26.33, layer X) or folding (layer Y) of the competent layers result depending on whether the layer suffers a positive or negative elongation in the shear zone. With certain orientations of layers (layer Z) it is possible to develop an initial phase of folding with later modifications as a result of stretching the previously formed folds, leading to unfolding or to the formation of boudinaged folds. With the progressive strains set up in shear zones it is not possible to find conditions of stretching (and boudinage formation) followed by shortening (folded boudinage).

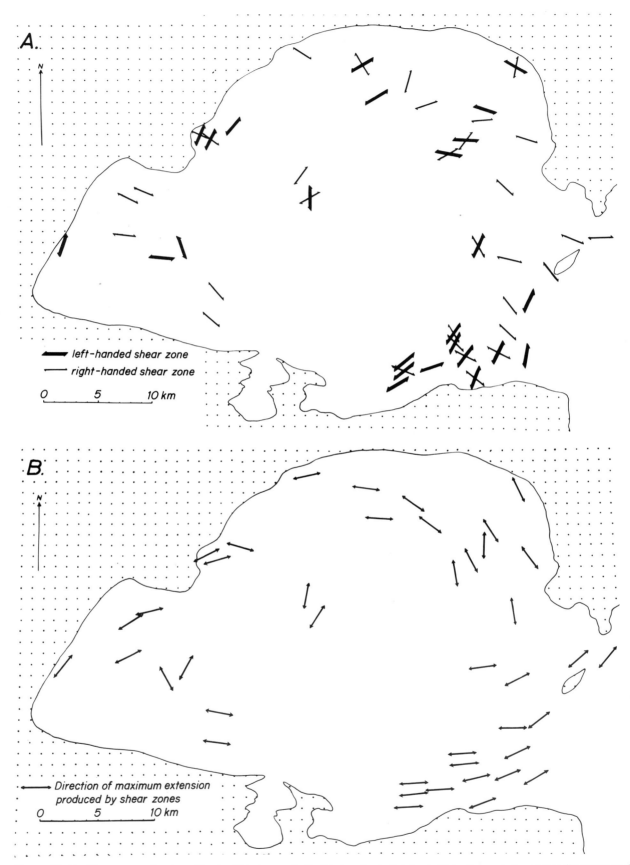

Figure 26.31. Shear zones developed in the Chindamora batholith (unstippled) intruded into Precambrian metasedimentary and metavolcanic rocks (stippled). The right-handed and left-handed shear zones allow the first consolidated exterior of the pluton to be tangentially elongated to make room for the input of more magma at the pluton centre.

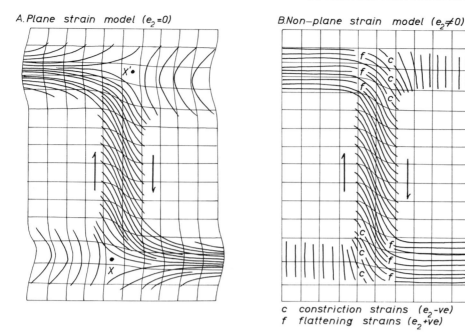

A. Plane strain model ($e_2 = 0$)

B. Non-plane strain model ($e_2 \neq 0$)

c   constriction strains (e_2 -ve)
f   flattening strains (e_2 +ve)

Figure 26.32. Strains developed at shear zone terminations. After Ramsay and Allison (1979).

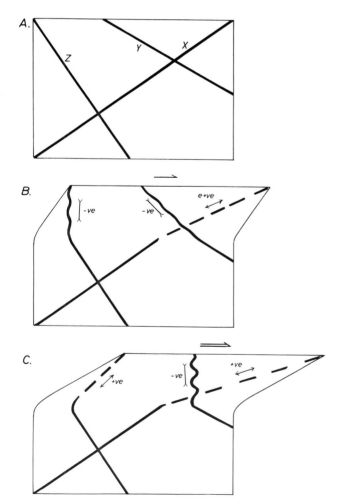

Figure 26.33. The effects of shearing on the lengths of competent layers X, Y and Z, and the structures likely to develop in these layers.

Sheath folds

In regions where fold forms exist prior to the development of shear zones, the geometry of these folds is often severely modified by the strong imposed strains. Figure 26.34 shows the effect of strong shear on an initial fold which had a slightly curving axis. After shear, the interlimb angle is generally considerably reduced, and the axial directions of the initial fold become much more strongly bowed than that of the initial fold: gently bowed domes and basins become transformed into very acute pointed structures termed **sheath folds** (Cobbold and Quinquis, 1980). This effect is only the result of the high strain in the rock and the structures of this geometry can be found in environments outside the shear zone terrain. Do not therefore regard the presence of sheath folds as automatically indicative of the proximity of a shear zone.

Shear zones with diverging walls

In some terrains one finds shear zones where the walls are not parallel. The compatibility constraints on these zones are much more complex that those we have so far examined. The geometric effects of producing a wedge-shaped shear zone are shown in Figure 26.35. The narrow end of the wedge generally implies a stronger shortening across the shear zone than that found at the wider end, and the strain variations are somewhat similar to those we encountered in our study of folds with deformation developed by tangential longitudinal strain (p. 458). If the walls are unstrained we will have a type of extrusive motion of the material in the shear zone which has similarities with that discussed above in shear zones with differential shortening components normal and parallel to the zone (Figure 26.24). The models that we have adopted for wedge-shaped shear zones have been "pinned" along a line sub-perpendicular to the zone and passing through the right-hand side of each model. The strained material is indicated either as block-like units with

A. Original fold   B. Sheath fold form after simple shear

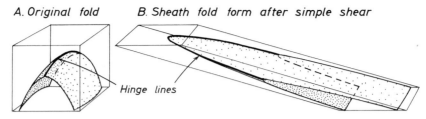

Hinge lines

Figure 26.34. *The development of sheath folds by strong shearing of an initial fold form. After Ramsay (1980).*

discontinuities along the shear zone walls (the dashed line grids), or with smoothly compatible solutions where the extruded material is held in contact with the walls (the solid line grids). As a result of the displacement of the material towards the left-hand side of each model, the vertical marker lines are folded, and show strongly induced shear strains

A. Wedge 'flattening' - weak

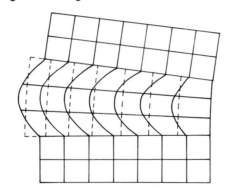

B. Wedge 'flattening' - strong

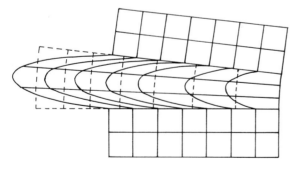

C. Wedge 'flattening' - heterogeneous

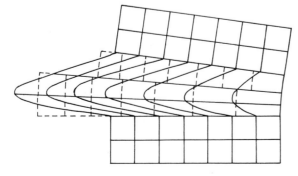

Figure 26.35. *The geometrical features of wedge-shaped shear zones developed by differential shortening across the shear zone.*

along the wall regions of the zones. Following the change of geometry of the folds from right to left it is found that, at first, the amplitudes increase markedly, but then stabilize (where the vertical dashed grid lines are of unchanged length). If we continue to trace these model folds to the left, the fold amplitudes start to decrease. We are of the opinion that some types of natural rock deformation shows these features. For example, in some orogenic zones it is not uncommon for the shortening in the basement to be rather uneven, and in these circumstances wedge-shaped zones between more rigid block like massifs can be tectonically extruded.

The models shown in Figure 26.24 have, for simplicity, only indicated the effect of differential squeezing between rigid walls. It is clearly possible to incorporate other types of shear strain by transporting, for example, the upper wall leftward relative to the lower wall. We suggest that students might reexamine the strain field of the western Helvetic nappes (Figure 11.10) in the light of these models, because there are certain similarities. In particular the base of the Morcles nappe where it comes in contact with the underlying slightly deformed Mesozoic sediments and undeformed Aiguilles Rouges basement massif shows very similar geometry to that of the lower part of the shear zone of Figure 26.24C. The large recumbent anticlinal fold of the Morcles nappe also shows very high strains in its root zone, becoming less strained towards the northwest concomitent with the deamplification of the fold.

Separation of shear and dilational components: 1. unstrained wall rocks

Answer 26.2★

The solutions of Equation 26.14 are set out in Table 26.7, one set using the dyke with initial orientation α, the other set using the dyke with orientation β. There are always two solutions for each pair of either $\alpha'\theta''$ or $\beta'\theta''$ values. The selection of the one appropriate to our problem is always simple. One solution pair always has a negative value of $1 + \Delta'_A$ and has a negative (right-hand) shear sense. As Δ_A can never be less than $-1 \cdot 0$ these solutions are geologically meaningless, while a right-hand shear sense is clearly false. The given results are comparable with those using α' and β' values set out in Table 26.5.

Separation of shear and dilational components: 2. strained wall rocks

Answer 26.3★

Mathematically the problem is not especially complicated

Table 26.7.
Answer 26.2★.

Locality	Solutions using α'				Solutions using β'			
	γ'	$1 + \Delta'_A$	γ''	$1 + \Delta''_A$	γ'	$1 + \Delta'_A$	γ''	$1 + \Delta''_A$
A	**0·20**	**0·97**	3·72	−4·89	−0·57	−1·02	**0·20**	**0·97**
B	0·75	−1·67	**0·80**	**0·95**	−1·37	−1·16	**0·81**	**0·95**
C	0·23	−1·29	**1·09**	**0·73**	−1·71	−0·89	**1·09**	**0·73**
D	0·12	−1·15	**1·20**	**0·72**	−1·83	−0·88	**1·17**	**0·71**
E	0·29	−1·29	**1·11**	**0·86**	−1·77	−1·08	**1·09**	**0·86**
F	**0·50**	**0·93**	1·58	−2·62	−0·93	−1·04	**0·50**	**0·93**

The solutions which are in bold type are those which are geologically realistic.

Table 26.8.
Answer 26.3★.

Locality	Solution pairs			
	γ'	$1 + \Delta'$	γ''	$1 + \Delta''$
A	−0·14	**1·01**	0·47	−0·65
B	0·82	−0·64	**−0·68**	**0·99**
C	1·08	−0·58	**−1·25**	**1·02**
D	1·33	−0·50	**−2·02**	**1·03**
E	1·63	−0·32	**−3·40**	**0·84**
F		No solutions		
F'	2·36	−0·17	**−95·47**	**7·53!**
F''	2·32	−0·16	**−48·43**	**3·83!**
F'''	2·28	−0·16	**−32·73**	**2·59!**
F^{IV}	2·12	−0·15	**−12·99**	**1·04**
F^V	2·10	−0·15	**−11·66**	**0·94**

but, because of the nature of the terms appearing in the quadratic equation 26.16, a programmable computer is required to find the solution. The two sets of solution pairs are set out in Table 26.8. The shear zone clearly has a negative (right-hand) sense and it is easy to select one of the answer pairs as being geologically realistic. These data show slight variations in the dilation factor, and we suggest that because these results are unsystematic across the shear zone there is probably little or no significant dilation with increase in shear strain. At locality F we could find no solution. We therefore changed the data input slightly to see how sensitive the final result was to very slight changes in the angular relationships. We therefore modified the original angle $\alpha' = 4·5°$ to 4·49, 4·48° and 4·47° (given as F', F'' and F''' in Table 26.8). Clearly such very small variations would be impossible to measure in practice. We did obtain solutions to the problem, and with progressive reduction of α' the very high positive dilation became smaller, together with a decrease in the rather absurdly high negative shear strain value. Continuing this angular reduction of α' to 4·42° and 4·41° (F^{iv} and F^v) we obtained a pair of solutions that had a Δ_A values which were very small (see our comments on data points $A–E$) and a resonable value for the shear strain. We think that, although not completely satisfactory, this gives a set of solutions that appears geologically realistic.

Calculation of shear zone components in sections other than profiles

Answer 26.4★

The basic geometric idea used for the solution of this prob-

lem is shown in Figure 26.36A. Three differently oriented planar surfaces always intersect to give an overall pyramidal form. If a shear zone cuts through this pyramid, triangular intersections are formed on each wall of the shear zone. If the distances between corresponding apices of the two triangles (XX', YY' and ZZ') are equal, and the two triangles are identical, the only displacements across the shear plane are those of simple shear parallel to the zone. The shear vector can be calculated from the orientation and shift of any of the apices of the triangles. It the triangles of intersection are not identical there must be a component of dilation acting perpendicular to the zone, probably (but not essentially) combined with a shear zone parallel shear component. If comparable triangle sides all show an identical angle of rotation a differential body rotation across the zone is present, and if the sides show unequal rotations a strain discontinuity exists across the shear zone. In our problem there are no differences in the orientations of comparable dykes, so we only have to seek dilation and shear components.

The first part of the solution requires the construction of the two intersection triangles made by the three dykes on the north and south walls of the shear zone. An equal area plot of the angular relationships of the dykes and shear zone enables the angular relationships between the triangle sides to be computed (Figure 26.36B). From the horizontal strike azimuth in the shear zone these were computed as: $A + 16°$, $B + 56°$ and $C + 49°$, the positive sign reflecting an anticlockwise rotation in the shear zone as viewed from the south. The two intersection triangles on the north and south walls of the zone were constructed using these angles together with the distance intersections along the outcrop surfacen trace on the shear zone. These triangles, as they appear on the shear zone surface, are shown in Figure 26.36C and D. The triangles are not identical and it follows that there is a dilation component across the shear zone. To make the evalution of dilation and shear components we must calculate the position of the pyramid apex on both sides of the zone. We need to construct the pyramid edge lines as projected from the apices of the two intersection triangles. The simplest way of doing this is to relate the three dyke orientations to the shear zone surfaces shown in the triangle constructions. In Figure 26.36E the pole of the shear zone πsz has been rotated to the centre of the net (to $\pi' sz$), and the poles to planes A, B and C have been similarly rotated to take up positions $\pi'A$, $\pi'B$ and $\pi'C$. From these new poles the great circles representing the dyke planes are constructed and the orientations of the apex lines of the pyramid AB, BC and AC relative to a horizontal shear zone computed (AB

A. Construction method

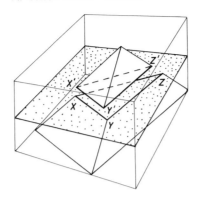

B. Equal area data plot

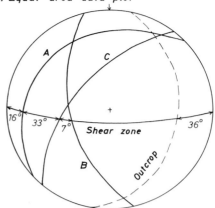

A

C

16° 33° 7° Shear zone 36°

B

Outcrop

C. Intersection triangle, N wall

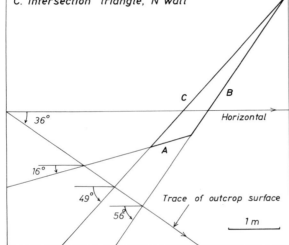

C   B

36°   Horizontal

A

16°

49°

56°   Trace of outcrop surface

1 m

D. Intersection triangle, S wall

36°

C B

16°   A

49°

56°

1 m

E. Data plot, horizontal shear zone

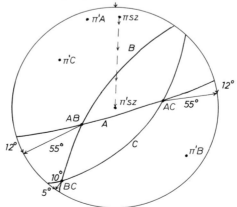

• π'A   • πsz

B

• π'C

12°

AC 55°

AB   A

12°   55°

C

• π'B

10°

5°   BC

F. Comparison of intersection triangles

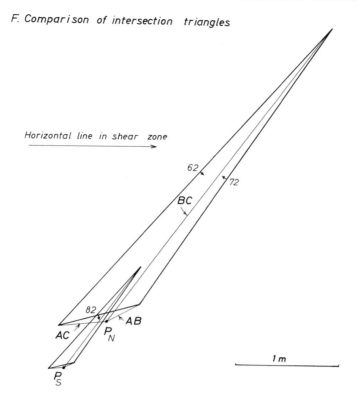

Horizontal line in shear zone →

62
72
BC
82
AC
P_N
AB
P_S

1 m

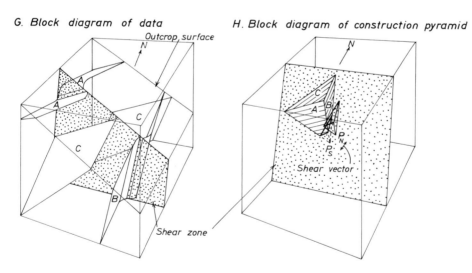

G. Block diagram of data

Outcrop surface
N
A
A
C
C
B
Shear zone

H. Block diagram of construction pyramid

N
C
B
A
A
P_N
P_S
Shear vector

Figure 26.36. Answer 26.4★.

55° in a direction 12° anticlockwise from triangle side *A*; *BC* 10° in a direction 5° anticlockwise from triangle side *C*; and *AC* 55° in a direction 12° clockwise from side *A*). The two triangles were then correctly drawn on the shear zone surface with their correct spatial relationships and the traces of the three pyramid edges added (Figure 26.36F). From this diagram we establish the apex and shifted apex of the pyramid, P_N and P_S on the north and south walls of the shear zone respectively. The join of P_N and P_S gives the shear vector of the movement in the zone: the shift is 0·6 m in a direction 257° azimuth and 49° plunge, the north side of the zone having been displaced upwards and to the east relative to the south side. This is almost parallel to the intersection

of the shear zone with dyke *C*, and clearly is the reason why this dyke shows hardly any deflection as a fold form in the ductile edge zone of the shear. This shear displacement vector crosses dykes *A* and *B* with opposite senses and this is the reason for the apparently contradictory sense of shear displacement in the ductile zone (Figure 26.36G). The differences of the normal distances of the apex points P_N and P_S from the shear surface give the dilation component. Because the pyramid closes towards the north of the zone, and because the intersection triangle on the north side is larger than that on the south side, it should be clear that there has been a gain in volume in the shear zone (Figure 26.36H). The normal distances of P_N and P_S have been

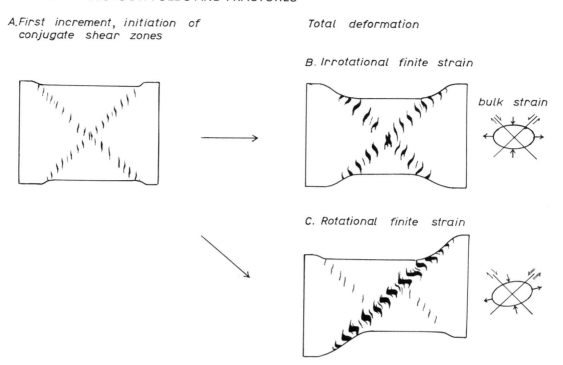

A.First increment, initiation of conjugate shear zones

Total deformation

B. Irrotational finite strain

bulk strain

C. Rotational finite strain

Figure 26.37. *Relationship of evenly and unevenly developed shear zones to the axes of bulk strain.*

calculated from the lengths and plunge angles of the edge lengths AB, AC and BC in the two intersection triangles and the difference gives an increase of distance of the two walls of 0·2 m. As this is taken up through a total width of the ductile part of the zone of about 0·7 m it implies an average volume gain of about 30%.

The figures that have been computed from the data are not absolutely accurate by any means. Because the two triangles are of pointed form it follows that any small errors in the orientations of dykes B and C are likely to lead to position errors that may be quite severe at the BC triangle apex. With a construction such as this the most accurate results will come about where the three surfaces are highly inclined to the shear zone and where the triangles of intersection show near equilateral forms. A further source of error in the example that we have investigated arises from the method used to determine the dyke orientations. Because the natural outcrop surface of Figure 26.10 is almost perfectly planar the dyke orientations were determined from the apparent dips on this surface and on two steeply inclined joint faces. Observational errors are bound to exist in all these data and will be incorporated into the primary orientation data that we used to solve the problem. Before leaving this problem we suggest that the student might profit from making a careful study of the complex three-dimensional geometry, particularly by comparing the actual geometry of the dykes in the outcrop (Figure 26.36G) with the construction of the displaced pyramid used to determine the displacement components (Figure 26.36H).

En-echelon vein arrays

Answer 26.5

When a conjugate shear zone system is initiated, it seems reasonable to assume that, if the rocks are fairly homo-

geneous and isotropic, the first formed shear zones will be symmetrically related to the first incremental principal strain directions and to the principal axes of stress at the time of initiation. If, relative to the orientation of the rock body, the stress and strain increment principal axes remain constantly oriented, it should be expected that the two orientations of shear zones would develop more or less equally, and that the overall geometry would be that of the symmetric zones of Figure 26.37B. The acute and obtuse bisectors of the dihedral angle between the shear zones will give the maximum shortening (minimum compression σ_3) and maximum extension (minimum compression σ_1) directions and the zonal intersection will give the intermediate strain and stress axis. If, however, the stress directions change their orientations relative to the rock mass, what generally seems to occur is that one of the two sets of initiated zones becomes more active than the other (Figures 26.14 and 26.37C). The bisectors of the shear zones will no longer give the bulk strain axes, but will probably lie close to the directions of maximum and minimum *initial* stress axes. In order to evaluate the bulk strain in the body it is necessary to make computations of the total shear displacement in each vein set, and to reconstruct the form of the bulk strain ellipse by reconstructing these data.

An equal area plot of the data derived from Figure 26.15 and Table 26.4 is shown in Figure 26.38. The two shear zones are indicated by the great circles P and Q. The angles between the zones are 41° and 139°. Because the shear zones are equally developed the maximum X and minimum Z bulk strains are found at the dihedral angle bisectors. The intermediate axis Y is located at the intersection of P and Q (construction method set out in Vol. 1, p. 158). The data from vein localities A, B and C come from sigmoidally folded veins in zone P and those from D and E in zone Q. The vein poles (black points in Figure 26.38) are spread along a great circle zone, the pole of which represents the

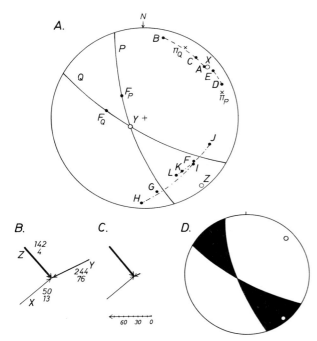

Figure 26.38. A: *Equal area plot of data from Table 26.4. Methods of plotting principal strains on a map are shown in* B, *full data;* C, *with scaled symbols; and* D, *with simplified projection.*

fold axis and coincides with the intermediate strain direction *Y*. Although the sets of en-echelon veins show S and Z forms in zones *P* and *Q* respectively, occasional linkages between adjacent veins produce geometric complications. The calcite veins which predate the shear zones are deflected in a way which indicates the ductile component of the deformation in the zones, and these deflections confirm the right- and left-handed shear senses deduced from the vein en-echelons. The early veins are folded in each zone: observations *F*, *G*, *H* and

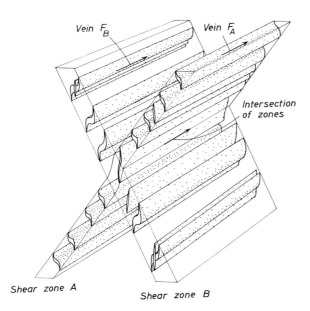

Figure 26.39. *Schematic diagram of the relationships of en-echelon vein arrays in conjugate shear zones.*

I relate to the fold in zone *Q* and *I*, *J*, *K* and *L* to the fold in zone *P*. The π-poles of these veins lie on partial great circles with poles F_P and F_Q. The fold axes have different directions in each zone controlled by the intersection of the shear zones with the pre-shear zone orientation of the early veins. These two directions of fold axes diverge downward into the outcrop surface and must have converged and intersected each other in the overlying rock now removed by erosion. The three-dimensional geometry is identical to that produced in certain types of conjugate kink folds (Figures 20.42D and E; 20.43).

A schematic diagram of the various components in sigmoidally folded en-echelon vein arrays is illustrated in Figure 26.39. Note the parallelism of a number of features: the zonal intersection line, intersection of any vein with the shear zone plane, intersections of cross cutting en-echelon veins and the fold axes of sigmoidally shaped veins.

The data which emerge from an analysis such as we have made are often fairly complex. It is very important to consider the most appropriate methods for recording the information that arises from field study in ways that enable the worker to appreciate the overall regional geometric patterns and to comprehend what may be the implications of the data. It is also vital to produce data syntheses in reports and publications which really will aid the reader, yet not in such a form as to oversimplify or lose valuable information. Each region will have its own special features which should guide the researcher in deciding on his illustration methods. We take it as obvious that data presented in numerical list form are only suited for appendices, because information presented in this way is generally incomprehensible to the general reader. We have also emphasized that, although equal area or stereographic projections can convey certain types of information very clearly, they should not be regarded as a universal panacea (see discussion in Vol. 1, p. 154).

The main information coming from an analysis of conjugate shear zones of all types relates to the orientations of the three principal bulk strain axes at each locality. The angles between the zones might also indicate important mechanical aspects of the rock deformation or constraints on kinematic problems. Figure 26.38 makes three suggestions by which these data might be incorporated on to a map or synoptic diagram. The axial direction data can be plotted in the form of three arrows with correctly directly azimuths meeting at the observation point (Figure 26.38B): the orientation data can be recorded by the side of each arrow and the axes defined by lettering or by using a line thickness convention. This method has the advantage of completeness, but the resulting map can look quite complicated. One problem with assimilating these data is that the azimuth must always be read in conjunction with the plunge values. This type of diagram can be simplified by using a scaled length for each arrow shaft to convey the different plunge angles: in Figure 26.38C it is immediately apparent that the *X*- and *Z*-directions have low angles of plunge, whereas the *Y*-direction is steeply inclined. A third method has been developed by using a simplified projection technique, the projection containing black and white sectors related to the crossing shear zones. This is a somewhat similar method to that used for the analysis of the first motions of earthquakes to evaluate the principal stress orientations (Session 25). The black sector records the shortening zone, and the white sector the stretching zone. The *Y*-direction is easily read off

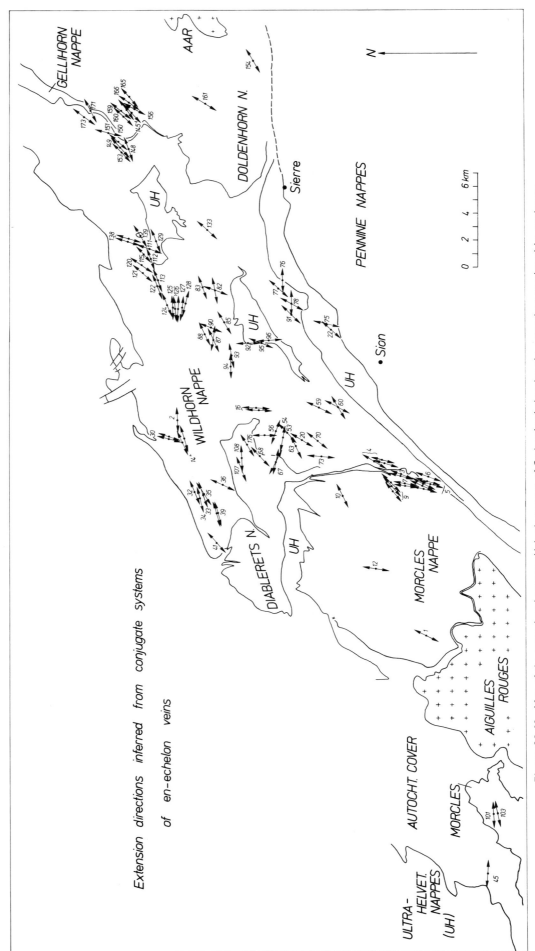

Figure 26.40. Map of the central and western Helvetic nappes of Switzerland showing the extensions produced by conjugate brittle–ductile shear zones. After Dietrich (1986).

from the zonal intersection. The maximum and minimum strains can be approximately "eyeballed-in", or indicated by symbols (Figure 26.38D). The advantage of this method over the other two is that it conveys the angular relationship between the crossing shear zones.

In order to illustrate what can be done with these methods we use the work of Dorothee Dietrich and Song Honglin in their studies of some of the conjugate vein array systems in the Helvetic nappes of central and west Switzerland, a region we have discussed in some detail in relation to finite strain and progressive strain history (Figures 11.9, 11.10, 14.21). Figure 26.40 shows the azimuths of the main stretching directions produced during the formation of the last sets of conjugate shear zones in this regions. The regional plan is very clear, indicating a stretch in a NE–SW direction sub-parallel to the fold axes of the main folding phase. The numbers refer to localities, and a more complete indication of the principal extensions can be placed in list form. Figure 26.41 shows the simplified projection technique. This map shows much more information that does Figure 26.40. Although at first sight it may look complex, further study will reveal that an enormous amount of useful information is recorded, and that it does not take too long to see the salient points.

1. The white extension sectors lie mostly, but not always in a NE–SW direction with a low plunge for X.
2. The black compression sectors generally occupy less area than the white sectors (2θ less than $90°$) but the Y- and Z-directions are generally less tightly constrained than are the X-axes.
3. The Y-axes tend to lie closer to the bedding or cleavage planes than do the Z-axes. This suggests that some of the initiation of the shear zones might be, in part, fabric controlled.
4. The shear zones cross the bedding or cleavage in such a way as to produce a complex array of fold axes the directions of which generally lie at a moderate to high angle to the NE–SW axes of the regional folds.

Angles between conjugate en-echelon shear zone vein arrays

The angles between conjugate shear zones can show quite a large variation (e.g. see the variations in axes of black sectors of the projections in Figure 26.41. In our discussion of the angular relationships of ductile shear zones we saw how the initial angular relationships might be modified as a result of ductile deformation in the inter-shear zone lozenges. In en-echelon brittle–ductile vein zones it is generally found that the strains in the regions between the zones are very small. This implies that the 2θ angles we measure in the field are likely to be close to those at initiation of the zones. Where the vein openings are strongly developed we generally find that the acute angle between the zones faces the direction of bulk shortening, but where the vein systems are accompanied by features suggestive of pressure solution effects, the 2θ angles become larger and may reach or even exceed $90°$. In Session 3 we made some suggestions to account for different orientations of the individual veins with respect to the shear zone walls using the concept of shearing acting together with dilation (Figures 3.21, 3.22, pp. 50–52). If we link these ideas with observations made

earlier in this session that the vein tips of both sets of en-echelon veins are often oriented sub-parallel to a regional extension set of veins, we arrive at a series of geometric models linking the zonal orientations with the orientations and types of regional bulk strain inducing the shear zones (Figure 26.42). This figure shows a complete spectrum of types of bulk two-dimensional deformation between uni-axial bulk stretching (A, with strong positive dilation), through various combinations of proportions of stretching and contraction (B, C and D), to uniaxial bulk shortening (E, with strong negative dilation). We think it reasonable to relate overall positive dilation in any system with the forma-tion of veins (and with part of the filling material of the veins being derived from outside the system under discussion), and to relate overall negative dilation with the development of pressure solution seams or stylolites (and with removal of material from the overall system). Putting this information together with the ideas that the actual veins (or their tips) are related to the regional extension veins or that sub-planar pressure solution features are related to regional contraction features, it follows that the characteristic angle θ between the veins and the shear zone walls relates directly to the 2θ angle facing the maximum regional contraction direction Z. The end members of this spectrum, **Models A and E**, show only veins or only stylolites respectively with no tendency for the alignment of the structures into zones.

The **positive dilational model B** shows strong stretching with some compensating shortening and leads to shear zones with well developed extensional veins making angles of less than $45°$ with the vein walls (unless sigmoidally folded by strong shearing) and with a 2θ angle of less than $90°$. Material to fill the veins needs to come, in part, from outside the region of vein formation (Figure 3.22). This model, incidentially, is an extremely common one in regions of en-echelon dyke swarm activity where the overall positive dilation is quite certain.

The **constant volume model C** shows both veins and pressure solution effects in the sets of shear zones, the two effects more or less compensating for each other so that the soluble material removed from the solution zones is directly transferred into neighbouring veins. Because the two effects are closely interlinked it is possible to find veins cutting pressure solution stripes and the stripes removing vein mate-rial; the system, in a manner of speaking, is eating up its own garbage (Figure 26.43). In zones formed in this way above the shears are fairly strong (more than about $\gamma = 1$) the pressure solution zones are sigmoidal as well as the vein systems because they are also planar markers which can undergo line rotation.

The **negative dilational model D** shows a prepon-derance of solution effects over veining. The individual shear zones show intensive pressure solution stripes or stylolites with the solution planes arranged at angles of less than $45°$ to the vein walls (Figures 26.44, 26.45). Where large shear strains have taken place these pressure solution zones are frequently sigmoidally shaped. In Figure 26.44 the zonal pressure solution stripes connect from one to the other and might relate to the channelways that are necessary in this model for the removal of rock material from the system. In these zones the angles θ between veins and zone walls exceed $45°$ and the 2θ values between conjugate zones therefore exceed $90°$.

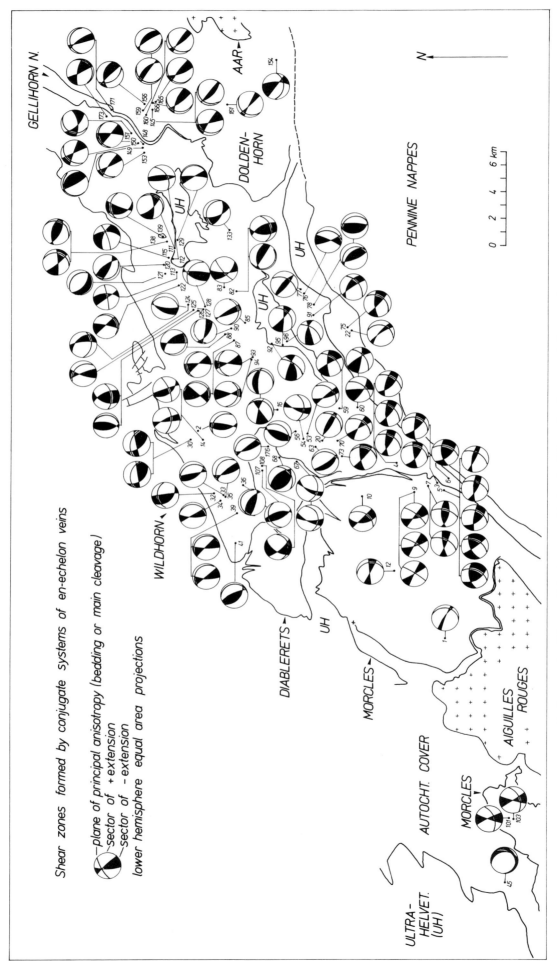

Figure 26.41 The central and Western Helvetic nappes of Switzerland illustrating the geometric features of conjugate vein systems. After Dietrich (1986).

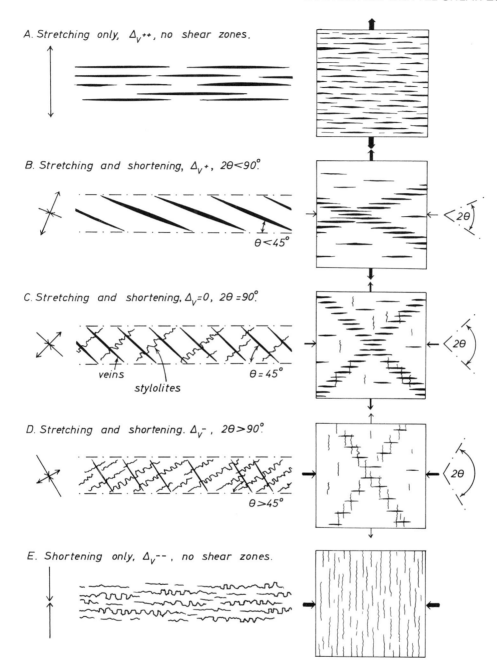

Figure 26.42. *Model spectrum for relating the geometric features of conjugate en-echelon vein arrays to the overall bulk shape changes and dilation taking place. The vein and pressure solution surfaces are located at their initiation angles. With high shear strains both take on curved sigmoidal forms.*

Superposed shear effects in brittle–ductile shear zones

We have seen that the geometric features of en-echelon vein arrays can be extremely complex. So far we have considered only those complications that have arisen from structural development related to a continuous sense of movement in the same shear zone. In many terrains, the geometric forms of the vein arrays do not appear to be explained in this way, but indicate changes of movement sense during shear zone development. It is possible for the shear vector to change

direction progressively during zone development. This effect could be appreciated by a range of en-echelon or sigmoidally folded veins showing systematic differences in the lines of intersection between veins and zone walls and in the direction of the axes of the folded veins. If deformation in the zone proceeds by simple shear (or simple shear and wall perpendicular dilation) the original orientations of these linear features are not deflected by later shears with differing vector orientations, and new veins will be generated such that their characteristic intersection and fold axial features

Figure 26.43. *Constant volume brittle–ductile shear zone with sigmoidally folded quartz veins and sigmoidally folded dark pressures solution stripes, cf. Figure 26.42C. The country rock is a rather homogeneous greywacke sandstone. Widemouth Bay, Cornwall, UK.*

Figure 26.44. *Negative dilational brittle–ductile shear zone in homogenous greywacke. The pressure solution stripes are strongly developed and have removed much of the material deposited in the veins, cf. Figure 26.42D. Millook Haven, Cornwall, UK.*

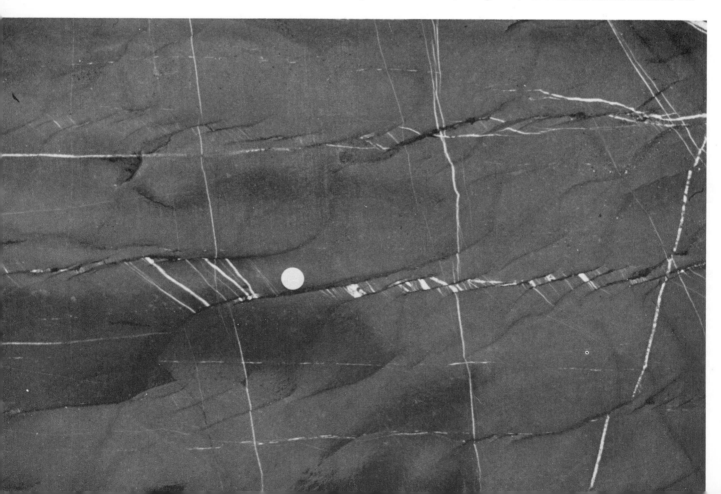

Figure 26.45. *Negative dilational brittle–ductile shear zone in Mesozoic limestone. Intense pressure solution has developed strongly marked stylolites, whereas the extensional vein systems are confined to tiny crack-seal calcite veins formed sub-perpendicular to the stylolites. This shear zone is one where the obtuse angle between conjugate sets faces the principal compression direction (cf. Figure 26.42D). Martigny, Valais, Switzerland.*

Figure 26.46. *Complex en-echelon vein systems with geometry indicative of change of shear sense in the zones. Gavarnie, West Pyrenees, France.*

will lie perpendicular to the new shear vectors. With extreme changes of movement vector in the zone it may be possible to find superposed small scale features indicating reversals of shear sense. Figure 26.46 illustrates such an example: the overall vein geometry, vein intersection relationships, vein fibre directions and two sets of crossing pressure solution stripes indicate an early right-hand shear sense was followed by displacements with left-hand sense.

Brittle shear zones: calculation of the principal stress directions

Answer 26.6

An equal area plot of the data is shown in Figure 26.47A. The pole of the fault surface was constructed (π_1) and the direction of the fibre orientation shown with a split circle symbol (L_1), this circle being split into two halves by a part of the great circle representing the fault. Because these data are real, they contain some slight measurement errors. This means that the lineation and the fault plane pole are not exactly perpendicular (they should be, because the lineation lies *on* the fault plane). To make the data plot poles truly perpendicular we assumed that both were subject to equal error, and moved them in opposite sense along the pole great circle join. This correction is not perfect, but it is all we can hope to do. One side of the split symbol of the lineation was then blacked-in, the side of the fault going *down* relative to the opposite side. The $L_1\pi_1$ plane was constructed: σ_1 was located at $\theta = 19°$ from the π_1 pole and σ_3 at 19° from the L_1 pole. It is most important to find these locations correctly using the relative motion across the fault surface; the σ_3 axis (maximum compression) always makes an angle θ with the white side of the motion vector. The principal stress axes derived from the other locations are indicated. The stress axes located in this way are not absolutely parallel, but do show a consistent grouping. In Figure 26.47B we show a projection of all the data we collected from a region of about 1 km^2 and the general sense of the stress directions is fairly consistent with a E–W maximum compression, and a steeply N–W plunging direction of minimum compression.

Determination of shear sense in shear zones

When the boundary of a ductile or brittle–ductile shear zone can be observed, the measured strain gradient or the forms of structures produced in this gradient can practically always be used to unambiguously define the shear sense. These features can be listed:

1. Change in **finite strain state**.
2. **Sigmoidal form of schistosity** induced during the deformation.
3. The orientations of **folded and boudinaged competent layers**, dykes, etc.
4. The geometry of **en-echelon extension vein arrays**.
5. **Fracture openings**, perhaps together with fibrous shear veins.

However, shear zones can be very wide, or the boundaries obscure. Under these circumstances the features listed above might be usable, but other small scale structures can also be

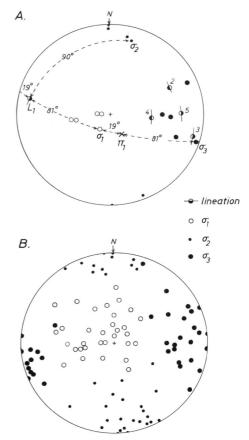

Figure 26.47. A: *Equal area data plot, Answer 26.6. The constructions to find the three principal stresses σ_1, σ_2 and σ_3 from data locality 1 are shown in dashed lines using $\theta = 19°$.* B: *Compilation of all available data.*

used to define the shear sense. Simpson and Schmid (1983) and Passchier and Simpson (1986) have discussed the types of criteria that might be established in outcrops, in hand specimens or in thin sections to evaluate the shear sense.

A. s–c band structure

Inside many shear zones the displacement and strain gradients may be heterogeneous. As a result the schistosity (s-structure) becomes uneven both in orientation and intensity. The zones of high and low strain run parallel to the shear zone and give a banded appearance to the rock (shear banding, *cisaillement* in French, hence c-band after Berthé *et al.*, 1979). The oblique relationship of the two structures gives a clear indication of the displacement sense (Figures 26.48A and 26.49).

B. Rotation of porphyroblasts or porphyroclasts

In many metamorphic rocks it is well known that large porphyroblasts can develop; albite, garnet, staurolite— being equidimensional—are particularly useful for the determination of shear sense. These crystals can roll like ball bearings and by matching the inclusion trails they contain inside the crystal (s_i) with those found outside (s_e), the rotation sense can be determined (Figure 26.48B). Such crystals can grow as porphyroblasts during the shear motion

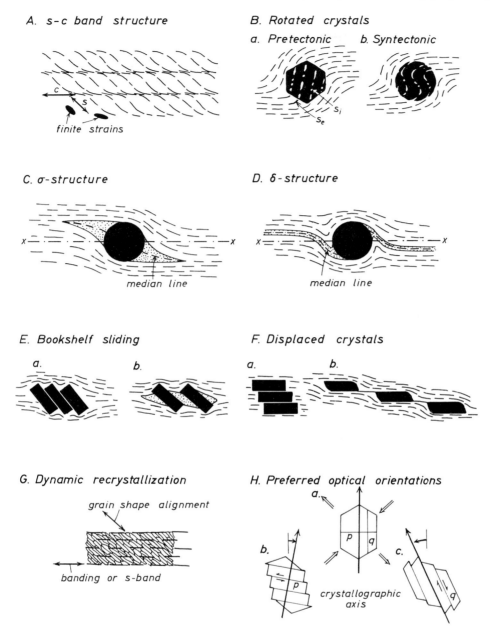

A. s-c band structure

B. Rotated crystals
  a. Pretectonic   b. Syntectonic

C. σ-structure

D. δ-structure

E. Bookshelf sliding

F. Displaced crystals

G. Dynamic recrystallization

H. Preferred optical orientations

Figure 26.48. *Different types of criteria used for the determination of shear sense in shear zones. All are with left-handed (positive) shear sense.*

when they produce a "snowball structure" (Figures 26.48B (b) and 26.50). The internal three-dimensional structures of the inclusion trails is complex (Powell and Treagus, 1967; Rosenfeld, 1970) and only a central section through the crystal normal to the rotation axis produces a simply interpreted pattern.

C. σ-structure

This structure is found around porphyroclasts, and relates to the form of the dynamically recrystallized pressure shadow tail in relation to the porphyroclast (Passchier and Simpson, 1986). In this structure the median line of the tail does not cross the trend of the average schistosity (Figures 26.48C, line *XX*, and Figure 26.51), and the structure appears to form because the rotation rate of the tail is higher

than that of the porphyroclast. This structure is characteristic of shear zones with *low shear strains* where the *recrystallization rates are higher than the rotation rates.*

D. δ-structure

This structure, like σ-structure, is developed around porphyroclasts, but where the recrystallized pressure shadow tail is strongly rotated by the porphyroclast (Figures 26.48D and 26.52). The descriptive name describes the extremely curved and often embayed nature of the generally narrow tail, and where the median line crosses the general trend of the schistosity. This type of structure is particularly characteristic of regions of *high shear strains* where the *recrystallization rates are lower than the rotation rates* (Passchier and Simpson, 1986).

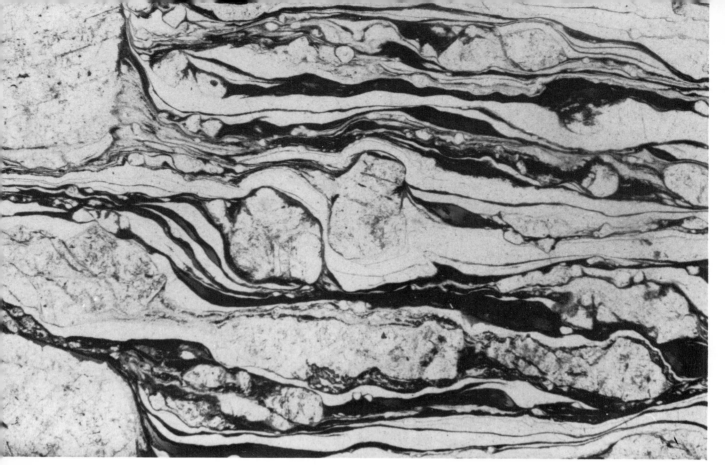

Figure 26.49. *Thin section of s–c structure in mylonite containing feldspar porphyroclasts in a quartz–mica ribbon matrix and showing left hand shear sense. The c-planes run sub-horizontally in the photograph. Santa Rosa, California. Enlargement × 20. From Passschier and Simpson (1986).*

E. Bookshelf sliding

When broken crystals are subjected to shear, the individual parts are rotated in the direction of shear, a feature which sets up a contrary shear motion between the fragments much like a collapsing set of books (Figure 26.48E). Further shearing motion can lead to separation of the individual fragments and the development of recrystallized pressure shadow zones between the fragments (Figures 26.48E(b) and 26.53).

F. Displaced crystals

Where crystals possessing a well developed crystal cleavage lie with this cleavage close to the shear plane, they often become internally detached by gliding on this surface (Figure 26.48F). In highly deformed mylonites, for example, it is not uncommon to find individual mica fragments connected by zonal films of phyllosilicates to form free "floating mica fish" (Figure 26.48F(b)).

G. Dynamic recrystallization

Rocks which have been very highly deformed by crystal plastic processes frequently show an intense banding and schistosity. Within the more deformed crystals (especially with quartz and calcite) the bands are crossed obliquely by small grains with a marked individual alignment (Figures 26.48 and 26.54). This fine grained structure appears to result from active dynamic recrystallization and readjustments of the crystal shape to conform with the last increments of deformation as recorded by the shape of the incremental strain ellipse (Schmid *et al.*, 1981).

H. Preferred optical orientations

Plastic flow of crystals leads to a variety of different types of preferred alignment patterns of the optical axes of the crystals in the rock aggregate. We do not have space here to go into the details of a subject that is quite complicated. Many of these complications of interpretation of preferred orientation patterns result from the nature of the environmental conditions during deformation, and the way this controls the different crystallographic glide systems. The principles are shown schematically in Figure 26.48H. Figure 26.48H(a) shows a crystal with optical axis aligned initially perpendicular to the shear direction. Left-handed (+) simple shear will produce a tendency for contraction in a direction at 45° to the shear plane and extension in a direction at −45° to the shear plane. The crystal can be plastically deformed in response to these axes, but what happens depends upon which glide system is active. We have shown two possibilities, (b) where glide system *p* operates, and (c) where glide system *q* operates. The original crystal boundary shape is modified to come to lie close to that of the incremental strain ellipse, either by left-hand glide on system *p*, or right-hand glide on system *q* (cf. the models F and E in Figure 26.48). the glide systems undergo rotation and, in response, the optic orientation also makes a rotation. In the case of activity by *p*-glide the optical axis moves towards the compression direction (and *against the shear sense*), whereas if *q*-glide is active the optic axis moves away from the compression direction (and *with the shear sense*). It should be clear from this model that the different types of preferred optic orientations are controlled by the nature of the predominant glide system. For further details the reader is referred to reviews by Schmid *et al.* (1981) for calcite, Schmid and Casey (1986) for quartz, and Boudez *et al.* (1983) for a general review.

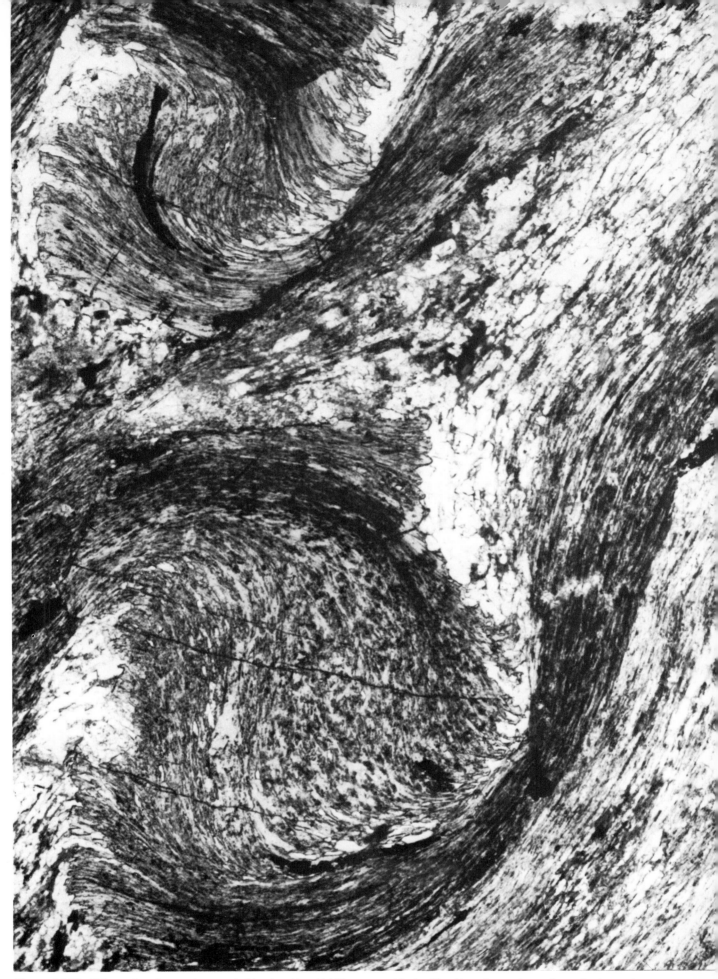

Figure 26.50. *Thin section of rotated garnets in Liassic schists, Val Piora, central Swiss Alps. Note the continuously curving internal inclusion trails (S$_i$) and the ragged porphyroblast boundaries—both features of syntectonic growth.*

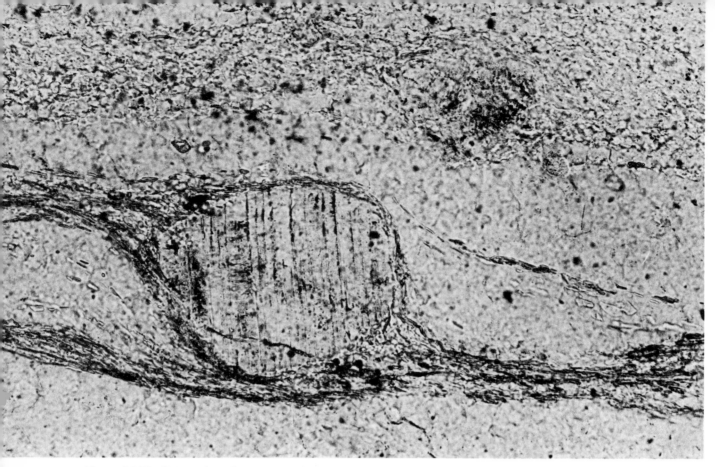

Figure 26.51. Thin section of σ-structure, showing a feldspar clast with recrystallized quartz–mica tail in mylonite, Santa Rosa, California. The shear sense is left handed. Enlargement ×200. After Passchier and Simpson (1986).

Figure 26.52. Thin section of δ-structure around a plagioclase porphyroclast in a mylonite, Saint Barthelelmy, French Pyrenees. The shear sense is left-handed. Enlargement ×200. After Passchier and Simpson (1986).

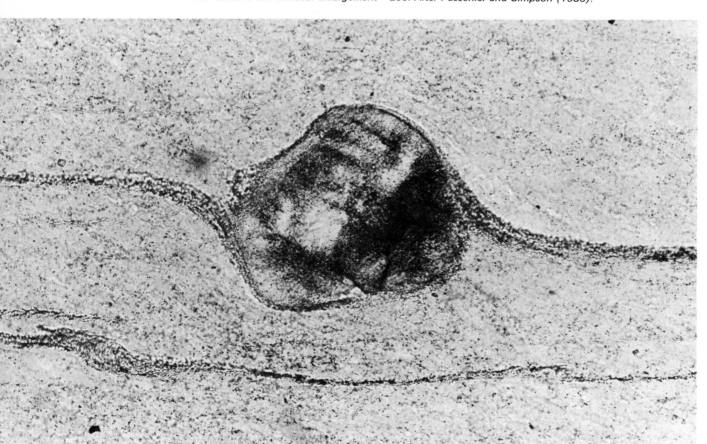

Figure 26.53. Thin section of broken and rotated plagioclase feldspar in mylonite, Lochcarron, NW Scotland. The shear sense is right-handed. Enlargement ×50.

Figure 26.54. Thin section of σ-structure in a feldspar porphyroclast together with dynamically recrystallized quartz crystals with long axes obliquely inclined to the trend of the main quartz ribbons. Santa Rosa mylonite, California. Enlargement ×200. After Passchier and Simpson (1986).

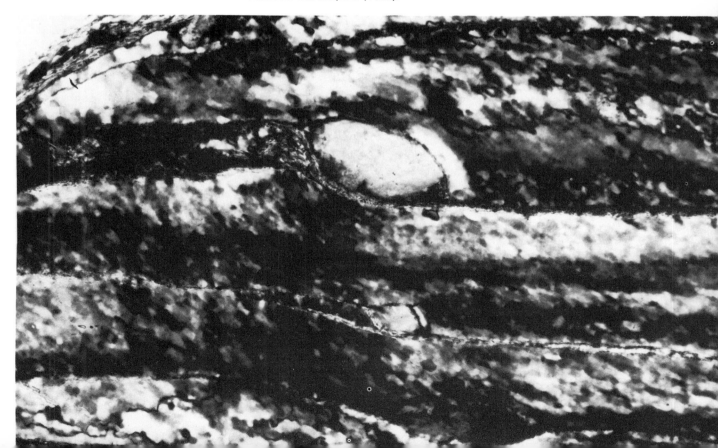

KEYWORDS AND DEFINITIONS

Internal boudinage

The geometric thinning of a marker layer where the layer is cut across by two ductile shear zones (Figure 26.28). This type of boudinage is independent of mechanical competence contrasts between the layer and its matrix.

Shear zone

A zone of high deformation which is characterized by a length to width ratio of more than 5:1. A **brittle shear zone** is marked by a fracture surface or **fault** across which the shear displacement sets up an abrupt geometric discontinuity (Figure 26.1A). In contrast, a **ductile shear zone** where the deformation across the zone is continuous, the zone being marked by a high ductile strain (Figures 26.1D, 26.9, 26.19). **Brittle–ductile shear zones** have characteristics of ductile shear, with elements of discontinuity in the form of faults or vein systems (Figures 26.1B and C, 26.11, 26.15). All types of shear zones can occur in **subparallel sets** or can be arranged in **crossing conjugate sets**. In profile, crossing conjugate shear zones have one set with **left-hand shear sense**, the other set with **right-hand shear sense**.

Sheath fold

A fold with tight or isoclinal profile and which shows variations in its hinge line of more than 90°. These folds show characteristically closed forms in section (Figure 15.5). They are formed by the superposition of folds of different phases or by the development of a very high strain state ($R_{XZ} > 10:1$) on folds with curving hinges. Ductile shear zones often show very high strains and sheath folds are commonly formed in these zones.

Strain factorization

A mathematical technique for finding separate deformation components or **factors** which describe a complex strain system in terms of the superposition (matrix multiplication) of two or more simple (mathematically speaking) strains (Figure 26.21). Strain factors may have the property of being **independently compatible**, each variable factor being able, by itself, to describe a perfectly compatible deformation (Figure 26.5). Factors are more usually **independently incompatible**; each variable factor does not describe a compatible strain (Figure 26.20). Many independently incompatible factors are **interlinked**, in that variation of one factor automatically gives rise to variations in other factors.

KEY REFERENCES

This is a vast subject with a very extensive literature. In our choice of key references we would first suggest that the student checks that the reference lists at the end of Session 2 and 3 have been covered. Our special choice for Session 26 has been based on a selection of papers which emphasise numerical methods and those which might lead the reader into new subject fields in geochemistry and istope chemistry.

Beach, A. (1976). The interrelations of fluid transport, deformation, geochemistry and heat flow in early Proterozoic shear zones in the Lewisian complex. *Phil. Trans. R. Soc. Lond.* **A280**, 569–604.

Casey, M., 1980. Mechanics of shear zones in isotropic dilatant materials. *J. Str. Geol.* **2**, 143–147.

This paper is of particular interest to show how mineralogical changes can take place locally in shear zones and how these changes might assist the development and propagation of shear zones.

In our session analysis we have concentrated on the strain features of shear zones. This paper gives a somewhat different although related view, emphasizing the control of fracture geometry by stress conditions. It shows how the mean stress condition might influence the angles between conjugate shear zones, a view which has analogies with the volumetric dilation effects discussed in the current session.

Cobbold, P. R. and Quinquis, H. (1980). Development of sheath folds in shear regions. *J. Str. Geol.* **2**, 119–126.

Theoretical and experimental methods are described to account for the strongly non-cylindrical folds often developed in shear zones, and it is shown how passive folds subjected to large shear strains can produce such structures.

Coward, M. P. (1980). Shear zones in the Precambrian crust of South Africa. *J. Str. Geol.* **2**, 19–27.

This paper describes shear zones developed on a continental scale which can be recognized as major crustal lineaments bounding orogenic zones or sub-sectors of orogenic belts.

Dietrich, D., McKenzie, J. A. and Honglin Song (1983). Origin of calcite in syntectic veins as determined from carbon–isotope ratios. *Geology* **II**, 547–551.

This paper is especially noteworthy in applying the techniques of stable isotope measurements to the calcite vein fillings of sets of en-echelon vein arrays in the Helvetic nappe region of Switzerland in an attempt to determine whether the carbonate is derived from local or distant sources. It is shown that the earliest veins derived their filling from some distant source when the fracture system was probably relatively open and interconnected, whereas the later vein fillings have isotope compositions which are identical to the local wall rock material.

Davies, F. B. (1982). Strain analysis of wrench faults and collision tectonics of the Arabian–Nubian Shield. *J. Geol.* **82**, 37–53.

This paper is an attempt to bridge the gap between an analysis of the features of individual shear zones and major plate tectonic geometry. The analytical data show that the shear zones have components other than those of simple shear.

Hageskov, B. (1985). Constrictional deformation of the Koster dyke swarm in a ductile sinestral shear zone, Koster Islands, S.W. Sweden. *Bull. Geol. Soc. Denmark* **34**, 151–197.

This describes a region in the South west part of the Baltic Shield where a very extensive swarm of dolerite dykes can be traced progressively into a mega shear zone where the original thickness and spacing is decreased, and the dykes become modified into schistose amphibolites. There are many interesting calculations of the deformations in the shear zone.

Kligfield, R., Crespi, J., Naruk, S. and Davis, G. H. (1984). Displacement and strain patterns of extensional orogens. *Tectonics* **3**, 577–605.

This paper discusses models for the development of shear zones at different levels in the crust and gives particular attention to extensional tectonics and the geometric relationships of high level domino faults and low level shear zones.

Knipe, R. J. and White, S. H. (1979). Deformation in low grade shear zones in the Old Red Sandstone, S.W. Wales, *J. Str. Geol.* **1**, 53–66.

This describes the development of en-echelon vein arrays associated with pressure solution zones over a range of scale from field geometry, microscopy fabric and mineral structure under the transmission electron microscope.

Ramsay, J. G. and Allison, I. (1979). Structural analysis of shear zones in an Alpinised Hercynain granit, Maggia Lappen, Pennine Zone, Central Alps. *Schweiz. Miner. Petrogr. Mitt.* **59**, 251–279.

Perhaps the most important feature of this paper is the large scale, very detailed coloured map of an exceptionally well exposed region in the Swiss Pennine Alps and the discussion of this field geometry. We recommend the student to look at the complexities of this real shear zone system and identify the geometric features that we have described in this current Session (conjugate zones, lozenge structure, sigmoidal schistosity, dyke deflections, strains at shear zone terminations, etc.).

Shear zone shear sense indicators

The following can be recommended as providing excellent reviews on the use of small scale and microscopic features to determine the shear sense.

Berthé, D., Choukroune, P. and Jegouzo, P. (1979). Othogneiss, mylonite and noncoaxial deformation of granites: the example of the South Armorican shear zone. *J. Str. Geol.* **1**, 31–42.

Bouchez, J. L., Lister, G. S. and Nicolas, A. (1983). Fabric asymmetry and shear sense in movement zones. *Geol. Rundschau* **72**, 401–419.

Passchier, C. W. and Simpson, C. (1986). Porphyroclast systems as kinematic indicators. *J. Str. Geol.* **8**, 831–843.

Simpson, C. and Schmid, S. 1983. An evaluation of criteria to deduce the sense of movement in sheared rocks. *Geol. Soc. Am. Bull.* **94**, 1281–1288.

SESSION 27

Joints

Different types of joints together with their possible origins are described. It is shown how joints may be classified following different schemes. This discussion leads to the analysis of two different geological situations: first an example of jointing in uniformly dipping sedimentary strata is studied, then the joint pattern in folded strata is analysed. Several techniques used to elaborate and present the orientation data are discussed. As additional information stylolitic and slickenside structures can help in the interpretation of the joint data. A brief discussion of surface features shows the importance of detailed and precise descriptions of joint characteristics apart from the record of their orientations. The session concludes with a discussion of the most important parameters which should be recorded and analysed in a well-operated joint survey.

INTRODUCTION

Joints are probably the most common structures exposed at the present day surface. Although they penetrate for some considerable distance into the uppermost crust, their frequency appears to decrease with depth. They are found in every rock type: unconsolidated and consolidated (lithified) sediments, igneous and metamorphic rocks. **Joints** are defined as fractures of geological origin along which *no appreciable displacement has occurred* (Figure 27.1). Of course, there probably has been some very slight movement along the fracture, but even where thin sections of joints are viewed with the microscope (and such thin sections are extremely difficult to prepare), no obvious displacement can be detected (Figure 27.2). Joints occur generally in parallel or sub-parallel **joint sets** (Figure 27.1), and where differently oriented sets cross the resulting geometric pattern is known as a **joint system**. The crossing sets that form a system may have developed at the same time, but contemporaneity is not a necessary requirement. Joints are sometimes further classified as **extension joints** or **shear joints**, a subdivision based on the angular relations of crossing sets. Because no movement normal or parallel to the joint walls can be observed, this classification cannot generally be checked. Some geologists are of the opinion that joints can only be generated by extensile displacement.

Although individual joint fractures may be quite short (1–10 m), in certain regions it is found that **master joints** run for very long distances. A study of air photographs often gives a very convenient method for evaluating the plans of a joint system because master joints often influence the geomorphological activity at the surface. Many of the striking **lineaments** seen on air photographs are master joints rather than major faults. Although joints are sometimes geometrically associated with sub-parallel fault planes in many terrains, particularly those of sub-horizontal cratonic sediments, the abundant joints are unassociated with faults (Figure 27.3).

The spacing distance between adjacent joints in a joint set varies from millimetres to tens of metres. This variation can often be related to rock type. Figure 27.3 illustrates the change from widely spaced joints in a massive sandstone horizon to a much closer spacing in the underlying silty and shaly sedimentary rocks. In some situations where interbedded lithologies occur one type may be jointed while in the other no joints can be found (Figure 27.4).

Although the open crack-like form of a joint exposed on an outcrop gives a false impression of the gape of the walls, being governed by weathering and near surface solution processes, there is no doubt that joints in the unweathered rock below surface do have sufficient space between their walls to allow preferential transport of fluids along the fracture. Joints are very important features controlling hydrogeological processes of groundwater flow, and many well known aquifers owe their water retaining properties to the presence of well developed joint systems. At deeper levels joints exert a control on the migration of geological fluids: water, petroleum and gas. The migration of these fluids through the fracture may lead to changes in the nature of the joint surface and in the neighbouring wall rocks. In porous sandstones water enriched in silica can migrate into the walls to form overgrowth cements on the original sand grains. Such impregnations strengthen the walls and give a rock more resistance to surface weathering processes (Figure 27.5). Joints which develop during the late stage cooling phases of igneous and metamorphic rocks can act as channels for hot, chemically active fluids, and frequently show "bleached"

Figure 27.1. *Joint sets cutting banded greenschist. West Ardara, Donegal, Eire.*

Figure 27.2. *Thin section of a joint passing through folded phyllite. Brigels, central Alps, Switzerland. Enlargement ×35.*

Figure 27.3. *Vertical joints, with spacing varying with lithology, cutting horizontal sandstones. Monument Valley, Colorado, USA.*

Figure 27.4. *Well bedded limestone and marl with joints developed only in limestone. Castellane, France.*

Figure 27.5. *Joints with walls impregnated with silica cutting cross bedded sandstones. Valley of Fire, Nevada, USA.*

Figure 27.6. *Joints with bleached walls, in granite gneiss, Lavertezzo, S Switzerland.*

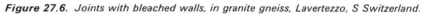

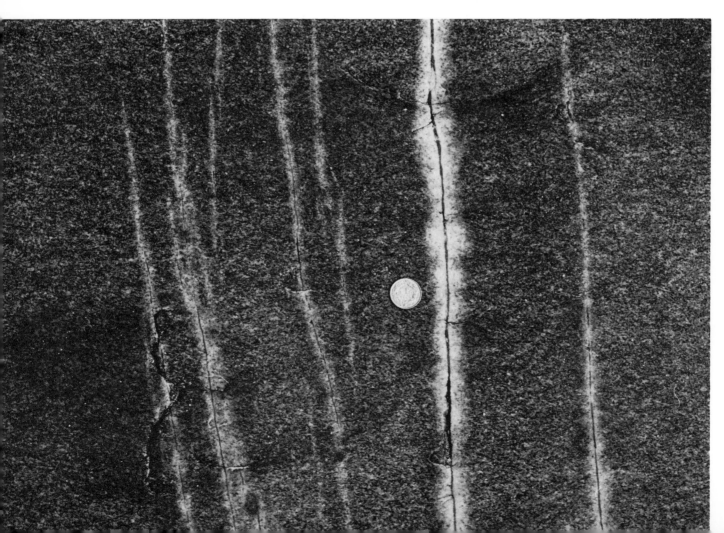

Figure 27.7. Unloading joints. Yosemite National Park, California, USA.

Figure 27.8. Arch structure in cross bedded aeolian sandstone. Zion park, Utah, USA.

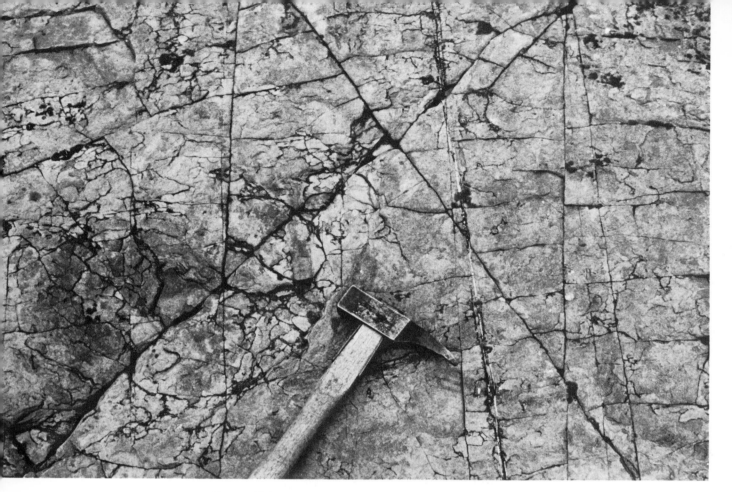

Figure 27.9. Systematic and non-systematic joints. St Tudwals, N Wales, UK.

walls. Figure 27.6 shows a rather homogeneous granite gneiss altered in this way. The biotite of the gneiss has been transformed to orthoclase feldspar in the bleached zone, and the iron and magnesium released by the mineral reaction has been accommodated by the formation of a thin chlorite vein in the centre of each fracture.

The reasons for the formation of regularly oriented joint systems are very obscure, and joints are probably the least well understood of any structures. In certain environments it appears that joints can form as a result of material contraction. The irregular polygonal **desiccation- or mud-cracks** and the often startlingly regular **columnar joints** found in lavas, dykes and ignimbrites have this origin. In some sedimentary environments mineral changes arising from diagenesis and which lead to volume changes can also form joint-like fractures. Certain types of discrete bedding plane parallel fractures in shales and marls give a **jointing fissility**, and are probably produced in this way. It is well known that joints can be produced by the removal of an overburden. The dome-like forms of the granite mountains of Yosemite National Park, California, are to a large extent controlled by erosion along **unloading joints** formed by stress relief when an ice overburden was removed by melting (Figure 27.7). Joints can also form by the stress changes taking place during rock fall activity. Many of the well known arch structures found in the sedimentary terrains of Arizona, Colorado and Utah were guided by joint fractures which, in part, arose in this way (Figure 27.8). Superficial, "skin deep" surface **exfoliation joints** can be developed by mineral changes inducing volumetric changes during weathering activity, and perhaps also by cyclic stress variations caused by alternate heating and cooling in regions subjected to large diurnal temperature changes. However, the joints which form large scale regionally consistent patterns are those which mostly interest structural geologists. The regularity of these joint systems implies that they must have formed under stress states which showed some consistency both in orientations and magnitudes of the principal stresses. Although it has been suggested that these regular patterns might be connected with worldwide rock fatigue resulting from tidal torsion of the upper crust, most geologists are of the opinion that these joints are related to regional tectonic stresses. The strains induced by joint formation are very small, probably much less than one per cent. It is therefore most unlikely that joints were formed during the periods of high activity of tectonic deformation. The fractures that do form during the major tectonic activity are those which go hand in hand with large shape changes in the crust, namely faulting and vein formation. The tectonic strains revealed by joint systems must either be those induced during the last phases of orogenic activity or by the release of stored orogenic elastic strains during periods of late uplift, perhaps millions of years after the main tectonism was closed. Regular joint systems are also formed in cratonic regions where no orogenic activity has occurred. In these cratons we have the possibility that the fracture patterns might have been induced from underlying fractured basement, or that they relate to regional stresses set up by regional warpings, gentle doming or basin subsidence.

Geometric classification of joints

Joints which form regular, planar, sub-parallel sets are known as **systematic**. In contrast those which are curved, conchoidal and non-parallel are **non-systematic** joints. It is not uncommon to find both types developed together with non-systematic joints forming irregular links between systematic joints (Figure 27.9). There are no universally

Figure 27.36. Complex en-echelon cracks, small scale plumose markings and conchoidal step fractures in the fringe zones of a major joint in sandstone. Giri valley, Simla, N India.

in the analysis of regional joint patterns. More important than a very complete record of the orientation of every single fracture in an outcrop is the need to collect different aspects of joint sets in many small areas which are connected together in a network of data stations. One should always bear in mind that hardly any outcrop will show all different joint sets.

Clearly every joint survey depends on the quality of outcrop, but if possible the following parameters should be studied and recorded:

> orientation
> spacing/frequency, length
> persistency
> "appearance"; relative planeness and relative roughness
> aperture, grade and material of filling
> interrelationships of different joint sets

Too often data collection in a study of regional jointing only concentrates on the seemingly most important parameter, the orientation of the joint planes. We have concentrated very much on the orientation aspect in this session, but we are fully aware that this is but a first step in a valid approach to the interpretation of a given joint pattern. As a complete investigation of one outcrop is very time consuming we generally do not exceed a total number of measurements of 100–200 in each outcrop, so that we have time for a detailed description of special features of the joint sets. Of course the number of measurements will be controlled by the quality of the outcrop.

Quantitatively the distribution of joints in a rock mass is described by **spacing** (or by **frequency**) and **length**. These parameters are not always measurable and in some instances they may be matter for estimation. The spacing in

three dimensions controls the size of individual blocks. A volumetric value is therefore the most meaningful term for description, but rarely are outcrops found that allow measurements in three dimensions. In most cases the spacing d (expressed in metres) or its reciprocal value $f = 1\,\mathrm{m}/d$ (frequency defined for 1 m) serves as term for spacing. The length of a joint set is recorded by taking the minimum and the maximum values (along strike *and* along dip) observed in the outcrop.

Persistency is difficult to describe if outcrop size is small compared to the length of the best developed joints, so that the real value often has to be guessed. The distinction may be made between persistent joints that extend outside the exposure, sub-persistent joints which visibly terminate in the exposed rock mass and non-persistent joints which terminate against other discontinuities in the outcrop.

The general appearance of a joint will be expressed by the terms **planeness** and **roughness**. Both terms depend on the scale of observation. Clearly we could not confound planeness of a joint set with its persistency. A set which is very persistent in outcrop can at the same time be curved. A surface view of an area, e.g. on air photographs, with seemingly absolutely linear traces of steep or sub-vertical joints, does not prove the planeness of the joints as the observation is only two-dimensional. Adjectives such as mirror-smooth, smooth, stepped, rough, striated, etc. describe the relative roughness of joint planes. No immediate genetic meaning should be given to these terms as has sometimes been done in the past ("shear joints are smooth, extensional joints are rough").

For the practical work with joints the **aperture**, the **grade** and the **material of filling** are very important parameters. On one hand these parameters influence the

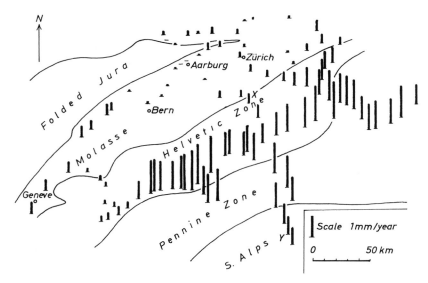

Figure 27.37. *Map of Switzerland showing measured uplift rates. After Gubler et al. (1981). Aarburg is the datum base to which the uplifts relate.*

fluid transport through the rock, on the other hand fillings may allow us to fix the age for the joint sets. The joint aperture will normally be described by adjectives such as closed, partly closed, open or wide. If widely open joints are present one should check if this gaping is the result of secondary gliding or near-surface rock creep. The description of the material which fills or coats the joints is important. This filling material may be unconsolidated sand or silt, soft clay or hard crystalline cement (e.g. calcite or quartz). The mineralogy of this crystalline material should be studied as it may assist the regional correlation of joint sets and perhaps determine the relative ages of the sets. Clay filling may prevent fluid transport even in gaping joints. The interpretation of the significance of joint filling may pose problems because weathering can very rapidly change the characteristic features of the filling material. During a joint survey the *nature and extent of the weathering* of the rocks should always be recorded.

An important part of any well conducted joint survey is a report on the **interrelationships of the different joint sets** and their relative ages. The evidence for the relative age comes mostly from the observation that younger joints come to an end (or abut) on older joints or are deflected where they are followed towards an older joint. This is often especially clear in simple orthogonal patterns, whereas in joint systems with, say, eight or more sets, it may be impossible to place them all in chronological order. Joints which are orthogonal are often not exactly contemporaneous. One model to explain the sequential development proposes systematic changes in the principal stress directions (Bock, 1980). One joint set first opens normal to the maximum tensile stress σ_1. With progressive formation of the joints of this set the value of the intermediate stress σ_2 becomes modified to become equal to, and eventually change over with, the maximum tensile stress. At this stage less well developed younger cross joints are formed perpendicular to the initial set of master joints (i.e. normal to the new σ_1 direction).

The presence of systematic oblique, diagonal conjugate joint sets generally seems to imply a common dynamic origin, although the sets may not be strictly contemporaneous. In such conjugate sets one of the two directions of joint planes often seems to be mechanically preferred, either as the result of some pre-existing fabric anisotropy or pre-existing discontinuities.

The origin of tectonic joints

The origin of tectonic joints is enigmatic. We have emphasized the low strain values inferred from joint development, a feature which precludes their formation during the development of the large strains arising during the main tectonic activity of a region. Although fractures are formed during the main deformation phases, these are opened during the activity to form extension vein systems or are subjected to strong shear displacement to form faults. The systematic orientation of joint sets within a joint system clearly implies the presence of a systematically oriented stress system. We are of the opinion that the best explanation for tectonic jointing is that proposed by Price (1966), that joints mostly arise from late uplift of a region to release the stored elastic strains contained in the essentially solid rock material. Because these elastic strains were imprinted during the last operative stress system the joint geometry will be linked in some way with the geometry of the previously formed structures. It is often apparent that uplift has occurred in regions of jointed rocks. In the plateau Jura and folded Jura discussed in this session, marine sediments of Upper Tertiary age are now elevated to levels several hundreds of metres above sea level datum, while throughout Switzerland accurate topographic leveling over many years shows that the Alpine region is currently being uplifted in terms of milimetres per year (Figures 27.37, 27.38). It seems reasonable to conclude that this uplift is the driving motion which enables the stored elastic strains to be released. Price has described the results of laboratory experiments that suggest that late tectonic strains may be stored elastically for tens or even hundreds of millions of years and therefore that joints may form a very long time after the imprinting of the stress state to which they relate.

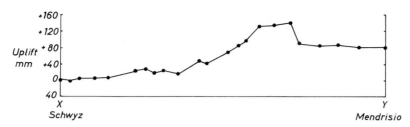

Figure 27.38. *Mean uplift curve for the section XY in Figure 27.37 for the central Alps over the period 1869–1972. After Kobold (1977).*

CONCLUDING REMARKS

Joint analysis is very time consuming, especially if one has decided to make an objective survey, where many fractures in a chosen area are measured and described. Usually the field geologist will carry out a subjective (biased) survey and describe only those discontinuities which seem to be important for the solution of his specific problem. In doing so he should resist the temptation to isolate the study of jointing from the rest of available geological data; i.e. every sound joint survey goes together with mapping and detailed description of the rock types, the major and minor structures, and so on. It is not enough to make an accurate data collection; it is of equal importance to choose a simple, clear way of representation that makes a complete record of the data. We have seen that many questions about joints and their origin are still unanswered today or the present answers are ambiguous. Therefore we should insist (and this not only in a joint survey!) on making sure that the information we collected in many tiresome hours will not be the object of only our interpretation, but that the data may easily be checked, reinterpreted and corrected by other geologists.

KEYWORDS AND DEFINITIONS

Joint A joint is a fracture of geological origin along which no appreciable displacement has occurred. A group of parallel or sub-parallel joints is called a **joint set** and where differently oriented sets cross they form a **joint system**. **Master joints** extend over very long distances. Regularly spaced, planar joints in sub-parallel sets are known as **systematic joints**; rather randomly oriented, non-planar and non-parallel joints are **non-systematic joints**.

Lineament Narrow linear fracture zones of regional scale which give rise to marked topographic features easily recognisable on air photographs and satellite images.

Stylolite (See also Session 3.) From the geometrical relations between individual stylolite columns and the fracture plane several types are distinguished: **normal stylolites** show interlocking columns which are oriented more or less normal to the surface. **Oblique stylolites** and **slickolites** have "teeth" oriented obliquely to the surface (slickolites having small angles). **Slickolite striae grooves** run parallel to the joint face. The term slickolite is a combination of the terms slickenside and stylolite and points to an origin which involves both shearing and solution (cf. Bretz, 1940, 1950) (Figure 27.27).

Slickenside A smoothed and polished surface that results from shearing.

KEY REFERENCES

Discussions of jointing are found in many textbooks and in numerous papers. The following choice is therefore somewhat subjective and we suggest that the reader consults the reference lists in the more recent papers and completes this reference list from them.

General

Hancock, P. L. (1985). Brittle microtectonics: principles and practice. *J. Str. Geol.* **7**, 437–457.

Meier, D. and Kronberg, P. (1987) (in German) "Klüftung in Sedimentgesteinen—Erscheinungsformen, Datenaufnahme, Datenbearbeitung, Interpretation." Enke Verlag, Stuttgart.

Price, N. J. (1966). "Fault and Joint Development in brittle and Semi-brittle Rock." 176 pp. Pergamon Press, London.

Secor, D. T. (1965). Role of fluid pressure in jointing. *Am. J. Sci.* **263**, 633–646.

Secor, D. T. (1965). Mechanics of natural extension fracturing at depths in the Earth's crust. *Geol. Surv. Pap. Can.* **68-52**, 3–48.

Examples of regional joint surveys

Babcock, E. A. (1973). Regional jointing in southern Alberta. *Can. J. Earth Sci.* **10**, 1769–1781.

Engelder, T. and Geiser, P. (1980). On the use of regional joint sets as trajectories of paleostress fields during the development of the Appalachian Plateau, New York, *J. Geophys. Res.* **85**, 6319–6341.

Maier, G. and Mäkel, G. H. (1982). The geometry of the joint pattern and its relation with fold structures in the Aywaille area (Ardennes, Belgium). *Geol. Rundschau* **71**, 603–616.

Nickelsen, R. P. and Hough, V. D. (1967). Jointing in the Appalachian Plateau of Pennsylvania. *Geol. Soc. Am. Bull.* **78**, 609–630.

Surface structures

Bankwitz, P. (1965 ff.) (in German) Ueber Klüfte. 1. Beobachtungen im Thüringischen Schiefergebirge. *Geologie* **14**, 241–253. 2. Die Bildung der Kluftfläche und eine Systematik ihrer Strukturen. (1966). *Geologie* **15**, 896–941. 3. Die Entstehung von Säulen Klüften. (1978). *Z. geol. Wiss.* **6**, 285–299. 4. Aspekte einer bruchphysikalischen Interpretation geologischer Rupturen. (1978). *Z. geol. Wiss.* **6**, 301–311.

Bankwitz, P. and Bankwitz, E. (1984). (in German) Die Symmetrie von Kluftoberflächen und ihre Nutzung für eine Paläospannungsanalyse. *Z. geol. Wiss.* **12**, 305–334.

Hodgson, R. A. (1961). Classification of structures on joint surfaces. *Am. J. Sci.* **259**, 493–502.

APPENDIX E
Force and Stress

The development of an understanding of the mechanical processes taking place in the Earth and of the development of tectonic structures necessitates a grasp of the concepts of force and stress. In the past some geological literature has tended to be unclear in the distinction between three fundamentally different but mechanically interrelated parameters: force, stress and strain. In this appendix we wish to develop the concepts of force and stress and, in particular, to develop the basic mathematical structure of these entities.

Force

A force is defined as that property which changes the state of rest of any material body or, if that body is in motion, which changes the speed or direction of linear motion. A good example is **gravitational force**: when we remove our hold on a stationary object held in our hands, the object will fall at an increasing speed as a result of gravitational force until it is brought to rest when it hits the ground surface. To describe this force we must define its point of application (e.g. centre of gravity of our hand-held object), its direction (vertical) and its magnitude. To define a unit of force we can choose a mass of 1 gram and a unit of acceleration of 1 cm per second; such a unit is known in the cm gram second (cgs) system as the **dyne**. For example, under gravitational force a body accelerates at 981 cm s^{-2} and, if the body has a weight of 1 gram the force exerted is 981 dynes. Force has units of mass multiplied by acceleration (dimensions MLT^{-2} where M is mass, L, is length and T is time). In many practical problems the dyne is rather small as a practical measuring unit, and in standard international (SI) units we measure force in **newtons**. A **newton** (N) is the force of a mass of 1 kilogram accelerating at a rate of 1 metre per second per second. Dynes divided by 10^5 give the force value in newtons.

A force can be represented as a vector (Figure E1A), and the effects of several forces acting together can be resolved by simple vector addition into a total resolved single force (Figure E.1B). Forces can be separated into components: for example, a force F acting at a point P on a surface at an angle of α degrees to the surface may be separated into components (Figure E.1A) acting perpendicular to the surface (normal force $F_n = F \sin \alpha$) and a force acting parallel to the surface (tractive or sliding force tending to slide objects parallel to the surface $F_s = F \cos \alpha$).

In geological processes the most important forces controlling the development of tectonic structures are first gravitational forces and, second, forces arising by thermal variations in the Earth and which produce thermal convection and mass movement in the ductile parts of the Earth.

Stress

The mechanical effect of a force depends upon the size of the area over which it acts. In everyday observation you will know that the effect of someone treading on your toe when they are wearing stiletto heels is a far more painful experience than when the same person is wearing flat heels. Although the force is the same the area over which it is concentrated is much less. The property **stress** is defined as the amount of force acting per unit area. Because the forces acting on a surface can be separated into normal (F_n) and sliding (F_s) components the stresses can also be separated into **normal stress** components acting perpendicular to the surface (σ) and **shear stress** component acting parallel to the surface (τ). If a force F acts on a surface of area A at an angle of α to the force vector then the normal shear stresses are

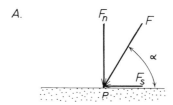

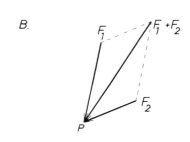

$$\sigma = F_n/A = F \sin \alpha/A \tag{E.1}$$

Figure E.1. A: *Force vector F resolved into a normal force F_n and a sliding force F_s. B: Resolution of two forces F_1 and F_2 using the parallelogram of force construction.*

667

$$\tau \;=\; F_s/A \;=\; F \cos \alpha /A \qquad \text{(E.2)}$$

The physical dimensions of stress are those of force divided by area ($ML^{-1}T^{-2}$ where M is mass, L length and T time). In the cgs system stresses are measured in units of dynes per square centimetere. In SI units, stress is measured in **pascals** (Pa), and a pascal in the stress exerted by 1 newton acting over an area of 1 square metre. In geological measurements we commonly use units of a **bar** or **kilobar** (1000 bars). The interrelationships are that one bar is approximately one **atmosphere**, and this is approximately the stress arising from the gravitational effect on 1 kilogram acting per square centrimetre.

$$1 \text{ bar} \;=\; 1 \times 10^6 \text{ dyne cm}^{-2} \;=\; 1 \times 10^5 \text{ Pa}$$

The sign conventions for normal stresses which are used here conform to that usual in engineering mechanics: tensile normal stresses are positive and compressional stresses are negative. A reverse notation is sometimes found in geological discussions because stress situations in the Earth are usually compressive (and negative in the engineering notation). We adopt the engineering notation because we are of the opinion that it is more logical to connect positive tensile stresses with positive strain elongations and negative compressional stresses with negative contractional elongations. The notation for shear strains is that positive shear strains are left-handed and negative are right-handed.

In stress analysis it is frequently necessary to describe the stresses in terms of some coordinate system. In two-dimensional situations with Cartesian coordinates x and y the normal stresses acting parallel to these directions are described as σ_x and σ_y respectively. Any shear stress acting parallel to x and with normal parallel to y is described as τ_{yx}, that acting parallel to the y-axis and with normal parallel to x as τ_{xy} (Figure E.2)

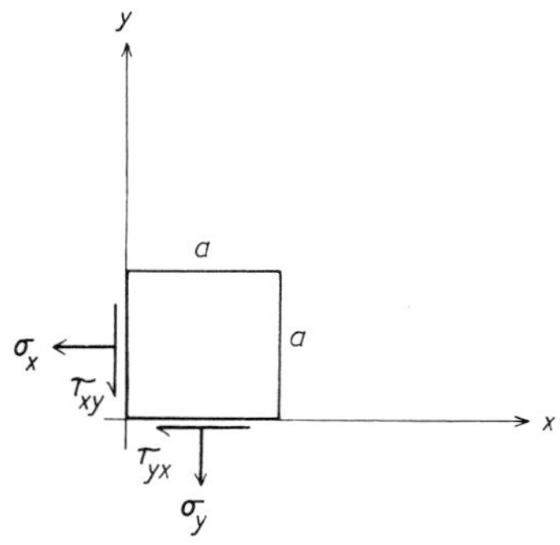

In any small cubic element with side lengths a in directions x and y and with no stresses acting perpendicular to the xy plane in direction z (a situation known as **plane stress**) *four stress components* act on the sides of this cube. For this analysis to hold the element must be small enough for the stress distribution on opposite faces of the cube to be the same (**homogeneous stress**) or, mathematically, for the cubic element to represent a point in a heterogeneously stressed body. If such an element is in static equilibrium and undergoing no accelerating spin about its centre point it follows that the moments which tend to cause rotation (shear stress × area of action × length of rotation arm) must be zero. This implies that the numerical values of τ_{xy} and τ_{yx} are equal. In equilibrium plane stress, therefore, only three of the four stress components are independent, σ_x, σ_y, $\tau_{xy} = \tau_{yx}$.

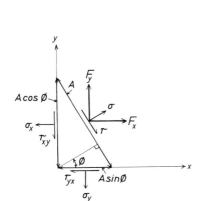

Figure E.2. *Stress components acting on the sides of a square element with sides of length a.*

Figure E.3. *Calculation of the normal stress σ and shear stress τ on a surface whose normal makes an angle φ with the x-direction.*

Analysis of plane stress

Taking the geometric relationships of Figure E.3 it is possible to determine the state of stress on any surface of area A whose **normal** makes an angle of ϕ with the x-direction. On this surface we have forces acting in the x- and y-directions. This surface cuts the x and y coordinate axes with length of $A \sin \phi$ and $A \cos \phi$ respectively. Remembering that force is calculated by stress multiplied by area we can find the force components acting in directions x and y on the oblique surface as a result of the stress components acting on the two faces parallel to x and to y.

$$F_x \;=\; \sigma_x A \cos \phi + \tau_{yx} A \sin \phi \qquad \text{(E.3)}$$

$$F_y \;=\; \tau_{xy} A \cos \phi + \sigma_y A \sin \phi \qquad \text{(E.4)}$$

These two forces may also be written in terms of normal force F_n and sliding force F_s with respect to the oblique surface

$$F_n \;=\; \sigma A \;=\; F_x \cos \phi + F_y \sin \phi \qquad \text{(E.5)}$$

$$F_s \;=\; \tau A \;=\; F_x \sin \phi - F_y \cos \phi \qquad \text{(E.6)}$$

The minus sign in Equation E.6 arises because the tractions along the surface caused by F_x is in the opposite sense to that caused by F_y. Combining Equations E.3–E.6 we obtain

$$\sigma \;=\; \sigma_x \cos^2 \phi + \sigma_y \sin^2 \phi + 2\tau_{xy} \sin \phi \cos \phi \qquad \text{(E.7A)}$$

or

$$\sigma = \frac{\sigma_x + \sigma_y}{2} + \frac{(\sigma_x - \sigma_y)}{2} \cos 2\phi + \tau_{xy} \sin 2\phi \qquad \text{(E.7B)}$$

and

$$\tau = (\sigma_x - \sigma_y) \sin \phi \cos \phi - \tau_{xy}(\cos^2 \phi - \sin^2 \phi) \qquad \text{(E.8A)}$$

or

$$\tau = \frac{(\sigma_x - \sigma_y)}{2} \sin 2\phi - \tau_{xy} \cos 2\phi \qquad \text{(E.8B)}$$

In general the normal stresses will vary across the surface depending upon the values taken by ϕ and will, for certain values of ϕ, attain maximum and minimum values. To discover these maximum and minimum values we differentiate Equation E.7B with respect to ϕ and equate to zero

$$\frac{d\sigma}{d\phi} = 0 = -(\sigma_x - \sigma_y) \sin 2\phi + 2\tau_{xy} \cos 2\phi$$

which gives directions

$$\tan 2\phi = \frac{2\tau_{xy}}{\sigma_x - \sigma_y} \qquad \text{(E.9)}$$

In a 360° range this gives two solutions (ϕ_1 and ϕ_2) with $2\phi_1 - 2\phi_2 = 180°$, implying that the two directions are perpendicular. These two directions are those of maximum and minimum tensile stress and are known as the **principal axes of stress** and the values of normal stress are the **principal stresses** σ_1 and σ_2. Along these two directions *no shear stresses exist* (the solution of Equation E.8B with $\tau = 0$ gives the same result as Equation E.9). Because of this absence of shear stress it follows that F_x and F_y may be resolved in terms of the principal stress σ_1 (or σ_2) and directions ϕ_1 (or ϕ_2).

$$F_x = A\sigma_1 \cos \phi_1 \qquad \text{(E.10)}$$

$$F_y = A\sigma_1 \sin \phi_1 \qquad \text{(E.11)}$$

combining these with Equations E.3 and E.4

$$\sigma_1 \cos \phi_1 = \sigma_x \cos \phi_1 + \tau_{xy} \sin \phi_1 \quad \text{or} \quad \tan \phi_1 = (\sigma_1 - \sigma_x)/\tau_{xy} \qquad \text{(E.12)}$$

$$\sigma_1 \sin \phi_1 = \tau_{xy} \cos \phi_1 + \sigma_y \sin \phi_1 \quad \text{or} \quad \tan \phi_1 = \tau_{xy}/(\sigma_1 - \sigma_y) \qquad \text{(E.13)}$$

from these equations

$$\sigma_1^2 - \sigma_1(\sigma_x + \sigma_y) + \sigma_x\sigma_y - \tau_{xy}^2 = 0$$

The two roots of this quadratic equation give the values of the principal stresses in terms of the stress components.

$$\sigma_1 \quad \text{or} \quad \sigma_2 = \tfrac{1}{2}[\sigma_x + \sigma_y \pm ((\sigma_x - \sigma_y)^2 + 4\tau_{xy}^2)]^{1/2} \qquad \text{(E.14)}$$

Stress in three dimensions

The discussion of plane stress above may be extended into three-dimensional general stress states (non-plane stress), and it can be shown (Ramsay, 1967, pp. 31–34) that, in general, a homogeneous stress can always be analysed in terms of three mutually perpendicular **principal axes of stress** ($\sigma_1 \geqslant \sigma_2 \geqslant \sigma_3$) which represent the greatest, the intermediate and the least tensile stress respectively.

In three dimensions we can define an average or **mean stress** (sometimes termed **hydrostatic stress**) as

$$\bar{\sigma} = (\sigma_1 + \sigma_2 + \sigma_3)/3 \qquad \text{(E.15)}$$

Note that the mean stress does not generally take the same value as that of the intermediate principal stress σ_2. The normal stresses in different directions generally differ from the mean stress by an amount known as the **deviatoric stress** σ':

$$\sigma' = \sigma - (\sigma_1 + \sigma_2 + \sigma_3)/3 \qquad \text{(E.16)}$$

The principal deviatoric stresses σ'_1, σ'_2 and σ'_3 show how the principal stresses vary from the mean stress value:

$$\sigma'_1 = \sigma_1 - (\sigma_1 + \sigma_2 + \sigma_3)/3 = (2\sigma_1 - \sigma_2 - \sigma_3)/3 \qquad \text{(E.17)}$$

In a general state of stress the principal stresses may be all tensile ($+$), all compressive ($-$), or any variation between these extremes. The principal deviatoric maximum stress σ'_1, however, is *always* positive, and the principal deviatoric minimum stress σ'_3 is *always* negative. The principal deviatoric intermediate stress σ'_2 can be either positive, zero or negative.

The relationships between stress states and the increment of deformation it causes depend upon the rheological properties of the material. However, in principal the mean stress $\bar{\sigma}$ relates to the volumetric dilatation in a material whereas the directions and values of the deviatoric stresses relate to the differential deformation taking place (as distinct from the overall volumetric changes). In the Earth's crust, where the principal stresses are practically always compressive, positive extensions can occur, even though the absolute stress is negative in directions where the deviatoric stress is positive.

Stresses acting on a surface inclined to the principal axes of stress in plane stress

For a detailed analysis of stress variations in different directions within a homogeneously stressed element it is often useful to be able to refer to the values of normal stress σ and shear stress τ with respect to the values of the principal stresses σ_1 and σ_2 and the angle the surface makes with the directions of principal stress.

To make this analysis we choose our coordinate axes x and y to coincide with the principal axes of stress. Consider now a plane of area A with its normal making an angle of ϕ to the x-axis (Figure E.4). The projections of the plane on the x- and y-axes are of areas $A \sin \phi$ and $A \cos \phi$, respectively. The forces acting on the inclined plane in the x and y directions are

$$F_x = \sigma_1 A \cos \phi$$

$$F_y = \sigma_2 A \sin \phi$$

Resolving these forces to determine the normal force acting across the surface we obtain

$$A\sigma = \sigma_1 A \cos \phi \cos \phi + \sigma_2 A \sin \phi \sin \phi$$

or

$$\sigma = \sigma_1 \cos^2\phi + \sin^2\phi \qquad \text{(E.18A)}$$

This can also be put into double angle form using $\cos^2\phi = (1 + \cos 2\phi)/2$ and $\sin^2\phi = (1 - \cos^2\phi)/2$.

$$\sigma = \frac{\sigma_1 + \sigma_2}{2} + \frac{(\sigma_1 - \sigma_2)}{2}\cos 2\phi \qquad \text{(E.18B)}$$

The shear force acting along the inclined surface can also be expressed in terms of forces F_x and F_y

$$A\tau = \sigma_1 A \cos \phi \sin \phi - \sigma_2 A \sin \phi \cos \phi$$

or

$$\tau = (\sigma_1 - \sigma_2) \sin \phi \cos \phi \qquad \text{(E.19A)}$$

using the relationship $\sin \phi \cos \phi = (\sin 2\phi)/2$

$$\tau = \frac{(\sigma_1 - \sigma_2)}{2}\sin 2\phi \qquad \text{(E.19B)}$$

You will note mathematical structure correspondence between these equations and those we evolved for finite strain (cf. Equations E.18A–E.19B with Equations D.7, D.8, D.16 and D.17 respectively). The mathematical structure is very similar because stress and strain are both second-order tensor quantities, but on no account confuse the two entities. It should be clear that the two tensor quantities have quite different physical characteristics; stress has dimensions $ML^{-1}T^{-2}$,

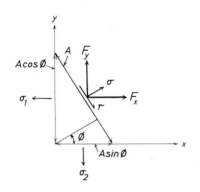

Figure E.4. *Calculation of the normal stress σ and shear stress τ on a surface whose normal makes an angle ϕ with the prinicipal stress direction σ_1.*

whereas strain is of scalar dimensions. Because Equations E.18B and E.19B both have the parametric form of a circle it follows that variations in normal and shear stress with ϕ can be represented in a **Mohr circle construction** (Figure E.5). The Mohr circle for stress is plotted slightly differently from that of the Mohr circle for finite strain discussed in Session 6 (Vol. 1, pp. 93–95). Because of the positive sign in the expression for $\cos 2\phi$ in Equation E.18B the angle 2ϕ is always measured clockwise from the positive direction of the x-axis, and because σ_1 is always greater in value than σ_2, the point representing σ_1, always lies to the right of σ_2 (cf. the Mohr circle for strain were $\lambda'_2 > \lambda'_1$). Another difference between the Mohr circle constructions for stress and strain is that stresses can take negative values when they are compressive, whereas the reciprocal quadratic extension λ' is always positive.

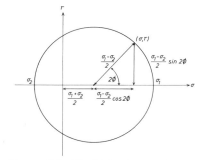

Figure E.5. *The Mohr diagram for representing stress states in two dimensions.*

The variations of normal and shear stresses with change in angle ϕ are shown graphically in Figure E.6. In plane stress ($\sigma_2 = 0$) the **mean stress** $\bar{\sigma}$ is defined from Equation E.16 as $(\sigma_1 + \sigma_2)/3$ and it is not equivalent to the **average normal stress** in the xy plane $(\sigma_1 + \sigma_2)/2$. The **maximum and minimum values of shear stress** are always found acting along two planes equally inclined to the principal axes of stress ($\phi = \pm 45°$), and the shear stress values are related to the differences in values of the principal stresses $[\pm(\sigma_1 - \sigma_2)/2]$. It is important to note that high numerical values of shear stresses are related to the stress differences, and not to the absolute values of the principal stresses. In three-dimensional stress states it can be shown that the numerically maximum values of shear stresses always occur on two conjugate surfaces intersecting along the intermediate stress axis σ_2, and making angles of $\pm 45°$ with the principal axes σ_1 and σ_3 (Ramsay 1967, pp. 36–38, and Figure E.7)

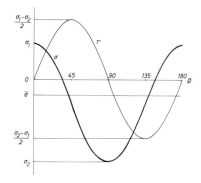

Figure E.6. *Graphical representation of variations of normal stress σ and shear stress τ on a surface which makes an angle ϕ with the principal stress direction σ_1.*

Stress variations in two dimensions

In the sections above we have focused attention on an analysis of stress in small elements which are under homogeneous states of stress. If we decrease the size of the element we progressively arrive at the concept of *stress at a point* in a body. In most geological situations the state of stress in a rock mass varies from point to point. This has very important implications because different stress states can lead to a variation in the stability of the material from locality to locality. For example, under certain states of stress a rock may behave elastically, and as these stress state changes in time or space the rock may reach its elastic limit and fracture. Because of the general conditions of equilibrium there are well defined constraints on the possible variations in stress which can be expressed in mathematical formulations known as the **stress equilibrium equations**.

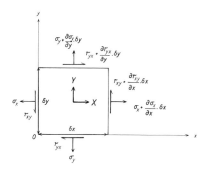

Figure E.7. *The planes of maximum and minimum shear stress τ in relation to the three principal axes of stress.*

Consider a small element in a body under conditions of plane stress with sides of length δx and δy (Figure E.8) and with thickness δ_z measured perpendicular to the plane of the figure. Along the faces which run along the x- and y-axes we can express the stresses in terms of the stress components σ_x, σ_y, τ_{xy}, τ_{yx}. Let us now envisage the possibility that the stress components change systematically in the directions x and y. For example the normal stress acting in the x-direction changes by an amount given by the expression $\partial\sigma_x/\partial x$. Then this relationship implies that the normal stress in the x-direction acting on the right-hand side of the block and at distance ∂x from the origin is given by σ_x multiplied by its change $\partial\sigma_x/\partial x$ and multiplied by the distance ∂x. Using similar reasoning we are able to write down all the stress components acting on the faces of the block (Figure E.8). We are now able to resolve all the forces derived from a consideration of each individual stress component multiplied by the area over which it acts. As well as the surface forces we might also have so-called body forces (e.g. gravity, centrifugal forces) which are forces per unit volume acting in the x-direction (X) and y-direction (Y). The total force acting in the x-direction per unit density is given by

$$F_x = \left(\sigma_x + \frac{\partial\sigma_x}{\partial x}\delta x\right)\delta y\delta z - \sigma_x\delta y\delta z$$

$$+ \left(\tau_{yx} + \frac{\partial\tau_{yx}}{\partial y}\delta y\right)\delta x\delta z - \tau_{yx}\delta x\delta z + X\delta x\delta y\delta z$$

For conditions of static equilibrium the acceleration of the body in the x-direction

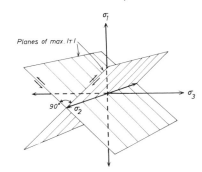

Figure E.8. *Variations of normal and shear stresses on the sides of a block with dimensions δx, δy and δz.*

is zero and this implies that

$$F_x \;=\; 0 \;=\; \frac{\partial \sigma_x}{\partial x} + \frac{\partial \tau_{yx}}{\partial y} + X \qquad \text{(E.20A)}$$

Similarly, by resolving the forces F_y acting in the y-direction and equating to zero

$$F_y \;=\; 0 \;=\; \frac{\partial \tau_{xy}}{\partial x} + \frac{\partial \sigma_y}{\partial y} + Y \qquad \text{(E.20B)}$$

APPENDIX F
Geological Mapping

In this appendix the main methods whereby a geologist records his observations of rock structures on a map and in a field note book will be outlined. The techniques employed for structural mapping do vary somewhat from person to person but there are a number of basic ground rules and these we wish to formulate here. We would emphasize that it is our firmly held opinion that field investigations and the production of a map record of the field observations form important keys to our understanding of the tectonic structure of the Earth. We can see exposed at the surface today a complete range of depths of erosion through the crust and even, at some localities, into the upper mantle. The field geologist is able to build from his observations a complete picture of the type of variation in structural style and the nature of the deformation that has occurred during the tectonic activity of the past.

Equipment

1. Topographic base map

Before starting any mapping project it is important to be clear about the scientific aims of the investigation, for this decision will guide the choice of map scale and control the nature of the techniques which are needed to cover the area in the detail necessary to resolve the problem. You may be fortunate to be able to choose one scale from a number of different topographic bases. For exploration work map scales of 1:100 000 to 1:25 000 are ideal. For more detailed geological work mapping is normally carried out at scales of 1:25 000 to 1:5000, and it will be possible to record most of the important geometric features of folds and fractures. For very detailed surface mapping, such as would be appropriate to investigating a mine claim, scales of 1:1000 to 1:50 may be the most convenient and these often require the geologist to make his own topographic base. In many regions no suitable topographic base maps which are of sufficient accuracy exist and the geologist will have to rely on air photographs or satellite images. The production of an accurate map from air photographs or by plane table or tape and compass techniques is very time consuming. Should such a problem come up it is usually best to have assistance from a map survey professional or photogrammetry expert. Even where a good topographic base is available **air photographs** or **satellite images** are especially useful in aiding a geological interpretation of the terrain. They are particularly useful in regions of folded and faulted sedimentary rocks and a geological interpretation of the ground morphology should be made before the field work is begun. In the field it is possible to make geological observations in ink directly on the photograph surface or, better still, on a transparent overlay. However, air photographs are not maps because they are often slightly tilted with respect to the vertical and, in hilly or mountainous terrain, the photographic image is falsely located because of parallax. This positional error increases from the centre towards the edge of each photograph. For a deeper discussion of the distortions contained in air photographs and discussion of the techniques of making photogeological interpretations the reader is advised to refer to Miller (1961); Kronberg (1974) and Curran (1985).

In some countries air photographs are compiled into **photographic mosaics**. These can be controlled or uncontrolled depending upon whether or not the centres of each photograph making up the mosaic are correctly related in scale and position. It is quite easy to make an uncontrolled mosaic oneself, but the production involves cutting away the edges of each photograph and pasting the prints on to a hard base board. If a second set of air photographs is not available to make a mosaic it is sometimes useful to make a loosely controlled **print lay-down** of the existing photographs and to prepare a photographic version of this assemblage.

In all mosaics and print lay-downs the detail between adjacent photographs will not line up because of variations of surface topography, air photo tilt and flying height.

The availability of maps and air photographs is, in certain countries, restricted for military or political reasons. This can be extremely frustrating because such restrictions generally serve no useful purpose. The geologist has sometimes to develop special techniques varying from careful diplomacy through various types of cunning subterfuge and even to the methods more appropriate to those of a professional spy in order to obtain copies or photographs of the basic topographic record he needs for his work! Remember, however, that Landsat images over a wide range of colour and infra red bands can be obtained of all of the Earth's surface, and that various high quality processing and enlarging techniques can transform these images into a practical working topographic base suitable for many types of regional geological investigation.

The topographic base map will generally be too large for direct work in the field. It should *never* be folded because information recorded near the folds will be subjected to excessive wear and tear during use, and will disappear. It is best to cut the base map into sheets of a size that is conveniently contained in your map case. Do not make these sheets too small, or you will always be frustrated by finding yourself near an edge or, worse still, near a corner, and you will constantly have to shuffle your map sheets like a card player. The size 20 × 30 cm is ideal. Each of the cut map sheets should be mounted with strong glue onto a base card, leaving a small margin around the map to protect its edges from wear. On the corner of each card number the sheets sequentially (A1, A2, A3, . . . , B1, B2, B3, . . . , etc.) so that you can quickly locate the next sheet as you move across the terrain. Indicate in ink the reference numbers of an existing (or personal) grid system so that you can refer exactly to individual map localities in your field note book (Figure F.1, lines 071 and 912). You will need some hard base (or map case) in order to be able to write effectively on your field map, together with some protection of the map from rain, dust or sweaty hands. Map cases can be highly sophisticated constructions with flaps, pockets for note books, air photographs, etc. and holders for pencils. The minimum requirements are a base of hard board, plastic or aluminium with a strong spring clip at one end to attach the map and a moveable elastic band at the other end to stop the map lifting in the wind, together with a plastic sheet for a cover (Figure F.1). Field techniques generally require constant use of a number of black graphite and coloured pencils and it is useful to have pencil holders on the map case or incorporated in a shoulder bag (where you can also have space for a note book, eraser and pencil sharpener and any other constantly used field equipment).

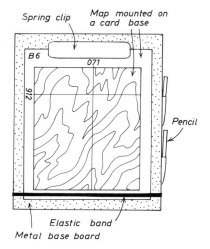

Figure F.1. *Example of field map with underlying base board and clip.*

2. Pencils

Black graphite pencils (H or 2H) are used to record orientation data and coloured pencils are used to record rock lithology on the field map. Coloured pencils should be of a waterproof type and should have a relatively thin coloured lead which has a strong consistency (e.g. in the UK Eagle Verithin range or Cumberland Derwent thin lead pencils are ideal). You will have to acknowledge that pencils will get lost during the course of your work. You will need at least one spare set of coloured pencils at the base camp and you will definitely need to carry spare black graphite pencils with you in the field. The method of solving the problem of pencil loss by tying each pencil on to the map case is not recommended: it can be guaranteed to produce a maypole effect and eventually a tangle that will keep your watchful assistant highly amused at the same time as it drives you into a frenzy.

Never record any field information directly on to the map in ink. Everyone makes mistakes at some time and it is very difficult to erase false inked data. It is best to record information with a sharp, relatively hard (2H) graphite pencil and to ink in this information at the end of every field day. A rule that must be observed without fail is *never* to go into the field with a map that contains data that has not been previously inked.

In using coloured pencils for recording lithology always keep your colour scheme simple, and the colours well contrasted. Choose a colour scheme for the various mapable formations that, wherever possible, conforms to a previously accepted set of standard colours. However, remember that the colour schemes that appear on published maps are not always the most practical schemes for the field map: five different shades of green might be perfect on a published map, but such a sophisticated scheme is unrealistic on a working field map subject to everyday wear and tear.

3. Compass and clinometer

These are absolutely essential instruments for recording azimuths and inclinations. The magnetic compass must be robust and simple to use. There are a great variety of instruments available today and many which are especially convenient for geologists combine a compass and clinometer in one instrument. Geological azimuth data need to be recorded to the nearest degree, so choose a compass with a scale which enables this to be done. It is occasionally necessary to find your location on the topographic base map by resection from distant landmarks, and this means that the compass must have some system of glass prism or mirror sight to be able to observe these landmarks and the compass scale at the same time. Another extremely important feature for a geologist's compass is that the magnetic needle or magnetic plate system should have a damping mechanism (an oil filled compass is best) so that the azimuth point comes quickly to rest. Many older models of the well known "Brunton" compass are very bad in this respect, the magnetic needle swinging like a yoyo before it is possible to take a reading. Remember that the compass North point relates to magnetic and not true North. It is essential to correct all direct compass readings to measurements with respect to true North (by the amount of local declination) before recording these in numerical and map form. Some compasses which can be especially highly recommended for geological work (e.g. Silva) have a method of adjusting the compass magnet so that the magnetic declination is automatically removed from the measurement, and every measurement can then be directly read off the compass scale in terms of true North. This compass also has a mounted internal grid system which enables the recorded true azimuth orientations to be directly plotted in the field with their correct orientation. A compass is such a key field instrument that its loss during field work could lead to scientific disaster. Keep your compass firmly attached either to your map case, field bag or your person with a strong nylon cord. If you plan to work in a very remote region you should take a spare compass to be kept in the base camp for emergencies.

A clinometer is most conveniently incorporated into the compass, and many equipment models do this. However, a simple, but perfectly accurate, clinometer can be made by mounting a protractor on to a wood or metal base with a free swinging plumb line (Figure F.2). Like azimuth directions, the angle of inclination must be measured to the nearest degree.

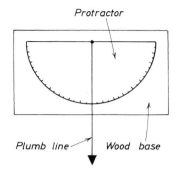

Figure F.2. *Simple home-made clinometer.*

4. Field notebook

Information that cannot be recorded on the geological map is written in the notebook. The most convenient size for such a book is about 20 × 10 cm, it must be strongly bound and made of the best quality paper you can find—economies in all of these respects are false. It is a personal choice whether the pages are unlined, lined or with a grid system. You will constantly be needing to find your next entry point in the book—some good quality notebooks have a tape marker embedded in the binding. If your notebook does not have this organize an elastic band to provide a moving book mark. A lost field book is a geological tragedy, and a brightly coloured notebook binding will reduce the risk of casual loss. Whether field notes are recorded in pencil or ink is a matter of personal choice. Soft pencil (B) produces a strong black line without problems of interpage smearing. Personally we do not favour direct recording of written or diagram data in ink. However, many geologists today use barrel and wire pens and record their field notes directly in ink. Although these pens are excellent for ink work in the base camp we have found a number of practical disadvantages in the field. Pens with a fine diameter (< 0·3 mm) often dry out and become blocked in the field, and all of these pens are extremely susceptible to changes in temperature and pressure and often extrude ink in a totally uncontrollable manner with atmospheric changes.

5. Hammer and soft steel chisel

A hammer is an important tool for obtaining rock specimens for laboratory work, and for chipping away weathered rock surfaces. In many classic localities the outcrops have been literally pulverized by the activity of geologists. We would request a code of hammer activity: never destroy interesting structures by uncontrolled hammering and never remove fossils from a natural outcrop unless there is a good scientific reason for doing so. Remember that, for most purposes, the information about rock types can just as well be obtained by observations on naturally fallen blocks. If you wish to obtain a particular block of rock from an outcrop it is best to use a soft steel rock chisel with the hammer. Never strike one hammer against another because hard steel shards are likely to break from one of the hammers and fly around like shrapnel.

Although a geologist should never lose his hammer, let us admit that many of us do. A good practical tip for lessening your chance of leaving your hammer at an outcrop, developed by the Greenland Geological Survey, is to paint the handle of your hammer some strikingly bright colour so that it can be easily distinguished from the colour of the local rock and vegetation.

6. Hand lens

A good hand lens with a moderate magnification ($\times 10$) is absolutely essential for the examination of a fresh rock surface to determine such features as mineral content, grain shape and micro fossils in a rock. It constantly amazes us how many geologists go into the field without a good hand lens and, in consequence, miss a great deal of important information. Tie your hand lens to your field bag with a piece of strong nylon cord or, even better, wear it round your neck where it is ideally positioned to make frequent observations.

7. Drafting materials

We will only suggest here the simple equipment that is necessary for base camp work after a day in the field. A firm board for providing a stable base for "inking-in" the maps is essential. Inks are essential for making permanent the pencil record made during the day. These should be waterproof (note that not all "drawing inks" are waterproof) and you will need black, red and perhaps also blue and green. You will need a soft rubber to remove the pencil lines and, because everyone makes errors, a razor blade and hard ink eraser. The type of pen used is a matter of personal taste. Although barrel pens (Rotring, Rapidograph, Faber-Castell, Staedler, Kern) are excellent, many people prefer the simplicity and flexibility of a drawing pen with a number of fine steel nibs and a fine cotton wiper for removing partially dried ink from the nib. A protractor and ruler will often be required at some stage in the redrafting, and a stereogram or equal area projection will be required to anayse orientation data.

8. Other equipment

Most geologists carry extra items of equipment in addition to those listed above, depending on their personal approach to field work and upon the special nature of the problem in hand. A **measuring tape** is useful if very detailed mapping is to be undertaken so that features such as bed thickness, joint spacing, etc. can be accurately recorded. A bottle of 10% concentration **hydrochloric acid** with a glass tube and rubber thumb dispenser is most useful in terrains where calcite limestone needs to be distinguished from dolomite. More elaborate chemicals can be carried where the geological problem in hand demands these. A small field **stereogram** or **equal area projection** can be used to make geometrical computations directly in the field. Although most interpretation of air photographs will be done in the base camp some geologists like to carry a **folding pocket stereoscope** to be able to make these observations in the field. If a good topographic base map with well controlled contours is available an **altimeter** is a great assistance to position location in the field. In mountainous terrain, where certain exposures may be inacessible cliffs, **binocular field glasses** are a great aid to observe structural detail.

Geological mapping procedures

The scientific aims of the investigation and the amount of time available for field work will dictate the data collection programme. In regional exploration mapping it is important to plan so that the distribution of available field time will enable a complete cover of the whole area. This will mean that you will generally have to rely on traverse mapping, and you should plan these traverses so that they cross the strike of the different rock units. With a more detailed mapping programme it may be possible to visit practically all the rock exposures. Whatever problem faces the geologist three basic contraints are applicable to every type of field mapping:

1. *The field map must be legible.* This means that it is essential to use the right type of equipment and that you must be accurate and neat. The field map must be a permanent record of your observations in ink.
2. *The field map must be readable by another geologist.* This requires that the observations must be plotted using consistent standard symbols rather than haphazard personal hieroglyphics and that, somewhere in the collection of field sheets, there must be a key to the symbols used and a legend to the colour code used to represent the different rock types.
3. *A field map must distinguish between observed facts and inferences drawn from these facts.* It is therefore necessary to indicate the differences between exposed and unexposed terrain, between observed and inferred contacts of different rock units. The limits of actual rock exposures need to be indicated on the map and a practical map technique to show exposed and inferred rock types is to colour outcrops with a strong colour, and regions where the same rocks are believed to exist in a weak colour of the same hue (Figure F.8)

At the start of the field investigation it is usually a good idea to spend at least a day making some preliminary traverse through the area to obtain a general impression of the nature of the geological problem, to decide on the main lithological subdivisions you might record on the map, and to decide on the types of small scale structural features that require mapping. It is always a good idea to spend a little time looking at the boulders in streams, because these can often give a very useful first impression of the rock types likely to be encountered in the catchment area of that particular stream. Rivers cutting through glacial moraine may not be so useful in this respect. It is best to commence mapping where the least weathered, best exposed and most accessible exposures are located.

Having arrived at a rock outcrop the first thing to do is to find your position on the map, either from simply identifiable features recorded on the topographic base or by resection of distant features using the compass to ray-in your position (two or better still three features are required for this). Determine the map coordinates of the locality (first east, then north) and note this reference in your field book together with the characteristic features of the lithological type and any general and special features of the rock (mineral composition, grain size, weathering). If the rock is of sedimentary origin look for diagnostic sedimentation structures and, if possible, determine their geometry so that later it will be possible to reconstruct the directions of transport of the sediment depositing current (cross bedding, ripple marks, bottom structures in turbidites). If the terrain is tectonically complex these primary sedimentation features can be used to determine the polarity of the beds (normal or overturned strata). In sedimentary rocks search for fossils which might give an indication of rock age (zone fossils) or indicate the environmental conditions of deposition (palaeoecology). Search for tectonically produced structures (cleavage, folds, lineations, joints, etc.) and for deformation markers that might be used to establish the principal strain directions and values. Note all these features in the field book and sketch any particularly important geometric relationships. You may decide that the outcrop shows certain features so well that it should be placed on photographic record. Remember to include a scale in the photograph, to note the exposure number in the film sequence and to make a quick sketch in the notebook of the field of view recording rock types and nature of any structures. Do not rely on a photographic record alone. For example, we know of at least one publication by a well known geologist in which all the diagrams were drawn from his photographs. Because he did not make notes of the rock lithology in his field book all the rock types which he called sandstone are, in fact, dolomite, and all those termed shale are marl. When making a photographic record always remember

Figure F.3. *Basic symbol for a planar structure.*

A.

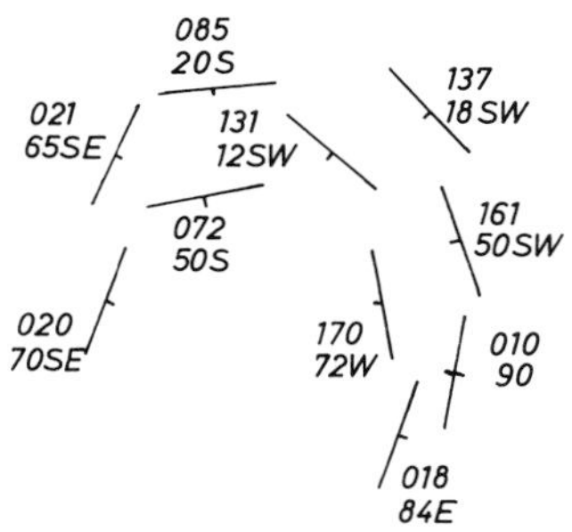

085
20 S

137
18 SW

021
65 SE

131
12 SW

072
50 S

161
50 SW

020
70 SE

170
72 W

010
90

018
84 E

B.

021 / 65 SE
170 / 72 W
131 / 12 SW
085 / 20 S
137 / 18 SW
020 / 70 SE
072 / 50 S
010 / 90
018 / 84 E
161 / 50 SW

Figure F.4. A: *Map representation of a series of bedding plane orientations.* B: *List form of the same data. Which is more understandable?*

254
15

74
15

Figure F.5. *Basic symbol for a linear structure.*

to get as close to the outcrop as possible while keeping the subject material in the field of view. So many published photographs are disappointing because the object of importance is subsidiary to a mass of vegetation or of rocks showing no features in particular. Most modern cameras can focus down to 0·5 m, and with a close-up lens, lens rings or a "macro" lens unit it is quite possible to take photographs at distances of a few centimetres from the object (see Figure 10.9). Remember that a greater range of depth of focus occurs where the *f*-number is high, but that high *f*-numbers require a longer time of exposure which may give movement problems if the camera is held in the hand. Before you take a photograph remember to clean up the outcrop surface by removing stones, leaves and distracting vegetation. In the field one may not notice the effect of a blade of grass or its shadow, but when the film is printed these geologically insignificant things may often detract from the features that are of geological importance.

Now to the field map. On the map indicate in pencil (H) the limits of the actual outcrop as accurately as possible and then, also in pencil, colour the mapped outcrop according to your lithological colour scheme. Measure any **planar structure** (bedding, cleavage, banding, jointing, etc.) first recording the *strike azimuth* with a compass to the nearest degree and second the *angle of dip* and its *directional sense*. If necessary, correct the magnetic azimuth reading to true north. You can record this information in your field book, but you should *record the most important geological readings directly on the map in the field*. With a protractor *accurately* draw the strike line and indicate with a side tick the dip sense. These data are recorded on the map in symbol form (Figure F.3) and the measurement made should be written next to the symbol. Two methods of recording the numerical data are in common use. The first (Figure F.3A) records the strike direction together with the angle and approximate direction sense of the dip; the second (Figure F.3B) describes the *direction* of the dip using a 360° compass notation followed by the angle of dip. With the first method the direction sense is not absolutely necessary, because it duplicates the information of the dip tick symbol, but it does give a useful check to avoid mistakes. Which of the measurements of planar structures you record directly on the map will depend upon the nature of the geological problem you are trying to solve but, in principle, record as many features as possible. Compare your reactions to mapped bedding orientation data of Figure F.4A to the list of data as would be recorded in the field book in Figure F.4B; which is more immediately understandable? The information recorded in map form immediately conveys the interrelationships between the observations. It shows a tight, overturned, synformal fold, with a rounded hinge zone, with axial surface dipping ESE at 75°–80° and with a fold axis plunging about 8°–10° to the SSW. Even an expert would be hard put to derive even a fraction of this information from the data in list form. Geological mapping is often a very stimulating exercise just because the geological significance of individual outcrops gradually emerges during the course of the work just like a jig-saw puzzle picture is gradually revealed as the separate pieces are assembled. A criticism that can be made of many field maps is that they do not contain enough measurements and records of structural information. A novice will clearly take quite some time to identify, measure and record a particular structure, but an experienced field geologist can be expected to make between 50 and 200 measurements each day. Such a large data bank is absolutely invaluable in providing a firm geometrically controlled base for a structural interpretation. Another important aspect of measuring orientations of structures is that it forces the geologist to observe carefully and identify features which, with a more casual investigation, might be overlooked. The second type of basic orientation information relates to the orientation of **linear structures** (fold axes, intersection lines of cleavage and bedding surfaces, stretching directions in deformed rocks, striae on fault planes, etc.). The **trend or azimuth** of the lineation is measured with the compass and the angle of inclination from the horizontal, the **angle of plunge**, with the clinometer. These data should be accurately recorded in pencil on the map in the field using an arrow symbol (Figure F.5). Remember that a linear structure plunging towards the WSW has a different arrow sense from that plunging ENE, and that this should be recorded using a 360° azimuth notation by the symbol (Figure F.5). All primary and secondary structures that can be defined by orientation numbers can be classified either as planar or linear features, and it is normal to denote different types of these features using variations in the basic notation or in

Primary structures (in black)

A. Planar: lithology, layering, banding in gneisses, flow banding and flow layering in igneous rocks.

1 strike and dip, basic notation
013
11NW

2 vertical
013
90

3 horizontal
hor.

4 polarity of layering

5 "normal" layering

6 "overturned" layering

7 exposed lithological contact

8 inferred (certain) lithological contact

9 inferred (uncertain) lithological contact

B. Linear: bottom markings in turbidities (flute casts, groove casts, bounce casts, etc.), ripple crests, flow lineations in igneous rocks.

1 azimuth and angle of plunge
203
16

2 horizontal
23
0

3 vertical
vert.

Secondary structures (in red): The double line symbol is recommended for use in the field, to be replaced by single line symbols when the field map is inked.

A. Planar: fold axial surfaces, cleavage, schistosity, fracture planes, shear zones, joints.

1 strike and dip, basic notation
012
45NW

2 suggested variations on basic symbol

3 vertical
012
90

4 horizontal
hor.

5 exposed fault contact (with dip tick)

6 inferred (certain) fault contact

7 inferred (uncertain) fault contact

8 normal fault (u upthrown side, d downthrown side)

9 reversed or thrust fault (triangle on hanging wall side)

10 strike slip fault (right-hand displacement)

11 antiform axial surface trace (vertical)

12 antiform axial surface trace (south dipping)

13 synform axial surface trace (vertical)

14 synform axial surface trace (south dipping)

B. Linear: fold hinge lines, mineral orientation lineation, long axes of elongated particles, cleavage–bedding intersection, stretching fabric, pencil structure, striae or fibrous crystals on fault surfaces.

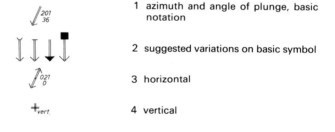

1 azimuth and angle of plunge, basic notation
201
36

2 suggested variations on basic symbol

3 horizontal
021
0

4 vertical
vert.

Figure F.6. Recommended list of symbols to be used on geological maps.

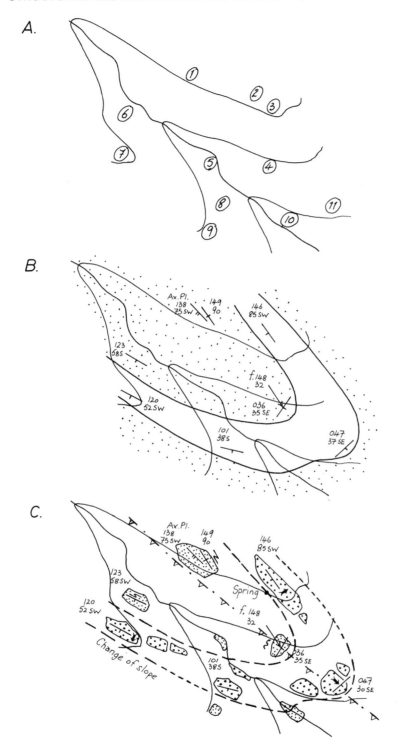

Figure F.7. *Types of map record of the same base data. A: records only locality numbers, B: is a solid geological map interpretation and C shows the mapping style recommended in this book clearly distinguishing facts and inferences from facts.*

colour of the symbols. Many geologists use colour coded symbols when they make a permanent record of their data in ink at the end of the day; black for basic features (bedding, polarity sense) and red for tectonic features (cleavage, fold axes). There are no generally agreed symbols for the various types of structures you will record, but a list of suggestions drawn from our practical experience is given in Figure E.6. Whatever you choose from this list, make a record somewhere on the field sheets of your notation.

The field map will develop day by day as information accumulates. Comparing the maps of different geologists, one sees a wide range of different techniques and approaches, some of which are to be discouraged whereas others can be recommended in helping to provide useful and high quality data collection. Beginners often produce a map which is just a series of locality numbers. All the data are placed in the field book and little or none on the map (Figure F.7A). Such a record is practically useless, and the geologist who uses such a technique is unlikely to see many of the major geological and structural problems of the region. These will only emerge when he begins to plot his data and he may then find it necessary to return to the terrain or leave the problem in an unsatisfactory and unresolved state. With this technique it is most unlikely that, during the data collection in the field, the geologist's mind has been actively working and thinking about the outcrop relationships from locality to locality. In our experience many of those geologists making such a data record understandably find field work boring and regard the activity as only necessary for supplying themselves with a series of hand specimens for later work in the laboratory.

Another technique which can be criticised is the production of a solid geological map of the bedrock formations directly in the field (Figure F.7B). With this technique it is not always clear what evidence there is behind the construction of geological boundary lines, what is reasonable inference and what is just guesswork.

In our opinion the best technique is one which clearly demarcates the observations and the inferences made from them (Figures F.7C, F.8). The differences between exposed and inferred boundaries are clear, and so are the differences of degree of certainty between different types of inference. Any geologist who later has to use such a map will know exactly where to find the exposures and where to seek additional information, should that be necessary. We have to admit that this technique may be time consuming and that, in well exposed terrain, neatness is imperative if the daytime drafting is to be interpretable during the evening inking-in period. With this technique *do* make interpretations in the field. For example it is often relatively easy to see where unexposed boundaries might be located during the course of the field work, especially if the topography is varied and complex. On such a field map it will be possible to record many features which may provide excellent indicators of sub-soil geology: slope changes, springs, soil colour, vegetation changes, etc. Without interfering with the solid geology record it is also possible to indicate information about Quaternary deposits (moraine, alluvium, rock falls, debris cones, etc.) using a variety of different point, circle and triangle symbols. After a day in the field it is often an excellent idea to make a full coloured pencil intepretation of the sub-surface geology using weaker colour intensities than those of exposed outcrops (Figure F.8)

Sample collection and record

During the course of a field investigation you will need to collect samples of rocks for later investigations of petrological features in the laboratory. In the field you will need to mark each rock specimen clearly with a sequential number using a waterproof felt tip marker pen and to record with grid coordinates the rock type and location in the note book. The sample number can also be recorded on the map. One convenient technique is to make a pin prick on the map at the sample locality and to record the specimen number on the back of the map sheet. For specimens which show important structural characteristics we strongly advise that the collected sample has an orientation mark so that any information resulting from further laboratory study can be directly related to geographical orientation. We recommend that all samples are oriented in this way. To collect an oriented sample a surface of the outcrop is selected which preferably shows some clearly marked planar surface. Often this surface will have some structural significance, being a bedding, cleavage or joint plane. The sample is then broken from the outcrop and then placed back on to its parent, perhaps with the assistance of sticky tape. The strike azimuth and dip orientation of the surface are then measured, the data recorded in the note book. The strike and dip directions are then marked on the rock sample with a felt tip pen, perhaps together with the orientation numbers (Figure F.9), before the specimen is detached from the outcrop and placed in the collection bag.

Concluding remarks

The maps produced by geologists vary from scruffy, torn, illegible pieces of paper

Figure F.8. Examples of field maps at scales of 1 inch to 1 mile (1:63,360).

Figure F.9. *Method of orienting a hand sample of rock with orientation of a marker plane shown with strike and dip symbol on the specimen.*

to near works of art. You will probably not make ideal field maps when you take your first steps in field mapping. However, with increasing experience and continuous self-criticism of your efforts, even those who are not born with a sense of neatness, or who have little or no inherent drafting ability, can produce good quality geological maps. Aim at the highest standard possible. If you are a student working in a group making a geological map compare your efforts critically with those of your colleagues, and especially with those of your instructor.

We personally find field mapping an enjoyable and stimulating intellectual activity. Remember, however, that it is not only an academic exercise. For the professional geologist a good accurate mapping technique provides one of the fundamental data documents necessary for much industrial research and development. The investment of large sums of money on major civil engineering, mining or petroleum development may partly depend upon the reliability of your field observations and your map. If your data collection and record was poorly made you will, at best, be unpopular with your project chief and, at worst, you may find yourself looking for new employment.

APPENDIX G
Mathematical Notations

1. Roman letters

A finite fold amplitude

$\dot{A}$ growth rate of fold amplitude ($\mathrm{d}A/\mathrm{d}t$)

a, b, c coordinate system to describe simple shear (ab shear plane, a shear direction)

a_0, a_1, a_2 coefficient of sine functions in a Fourier series

b_1, b_2 coefficient of cosine functions in a Fourier series

c curvature

D parameter expressing the intensity of distortion in a logarithmic strain plot

d, d_1, d_2 bed thickness

d parameter expressing the intensity of distortion in a Flinn diagram

E Young's modulus of elasticity

e, e_x extension, extension parallel to the x-coordinate direction $(l_1 - l_0)/l_0$

e_i incremental extension

e_1, e_2, e_3 principal finite extensions ($e_1 \geqslant e_2 \geqslant e_3$)

e_{i1}, e_{i2}, e_{i3} principal incremental extensions

$\dot{e}, \dot{e}_x$ rate of extension, rate of extension parallel to the x-coordinate direction ($\mathrm{d}e_x/\mathrm{d}t$)

$\dot{e}_1, \dot{e}_2, \dot{e}_3$ principal rates of extension ($\dot{e}_1 \geqslant \dot{e}_2 \geqslant \dot{e}_3$)

F force

F_n, F_s normal force, shear force

G rigidity modulus or shear modulus in elastic materials

K parameter expressing the shape of a strain ellipsoid in a logarithmic strain plot, bulk modulus of elasticity

k parameter expressing the shape of a strain ellipsoid in a Flinn diagram

l, l_0, l_1 length, original length and final length

l, m, n direction cosines with respect to coordinate directions x, y and z

m mass

R ellipticity of a finite strain ellipse

R_{xy} ellipticity of a principal XY section of a strain ellipsoid

T absolute temperature degrees Kelvin

t	time
W	finite fold wavelength
W_i	initial wavelength of a folded surface
W_a	fold arc-length measured between corresponding points in a wave train
$\bar{W}$	fold wavelength of an antiform–synform pair
w	line rotation
X, Y, Z	directions of the maximum, intermediate, and minimum axes of the finite strain ellipsoid
X_0, Y_0, Z_0	initial directions of lines which become the X, Y and Z directions of the finite strain ellipsoid
X_i, Y_i, Z_i	directions of the maximum, intermediate and minimum axes of the incremental strain ellipsoid

2. Greek letters

γ	shear strain
Δ_A	area change
Δ_v	volume change
$\varepsilon, \varepsilon_x$	logarithmic extension, logarithmic extension parallel to x-direction
$\varepsilon_1, \varepsilon_2, \varepsilon_3$	principal logarithmic extension
η, η_1, η_2	extensional viscosity coefficients
λ, λ_x	quadratic extension, quadratic extension parallel to x-direction
λ'	reciprocal quadratic extension ($\lambda' = 1/\lambda$)
$\lambda_1, \lambda_2, \lambda_3$	principal quadratic extensions ($\lambda_1 \geqslant \lambda_2 \geqslant \lambda_3 > 0$)
μ, μ_1, μ_2	shear viscosity coefficients
μ_d	coefficient of dynamic friction
μ_i	coefficient of internal friction
μ_s	coefficient of static friction
v	Poisson's ratio of elasticity; Lode's parameter of finite strain
ϱ	density
σ, σ_x	normal stress, normal stress parallel to x-direction
$\sigma_1, \sigma_2, \sigma_3$	principal normal stress ($\sigma_1 \geqslant \sigma_2 \geqslant \sigma_3$)
σ'	deviatoric stress
$\bar{\sigma}$	mean stress
τ, τ_{xy}	shear stress, shear stress parallel to y-direction with normal parallel to x-direction
ψ	angular shear strain
ω, ω_x	rotation, rotation around the x-axis
$\dot{\omega}$	vorticity or rate of rotation ($d\omega/dt$)

Source List

This list of papers, books and articles has been compiled so as to give further references to the topics covered in the Sessions of Volume 2. As with the source list of Volume 1, it is by no means a complete list: the titles have been selected so as to be of special relevance to the Session materials discussed here and within the framework of the approach of the book. Further references under the general headings below can be found in the key references of each session.

Fold geometry and folding mechanisms
(Sessions 15, 17, 19, 20, 21)

Bayly, M. B. (1964). A theory of similar folding in viscous materials. *Am. J. Sci.* **262**, 753–766.

Bayly, M. B. (1969). Anisotropy of viscosity in suspensions of parallel flakes. *J. Compos. Mater.* **3**, 705–708.

Bayly, M. B. (1970). Viscosity and anisotropy estimates from measurements on chevron folds. *Tectonophysics* **9**, 459–474.

Bayly, M. B. (1971). Similar folds, buckling and great circle patterns. *J. Geol.* **79**, 110–118.

Biot, M. A. (1961). Theory of folding of stratified viscoelastic media and its implications in tectonics and orogenesis. *Geol. Soc. Am. Bull.* **72**, 1595–1620.

Biot, M. A. (1965). Similar folding of the first and second kinds. *Geol. Soc. Am. Bull.* **76**, 251–258.

Biot, M. A., Ode, H. and Roever, W. L. (1961). Experimental verification of the theory of folding of stratified viscoelastic media. *Geol. Soc. Am. Bull.* **72**, 1621–1632.

Borradaile, G. J. (1978). Transected folds: a study illustrated with examples from Canada and Scotland. *Geol. Soc. Am. Bull.* **89**, 481–493.

Borradaile, G. J. (1972). Variably oriented co-planar primary folds. *Geol. Mag.* **109**, 89–98.

Boyer, S. E. (1986). Styles of folding within thrust sheets. *J. Struct. Geol.* **8**, 325–339.

Brühl, H. (1969). Boudinage in den Ardennen und in der Nordeifel als Ergebnis der inneren Deformation. *Geol. Mitt. Aachen* **8**, 263–308.

Campell, J. D. (1958). En echelon folding. *Econ. Geol.* **53**, 448–472.

Carey, S. W. (1962). Folding *J. Alberta Soc. Petrol. Geol.* **10**, 95–144.

Caron, J. M. and Jeanette, D. (1975). Cisaillements et cristallisations dans les roches métamorphiques: interprétation de quelques microstructures. *Geol. Soc. France Bull.* **17**, 708–720.

Casey, M. and Huggenberger, P. (1985). Numerical modelling of finite-amplitude similar folds developing under general deformation histories. *J. Struct. Geol.* **7**, 103–114.

Chapple, W. M. (1970). The finite-amplitude instability in the folding of layered rocks. *Canad. J. Earth Sci.* **7**, 457–466.

Chapple, W. M. and Spang, J. H. (1974). Significance of layer-parallel slip during folding of layered sedimentary rocks. *Geol. Soc. Am. Bull.* **85**, 1523–1534.

Chaudhuri, A. K. (1972). Concise descriptions of fold orientations. *Geol. Mag.* **109**, 231–233.

Cobbold, P. R. and Quinquis, H. (1980). Development of sheath folds in shear regimes. *J. Struct. Geol.* **2**, 119–126.

Cobbold, P. R. and Watkinson, A. J. (1981). Bending anisotropy: a mechanical constraint on the orientation of fold axes in an anisotropic medium. *Tectonophysics* **72**, TI–T10.

Curie, J. B., Patnode, H. W. and Trump, R. P. (1962). Development of folds in sedimentary strata. *Geol. Soc. Am. Bull.* **73**, 655–674.

Donath, F. A. (1962). Role of layering in geological deformations. *N. Yk. Acad. Sci. Trans.* **24**, 236–249.

Donath, F. A. and Parker, R. B. (1974). Folds and folding. *Geol. Soc. Am. Bull.* **75**, 45–62.

Dubey, A. K. and Cobbold, P. R. (1977). Noncylindrical flexural slip folds in nature and experiment. *Tectonophysics* **38**, 223–239.

Elliott, D. (1965). The quantitative mapping of directional minor structures. *Jour. Geol.* **73**, 865–880.

Elliott, D. (1968). Interpretation of fold geometry from lineation isogonic maps. *J. Geol.* **76**, 171–190.

Escher, A. and Watterson, J. (1974). Stretching fabrics, folds and crustal shortening. *Tectonophysics* **22**, 223–231.

Fletcher, R. C. (1974). Wavelength selection in the folding of a single layer with power law rheology. *Am. J. Sci.* **274**, 1029–1043.

Fyson, W. K. (1971). Fold attitudes in metamorphic rocks. *Am. J. Sci.* **270**, 373–382.

Ghosh, S. K. (1966). Experimental tests of buckling folds in relation to strain ellipsoid in simple shear deformations. *Tectonophysics* **3**, 169–185.

Ghosh, S. K. (1968). Experiments on buckling of multilayers which permit interlayer gliding. *Tectonophysics* **6**, 207–249.

Goguel, J. (1984). Introduction à l'étude mécanique des déformations de l'Écorce Terrestre. *Mem. Serv. Carte Geol. France*, 1–530.

Hara, I. (1966). Strain and movement pictures in competent layers in flexural folding, *J. Geol. Soc. Japan* **72**, 413–425.

Hobbs, B. E. (1971). The analysis of strain in folded layers. *Tectonophysics* **11**, 329–375.

Huddleston, P. J. (1977). Similar folds, recumbent folds and gravity tectonics in ice and rocks. *Jour. Geol.* **85**, 113–122.

Hudleston, P. J. and Stephansson, O. (1973). Layer short-ening and fold shape development in the buckling of single layers. *Tectonophysics* **17**, 299–321.

Johnson, A. M. (1969). Develoment of folds within Carmel formation, Arches National Monument, Utah. *Tectonophysics* **8**, 31–77.

Lacassin, R. and Mattauer, M. (1985). Kilometre-scale sheath folds at Mattmark and implications for transport direction in the Alps. *Nature* **316**, 739–742.

Latham, J. P. (1979). Experimentally developed folds in a material with a planar mineral fabric. *Tectonophysics* **57**, TI-T8.

Latham, J. P. (1985). The influence of non-linear material properties and resistance to bending on the development of internal structures. *J. Struct. Geol.* **7**, 225–236.

Latham, J. P. (1985). A numerical investigation and geological discussion of the relationship between folding, kinking and faulting. *J. Struct. Geol.* **7**, 237–250.

Laubscher, H. P. (1976). Geometrical adjustments during rotation of a Jura fold limb. *Tectonophysics* **36**, 347–365.

Laubscher, H. P. (1978). Foreland folding. *Tectonophysics* **47**, 325–337.

Matthews, P. E., Bond, R. A. B. and Van den Berg, J. J. (1971). Analysis and structural implications of a kinematic model of similar folding. *Tectonophysics* **12**, 129–154.

McBirney, A. R. and Best, M. G. (1961). Experimental deformation of viscous layers in oblique stress fields. *Geol. Soc. Am. Bull.* **72**, 495–498.

Minnigh, L. D. (1979). Structural analysis of sheath folds in a mega chert from the western Italian Alps, *J. Struct. Geol.* **1**, 275–282.

Moseley, F. (1968). Conical folding and fracture patterns in the Pre-Betic of southwest Spain. *Geol. J.* **6**, 97–104.

Mukhopadhyay, D. (1965). Effects of compression on concentric folds and mechanisms of similar folding. *J. Geol. Soc. India* **6**, 27–41.

Mukhopadhyay, D. (1972). A note on the mullion structures from the Ardennes and North Eifel. *Geol. Rundschau* **61**, 1037–1049.

Price, N. J. (1967). The initiation and development of asymmetric buckle folds in non-metamorphosed competent sediments. *Tectonophysics* **4**, 173–201.

Ramberg, H. (1961). Relationships between concentric longitudinal strain and concentric shearing strain during folding of homogeneous sheets of rock. *Am. J. Sci.* **259**, 382–390.

Ramberg, H. (1963). Strain distribution and geometry of folds. *Geol. Inst. Univ. Uppsala Bull.* **42**, 1–20.

Ramberg, H. (1963). Evolution of drag folds. *Geol. Mag.* **100**, 97–110.

Rickard, M. J. (1971). A classification diagram of fold orientations. *Geol. Mag.* **108**, 23–26.

Roberts, D. and Strömgard, K. (1972). A comparison of natural and experimental strain patterns around fold hinge zones. *Tectonophysics* **14**, 105–120.

Sanderson, D. J. (1973). The development of fold axes oblique to the regional trend. *Tectonophysics* **16**, 55–70.

Sanderson, D. J. (1979). The transition from upright to recumbent folding in the Variscan fold belt of southwest England: a model based on the kinematics of simple shear. *J. Struct. Geol.* **1**, 171–180.

Shimamoto, T. and Hara, I. (1976). Geometry and strain distribution of single layer folds. *Tectonophysics* **30**, 1–34.

Smith, R. B. (1969). The folding of a strongly non-Newtonian layer. *Am. J. Sci.* **279**, 272–287.

Smythe, D. K. (1971). Viscous theory of angular folding by flexural flow. *Tectonophysics* **12**, 415–430.

Sorby, H. C. (1853). On the origin of slaty cleavage. *New Philosoph J. Edinburgh* **55**, 137–148.

Treagus, S. H. (1973). Buckling instability of a viscous single-layer system, oblique to the principle compression. *Tectonophysics* **19**, 271–289.

Treagus, S. H. (1981). A theory of stress and strain variations in viscous layers and its geological implications. *Tectonophysics* **72**, 75–103.

Treagus, S. H. (1983). A theory of finite strain variation through contrasting layers, and its bearing on cleavage refraction. *J. Struct. Geol.* **5**, 351–368.

Treagus, J. E. and Treagus, S. H. (1981). Folds and the strain ellipsoid: a general model. *J. Struct. Geol.* **3**, 1–17.

Van Hise, C. R. (1894). Principles of North American Precambrian geology. *U.S. Geol. Surv. Ann. Rep.* **16**, 581–843.

Watkinson, A. J. (1975). Multilayer folds initiated in bulk plane strain, with the axis of no change perpendicular to the layering. *Tectonophysics* **28**, T7-T11.

Williams, G. D. (1977). Rotation of contemporary folds in the X-direction during overthrust processes in Lakse fjord, Finnmark. *Tectonophysics* **48**, 29–40.

Williams, J. R. (1980). Similar and chevron folds in multilayers using finite element and geometric models. *Tectonophysics* **65**, 323–338.

Kink folds and crenulation cleavage (Session 20)

Anderson, T. B. (1964). Kink-bands and related geological structures. *Nature* **202**, 272–274.

Anderson, T. B. (1969). The geometry of a natural orthorhombic system of kink-bands. *Geol. Surv. Can. Pap.* **69–52**, 200–228.

Baer, A. J. and Norris, D. K. (editors) (1969). Research in Tectonics, kink bands and brittle deformation. *Geol. Surv. Can. Pap.* **68–52**, 1–373.

Collier, M. (1978). Ultimate locking angles for conjugate and monoclinal kink bands. *Tectonophysics* **48**, T1–T6.

Donath, F. A. (1961). Experimental study of shear failure in anisotropic rocks. *Geol. Soc. Am. Bull.* **72**, 985–990.

Donath, F. A. (1969). The development of kink bands in brittle anisotropic rocks. *Geol. Soc. Am. Memoir* **115**, 453–493.

Faill, R. T. (1973). Kink band folding, Valley and Ridge province, Pennsylvania. *Geol. Soc. Am. Bull.* **84**, 1289–1314.

Frank, F. C. and Stroh, A. N. (1952). On the theory of kinking. *Proc. Phys. Soc. London* **65B**, 811–821.

Gay, N. C. and Weiss, L. E. (1974). The relationship between principal stress directions and the geometry of kinks in foliated rocks. *Tectonophysics* **21**, 287–300.

Gray, D. R. (1977). Differentiation associated with discrete crenulation cleavages. *Lithos* **10**, 89–101.

Gray, D. R. (1977). Morphological classification of crenulation cleavage. *J. Geol.* **85**, 229–235.

Greenly, E. (1919). The geology of Anglesey. *Geol. Surv. G.B. Mem.* 1–340.

Hammer, S. K. (1979). The role of discrete homogeneities

and linear fabrics in the formation of crenulations. *J. Struct. Geol.* **1**, 81–91.

Hara, I. and Paulitsch, P. (1971). Strain distribution in conjugate kink bands. *N. Jb. Geol. Paläont. Abh.* **139**, 346–368.

Honen, E. and Johnson, A. M. (1976). A theory of concentric, kink and sinusoidal folding and of monclinal flexuring of compressible elastic multilayers. *Tectonophysics* **30**, 197–239.

Johnson, M. R. W. (1956). Conjugate fold systems in the Moine thrust zone in the Lochcarron and Coulin Forest areas in Wester Ross. *Geol. Mag.* **93**, 345–350.

Johnson, A. M. and Ellen, S. D. (1974). A theory of concentric kink and sinusoidal folding and of monoclinal flexuring of compressible elastic multilayers. I. Introduction. *Tectonophysics* **21**, 301–339.

Knill, J. L. (1961). Joint drags in mid-Argyllshire. *Proc. Geol. Assoc. London* **72**, 13–39.

Marlow, P. C. and Etheridge, M. A. (1977). Development of a layered crenulation cleavage in mica schists of the Kanmantoo Group near Macclesfield, S. Australia. *Geol. Soc. Am. Bull.* **88**, 873–882.

Marshall, B. (1966). Kink-bands and related geological structures. *Nature* **210**, 1249–1250.

Matte, Ph. (1969). Les kink-bands: example de déformation tardive dans l'Hercynien du Nord-Quest de l'Espagne. *Tectonophysics* **7**, 309–322.

Naha, K. and Halyburton, R. V. (1974). late stress systems deduced from conjugate folds and kink bands in the Main Raialo syncline, Rajasthan, India. *Geol. Soc. Am. Bull.* **85**, 251–256.

Nicholson, R. (1966). Metamorphic differentiation in crenulated schists. *Nature* **209**, 68–69.

Paterson, M. S. and Weiss, L. E. (1966). Experimental deformation and folding in phyllite. *Geol. Soc. Am. Bull.* **77**, 343–374.

Ramsay, J. G. (1962). The geometry of conjugate fold systems. *Geol. Mag.* **99**, 516–526.

Roberts, D. (1971). Stress regime and distribution of a conjugate fold system from the Trondheim region, central Norway. *Tectonophysics* **12**, 155–165.

Rondeel, H. E. (1969). On the formation of kink bands. *Proc. K. Ned. Akad. Wet.* **72**, 317–329.

Tobisch, O. T. and Fiske, R. S. (1976). Significance of conjugate folds and crenulations in the central Sierra Nevada, California. *Geol Soc. Amer. Bull.* **87**, 1411–1420.

Turner, F. J. and Weiss, L. E. (1965). Deformation kinks in brucite and gypsum. *Proc. U.S. Nat. Acad. Sci.* **54**, 359–364.

Van Loos, A. J., Brodzikowski, K. and Gotowala, R. (1983). Structural analysis of kink bands in unconsolidated sands. *Tectonophysics* **104**, 351–374.

Weiss, L. E. and Turner, F. J. (1972). Some observations on translation gliding and kinking in experimentally deformed calcite and dolomite. *Am Geophy. Union Monogr.* **16**, 95–107.

Construction of profiles and balanced cross section (Sessions 18 and 24)

Argand, E. (1911). "Les nappes de recouvrement des Alpes Pennines et leurs prolongements structuraux." Francke, Bern.

Bally, A. W., Gordy, P. L. and Stewart, G. A. (1966). Structure, seismic data, and orogenic evolution of southern Canadian Rocky Mountains. *Canad. Petrol. Geol. Bull.* **14**, 337–381.

Buxtorf, A. (1907). Geologische Beschreibung des Weissenstein-Tunnels und seiner Umgebung. *Beitr. geol. Karte Schweiz (NF)* **21**.

Buxtorf, A. (1916). Prognosen und Befunde beim Hauensteinbasin—und Grenchenbergtunnel und die Bedeutung der letzteren für die Geologie des Juragebirges. *Verhandl. Naturforsch. Gesellsch. Basel* **27**, 184–254.

Chamberlin, R. T. (1910). The Appalachian folds of central Pennsylvania. *J. Geol.* **18**, 288–251.

Coates, J. (1945). The construction of geological sections. *J. Geol. Min. Met. Soc. India* **17**, 1–20.

Cooper, M. A. and Trayner, P. M. (1986). Thrust-surface geometry: implications for thrust belt evolution and section-balancing techniques. *J. Struct. Geol.* **8**, 305–312.

Coward, M. P. (1983). Thrust tectonics, thin skinned or thick skinned, and the continuation of thrusts to deep in the crust. *J. Struct. Geol.* **5**, 113–124.

Gwinn, V. E. (1964). Thin-skinned tectonics in the Plateau and northwestern Valley and Ridge provinces of the central Appalachians. *Geol. Soc. Amer. Bull.* **75**, 863–900.

Gwinn, V. E. (1970). Kinematic patterns and estimates of lateral shortening. *In* "Studies of Appalachian Geology" (G. W. Fisher *et al*, eds) 127–146. Wiley, New York.

Heim, A. (1921). "Geologie der Schweiz." Bern. Tauchnitz Verlag GmbH, Stuttgart.

Hossack, J. R. (1978). The correction of stratigraphic sections for tectonic finite strain in the Bygdin area, Norway. *J. Geol. Soc. London* **135**, 220–241.

Hossack, J. R. (1983). A cross section through the Scandinavian Caledonides constructed with the aid of branch-line maps. *J. Struct. Geol.* **5**, 103–112.

Hsü, K. J. (1979). Thin skinned plate tectonics during neo-Alpine orogenesis. *Am. J. Sci.* **279**, 353–366.

Huggenberger, P. (1985). Faltenmodelle und Verformungsverteilung in Deckenstrukturen am Beispiel der Morcle Decke (Helvetikum der Westschweiz). *Diss. ETH Zürich* **7861**, 1–326.

Julivert, M. and Arboleya, M. L. (1986). Areal balancing and estimates of areal reduction in a thin-skinned fold-and-thrust belt (Cantabrian zone, N.W. Spain): constraints on its emplacement mechanism. *J. Struct. Geol.* **8**, 407–414.

Mertie, J. B. (1959). Classification, delineation and measurements of non-parallel folds. *U.S. Geol. Surv. Prof. Paper* **314-E**, 91–124.

Müller, St., Egloff, R. and Ansorge, J. (1976). Struktur des tieferen Untergrundes entlang der Schweizer Geotraverse. *Schweiz. Mineral. Petrogr. Mitt.* **56**, 685–692.

Mugnier, J. L. and Vialon, P. (1986). Deformation and displacement of the Jura cover on its basement. *J. Struct. Geol.* **8**, 373–387.

Phillips, W. E. A. and Byrne, J. G. (1969). The construction of sections in areas of highly deformed rocks. *In* "Time and Place in Orogeny" (P. E. Kent *et al*. eds.). *Geol. Soc. London Spec. Pub.* **3**, 81–93.

Rich, J. L. (1934). Mechanics of low-angle overthrust faulting as illustrated by Cumberland thrust block, Virginia, Kentucky and Tennessee. *Am. Assoc. Petrol. Geol. Bull.* **18**, 1584–1596.

Rubey, W. W. (1926). Determination and use of thicknesses of incompetent beds in oil-field mapping and general structural studies.

Stockwell, C. H. (1950). The use of plunge in the construction of cross-sections of folds. *Proc. Geol. Assoc. Canada* **3**, 97–121.

Williams, G. D. and Fischer, M. W. (1984). A balanced section across the Pyrenean orogenic belt. *Tectonics* **3**, 773–780.

Woodward, N. B., Gray, D. R. and Spears, D. B. (1986). Including strain data in balanced cross sections. *J. Struct. Geol.* **8**, 313–324.

Superposed folding (Session 22)

Allenbach, B. and Carron, J. M. (1985). Modélisation de la réorientation des linéations autour des plis concentriques aplatis. *Sci. Géol. Bull. Strasbourg* **37**, 197–212.

Clifford, P., Fleuty, M. J., Ramsay, J. G., Sutton, J. and Watson, J. (1957). The development of lineation in complex fold systems. *Geol. Mag.* **94**, 1–24.

Fleuty, M. J. (1961). The three fold systems in the metamorphic rocks of Upper Glen Orrin, Ross-shire. *J. Geol. Soc. London* **117**, 447–479.

Ghosh, S. K. and Ramberg, H. (1968). Buckling experiments on intersecting fold patterns. *Tectonophysics* **5**, 89–105.

Ghosh, S. K. (1974). A theoretical study of intersecting fold patterns. *Tectonophysics* **9**, 559–569.

Ghosh, S. K. (1974). Strain distribution in superposed buckling folds and the problem of reorientation of early lineations. *Tectonophysics* **21**, 249–272.

Ghosh, S. K. and Chatterjee, A. (1985). Patterns of deformed early lineations over later folds formed by buckling and flattening. *J. Struct. Geol.* **7**, 651–666.

Holdsworth, R. E. and Roberts, A. M. (1984). Early curvilinear fold structures and strain in the Moine of the Glen Garry region, Inverness-shire. *J. Geol. Soc. London* **141**, 327–338.

Jacobson, C. E. (1984). Petrological evidence for the development of refolded folds during a single deformation event. *J. Struct. Geol.* **6**, 563–570.

O'Driscoll, E. S. (1964). Interference patterns from inclined shear fold systems. *Canad. Petrol. Geol. Bull.* **12**, 279–310.

Mukhopadhyay, D. (1965). Some aspects of coaxial refolding. *Geol. Rundsch.* **55**, 819–825.

Ramsay, J. G. (1958). Moine-Lewisian relations at Gleneig, Inverness-shire. *J. Geol. Soc. London* **113**, 487–520.

Reynolds, D. L. and Holmes, A. (1954). The superposition of Caledonoid folds on an older fold system in the Dalradians of Malin Head, Co. Donegal. *Geol. Mag.* **91**, 417–433.

Sitter, L. U. de (1960). Crossfolding in non-metamorphic rocks of the Cantabrian mountains and in the Pyrenees. *Geol. en. Mijnb.* **39**, 189–194.

Tobisch, O. T. (1966). Large-scale basin and dome pattern resulting from the interference of major folds. *Geol. Soc. Am. Bull.* **77**, 393–400.

Williams, P. F. (1985). Multiply deformed terrains—problems of correlation *J. Struct. Geol.* **7**, 269–280.

Fault geometry and mechanisms (Session 23 and 25)

Anderson, E. M. (1951). "The Dynamics of Faulting", 206 pp. Oliver and Boyd, Edinburgh.

Aydin, A. and Johnson, A. M. (1983). Analysis of faulting in porous sandstones. *J. Struct. Geol.* **5**, 19–31.

Aydin, A. and Reches, Z. (1982). Number and orientation of fault sets in the field and in experiments. *Geology* **10**, 107–112.

Bahat, D. (1983). New aspects of rhomb structures. *J. Struct. Geol.* **5**, 591–602.

Brock, W. G. and Engelder, T. (1977). Deformation associated with the movement of the Muddy Mountain overthrust in the Buffington window, southeast Nevada. *Geol. Soc. Amer. Bull.* **88**, 1667–1677.

Butler, R. W. H. (1982). A structural analysis of the Moine Thrust zone between Loch Eriboll and Foinaven, N.W. Scotland. *J. Struct. Geol.* **4**, 19–30.

Butler, R. W. H. (1982). The terminology of structures in thrust belts. *J. Struct. Geol.* **4**, 239–246.

Chapple, W. M. (1978). Mechanics of thin-skinned fold and thrust belts. *Geol. Soc. Am. Bull.* **89**, 1189–1198.

Cloos, E. (1955). Experimental analysis of fracture patterns. *Geol. Soc. Am. Bull.* **66**, 241–256.

Coward, M. P. (1980). The Caledonian thrust and shear zones of N.W. Scotland. *J. Struct. Geol.* **2**, 11–17.

Coward, M. P. (1982). Surge zones in the Moine thrust zone of N.W. Scotland. *J. Struct. Geol.* **4**, 247–256.

Crowell, J. C. (1974). Origin of late Cenozoic basins in southern California. *In* "Tectonics and Sedimentation", (W. R. Dickinson, ed.). *Soc. Econom. Palaeo. Mineral.* **22**, 190–204.

Davis, G. H. (1979). Laramide folding and faulting in southeastern Arizona. *Am. J. Sci.* **279**, 543–569.

Diegel, F. A. (1986). Topological constraints on imbricate thrust networks, examples from the Mountain City window, Tennessee, U.S.A. *J. Struct. Geol.* **8**, 269–279.

Elliott, D. (1976). The motion of thrust sheets. *J. Geophys. Res.* **81**, 949–963.

Elliott, D. (1976). The energy balance and deformation mechanisms of thrust sheets. *Phil. Trans. Roy. Soc. London* **A283**, 289–312.

Elliott, D. and Johnson, M. R. W. (1980). Structural evolution in the northern part of the Moine thrust belt, N.W. Scotland. *Trans. Roy. Soc. Edinburgh* **71**, 69–96.

Freund, R. (1970). Rotation of strike slip faults in Sistan, southeastern Iran. *J. Geol.* **78**, 188–200.

Gammond, J. F. (1983). Displacement features associated with fault zones: a comparison between observed examples and experimental models. *J. Struct. Geol.* **5**, 33–45.

Garfunkel, Z. (1972). Transcurrent and transform faults: a problem of terminology. *Geol. Soc. Amer. Bull.* **83**, 3491–3496.

Gay, N. C. (1970). The formation of step structures on slicksided shear surfaces. *J. Geol.* **78**, 523–532.

Gretner, P. E. (1972). Thoughts on overthrust faulting in a layered sequence. *Canad. Pet. Geol. Bull.* **20**, 583–607.

Grocott, J. (1981). Fracture geometry of pseudotachylite generation zones: a study of shear fractures formed during seismic events. *J. Struct. Geol.* **3**, 169–178.

Hamblin, W. K. (1965). Origin of "reverse drag" on the downthrow side of normal faults. *Geol. Soc. Amer. Bull.*

76, 1145–1164.

Hancock, P. L. and Kadhi, A. (1978). Analysis of mesoscopic fractures in the Dhruma-Nisah segment of the central Arabian graben system. *J. Geol. Soc. London* **135**, 339–347.

Hancock, P. L. and Barka, A. A. (1981). Opposed shear senses inferred from neotectonic mesofracture systems in the North Anatolian fault zone. *J. Struct. Geol.* **3**, 383–392.

Harland, W. B. (1971). Tectonic transpression in Caledonian Spitzbergen. *Geol. Mag.* **108**, 27–42.

Hoeppener, R., Kalthoff, E. and Schrader, P. (1969). Zur. Physikalischen Tektonik: Bruchbildung bei verschiedenen Deformationen im Experiment. *Geol. Rundsch.* **59**, 179–193.

Horsfield, W. T. (1980). Contemporaneous movement along crossing conjugate normal faults. *J. Struct. Geol.* **2**, 305–310.

Hutchinson, D. R., Klitgord, K. D. and Detrick, R. S. (1986). Rift basins of the Long Island platform. *Geol. Soc. Am. Bull.* **97**, 688–702.

Illies, J. H. (1977). Ancient and Recent rifting in the Rheingraben. *Geol. Mijnb.* **56**. 329–350.

Illies, J. H. and Greiner, G. (1978). Rheingraben and the Alpine system. *Geol. Soc. Am. Bull.* **89**, 770–782.

Jackson, J. and McKenzie, D. (1983). The geometrical evolution of normal fault systems. *J. Struct. Geol.* **5**, 471–482.

Johnson, M. R. W. (1981). The erosion factor in the emplacement of the Keystone thrust sheet (SE Nevada) across a land surface. *Geol. Mag.* **118**, 501–507.

Julivert, M. and Arboleya, M. L. (1984). A geometrical and kinematic approach to the nappe structure in an arcuate fold belt: the Cantabrian nappes (Hercynian chain, NW Spain). *J. Struct. Geol.* **6**, 499–519.

Knipe, R. J. (1985). Footwall geometry and the rheology of thrust sheets. *J. Struct. Geol.* **7**, 1–10.

Korn, H. and Martin, H. (1959). Gravity tectonics in the Naukluft Mountains of South West Africa. *Geol. Soc. Amer. Bull.* **70**, 1047–1078.

Lensen, G. J. (1958). Note on the compressional angle between intersecting transcurrent clockwise and anticlockwise faults. *N. Zealand J. Geol. Geophys.* **1**, 533–540.

Lindström, M. (1974). Steps facing against the slip direction: a model. *Geol. Mag.* **111**, 71–74.

McCross, A. M. (1986). Simple construction for deformation in transpression/transtension zones. *J. Struct. Geol.* **8**, 715–718.

McKenzie, D. P. and Morgan, W. J. (1969). The evolution of triple junctions. *Nature* **224**, 125–133.

Mitra, S. (1986). Duplex structures and imbricate thrust systems: Geometry, structural position, and hydrocarbon potential. *Am. Assoc. Pet. Geol. Bull.* **70**, 1087–1112.

Mitra, G. and Boyer, S. E. (1986). Energy balance and the deformation mechanisms of duplexes. *J. Struct. Geol.* **8**, 291–304.

Moody, J. D. and Hill, M. J. (1956). Wrenchfault tectonics. *Geol. Soc. Am. Bull.* **67**, 1207–1247.

Muraoka, H. and Kamata, H. (1983). Displacement distribution along minor fault traces. *J. Struct. Geol.* **5**, 483–495.

Nicholson, R. and Pollard, D. D. (1985). Dilation and linkage of echelon cracks. *J. Struct. Geol.* **7**, 583–590.

Oertel, G. (1965). The mechanism of faulting in clay experiments. *Tectonophysics* **2**, 343–393.

Paterson, M. S. (1958). Experimental deformation and faulting in Wombeyan marble. *Geol. Soc. Am. Bull.* **69**, 465–476.

Peach, B. N., Horne, J., Gunn, W., Clough, M. A. and Hinxman, L. W. (1907). The geological structure for the northwest Highlands of Scotland. *Geol. Surv. G. Brit. Mem.*, 1–668.

Price, R. A. (1986). The southeastern Canadian Cordillera: thrust faulting, tectonic wedging, and delamination of the lithosphere. *J. Struct. Geol.* **8**, 239–254.

Raleigh, C. B. and Griggs, D. T. (1963). Effect of the toe in the mechanics of overthrust faulting. *Geol. Soc. Am. Bull.* **74**, 819–830.

Riecker, R. F. (1965). Fault plane features: an alternative explanation. *J. Sea. Petrol.* **35**, 746–748.

Riedel, W. (1929). Zur Mechanik geologischer Brucherscheinungen. *Zent. Min. Geol. Pal.* **1929**, 354–368.

Rippen, J. H. (1985). Contoured patterns of the throw and hade of normal faults in the Coal Measures of northeast Derbyshire. *Yorkshire Geol. Soc. Proc.* **45**, 147–161.

Sanderson, D. J. and Marchini, W. R. D. (1984). Transpression. *J. Struct. Geol.* **6**, 449–458.

Scholz, C. H. (1977). Transform fault systems of California and New Zealand: similarities in their tectonic and seismic styles. *J. Geol. Soc. London* **133**, 215–229.

Segall, P. and Pollard, D. D. (1983). Nucleation and growth of strike slip faults in granite. *J. Geophys. Res.* **88**, 555–568.

Sibson, R. H. (1986). Earthquakes and lineament infrastructure. *Phil. Trans. Roy. Soc. London* **317**, 63–79.

Suppe, J. (1985). "Principles of Structural Geology", 1–537. Prentice Hall, N. Jersey.

Tchalenko, J. S. and Ambraseys, N. N. (1970). Structural analysis of the Dasht-e-Bayaz (Iran) earthquake fractures. *Geol. Soc. Am. Bull.* **81**, 41–60.

Wellman, H. W. (1954). Angle between principal horizontal stress and transcurrent faults. *Geol. Mag.* **91**, 407–408.

Wellman, H. W. (1971). Reference lines, fault classification, transform systems, and ocean-floor spreading. *Tectonophysics* **12**, 199–210.

White, R. S. and Williams, C. A. (1986). Oceanic fracture zones. *J. Geol. Soc. London* **143**, 737–741.

Willemin, J. H. (1984). Erosion and the mechanics of shallow foreland thrusts, *J. Struct. Geol.* **6**, 425–432.

Williams, G. D. (1985). Thrust tectonics in the south central Pyrenees, *J. Struct. Geol.* **7**, 11–17.

Wilson, J. T. (1965). A new class of faults and their bearing on continental drift. *Nature* **207**, 343–347.

Wojtal, S. (1986). Deformation within foreland thrust sheets by population of minor faults. *J. Struct. Geol.* **8**, 341–360.

Woodcock, N. H. and Fischer, M. (1986). Strike-slip duplexes. *J. Struct. Geol.* **8**, 725–736.

Woodward, N. B. (1986). Thrust fault geometry of the Snake River Range, Idaho and Wyoming, *Geol. Soc. Amer. Bull.* **97**, 178–193.

Crustal stresses, general mechanical principles (Session 25)

Adams, F. D. and Bancroft, J. A. (1917). On the amount of

internal friction developed in rocks during deformation and on the relative plasticity of different types of rocks. *J. Geol.* **25**, 597–637.

Alvarez, L. W., Alvarez, W., Asaro, F. and Michel, H. V. (1980). Extraterrestrial cause for the Cretaceous-Tertiary extinction. *Science* **208**, 1095–1107.

Beach, A. (1982). Chemical processes in deformation at low grades: pressure solution and hydraulic fracturing. *Episodes* **4**, 22–25.

Berry, F. A. F. (1973). High fluid potentials in California Coast Ranges and their tectonic significance. *Am. Assoc. Petrol. Geol. Bull.* **57**, 1219–1249.

Bieniawski, Z. T. (1967). Mechanism of brittle failure of rock. *Int. J. Rock. Mech. Min. Sci.* **4**, 395–430.

Brace, W. F. (1960). An extension of the Griffith theory of fracture to rocks. *J. Geophys. Res.* **65**, 3477–3480.

Byerlee, J. D. and Brace, W. F. (1968). Stick slip, stable gliding and earthquakes-effect of rock type, pressure, strain rate and stiffness. *J. Geophys. Res.* **73**, 6031–6037.

Chapman, R. W. (1979). Mechanics of unlubricated sliding. *Geol. Soc. Am. Bull.* **90**, 19–28.

Fletcher, R. C. and Pollard, D. D. (1981). Anticrack model for pressure solution surfaces. *Geology* **9**, 419–424.

French, B. M. and Short, N. M. (1968). "Shock Metamorphism. Mono Books N. Yk.

Friedman, L. (1972). Residual elastic strain in rocks. *Tectonophysics* **15**, 297–330.

Geiser, P. A. and Sansone, S. (1981). Joints, microfractures and the formation of solution cleavage in limestone. *Geology* **9**, 280–285.

Gretener, P. E. (1972). Thoughts on overthrust faulting in a layered sequence. *Canad. Petrol. Geol. Bull.* **20**, 583–607.

Gretener, P. E. (1969). Fluid pressure in porous media, its importance in geology: a review. *Canad. Petrol. Geol. Bull.* **17**, 255–295.

Grieve, R. A. F. (1982). The record of impact on Earth: implications for a major Cretaceous/Tertiary impact event. *Geol. Soc. Am. Spec. Paper* **190**, 25–55.

Griffith, A. A. (1920). The phenomenon of rupture and flow in solids. *Phil. Trans. Roy. Soc. London* **221**, 163–197.

Gubler, E., Kahle, H. G., Klingelé, E., Mueller, St. and Olivier, R. (1981). Recent crustal movements in Switzerland and their geophysical interpretation. *Tectonophysics* **71**, 125–152.

Gubler, E. (1976). Beitrag des Landesnivellements zur Bestimmung vertikaler Krustenbewegungen in der Gotthard-Region. *Schweiz. Mineral. Petrog. Mitt.* **56**, 675–678.

Hafner, W. (1951). Stress distribution and faulting. *Geol. Soc. Am. Bull.* **62**, 373–391.

Handin, J. (1969). On the Coulomb-Mohr failure criterion. *J. Geophys. Res.* **74**, 5343–5348.

Hodgson, J. H. (1957). Nature of faulting in large earthquakes. *Geol. Soc. Am. Bull.* **68**, 611–652.

Hsü, K. J. (1968). Role of cohesive strength in the mechanics of overthrust faulting and landsliding. *Geol. Soc. Am. Bull.* **80**, 927–952.

Hubbert, M. K. and Rubey, W. W. (1959). Role of fluid pressure in mechanics of overthrust faulting. *Geol. Soc. Am. Bull.* **70**, 115–166.

Lajtai, E. Z. (1969). Mechanisms of second order faults and tension gashes. *Geol. Soc. Am. Bull.* **80**, 2253–2272.

Lisle, R. J. and Strom, C. S. (1982). Least squares fitting of the linear Mohr envelope. *J. Eng. Geol. London* **15**, 55–56.

McClay, K. R. (1977). Pressure solution and Coble creep in rocks and minerals: a review. *J. Geol. Soc. London* **134**, 57–70.

McGarr, A. (1980). Some constraints on levels of shear stress in the crust from observations and theory. *J. Geophys. Res.* **85**, 6231–6238.

McGarr, A. and Gay, N. C. (1978). State of stress in Earth's crust. *Rev. Earth Planet. Sci.* **6**, 405–436.

Mercier, J. C., Anderson, D. A. and Carter, N. L. (1977). Stress in the lithosphere. Interferences from steady state flow of rocks. *Pure Appl. Geophysics* **115**, 119–226.

Müller, W. H. and Briegel, U. (1980). Mechanical aspects of the Jura overthrust. *Eclogae Geol. Helv.* **73**, 239–250.

Müller, W. H. and Hsü, K. J. (1980). Stress distribution in overthrust slabs and Mechanics of Jura deformation. *Rock Mech.* **9**, 219–232.

Orowan, E. (1960). Mechanisms of seismic faulting. *Geol. Soc. Am. Memoir* **79**, 323–345.

Pavoni, N. (1977). Erdbeben im Gebiet der Schweiz. *Eclogae Geol. Helv.* **70**, 351–370.

Phillips, W. J. (1972). Hydraulic fracturing and mineralisation. *J. Geol. Soc. London* **128**, 337–359.

Plessman, W. von (1964). Gesteinslösung, ein Hauptfaktor beim Schieferungsprozess. *Geol. Mitt.* **4**, 69–82.

Price, N. J. (1963). The influence of geological factors on the strength of coal measure rocks. *Geol. Mag.* **100**, 428–443.

Roy, D. W. and Hansman, R. H. (1971). Two occurrences of shatter-cone-like fractures. *Geol. Soc. Am. Bull.* **82**, 3183–3188.

Rutter, E. H. (1986). On the nomenclature of mode of failure transitions in rocks. *Tectonophysics* **122**, 381–387.

Rutter, E. H. (1976). The kinetics of rock deformation by pressure solution. *Phil. Trans. Roy. Soc. London* **283**, 203–219.

Schmid, S. M. (1983). Microfabric studies as indicators of deformation mechanisms and flow laws operative in mountain building. In "Mountain Building Processess" (K. J. Hsü, ed.) 95–110. Academic Press, London and Orlando.

Schmitt, T. L. (1981). The west European stress field: new data and interpretation. *J. Struct. Geol.* **3**, 309–315.

Sibson, R. H. (1983). Continental fault structure and the shallow earthquake source. *J. Geol. Soc. London* **140**, 741–767.

Sibson, R. H. (1973). Interactions between temperature and pore pressure during earthquake faulting. *Nature* **243**, 66–68.

Sibson, R. H. (1980). Power dissipation and stress levels on faults in the upper crust. *J. Geophys. Res.* **85**, 6239–6247.

Sibson, R. H. (1981). Controls on low stress hydro-fracture dilatancy in thrust, wrench and normal fault terrains. *Nature* **289**, 665–667.

Sibson, R. H., Moore, J. M. and Rankin, A. H. (1975). Seismic pumping—a hydrothermal fluid transport mechanism. *J. Geol. Soc. London* **131**, 653–659.

Silver, L. T. and Schultz, P. H. (1982). Geological implications of impacts of large asteroids and comets on the earth. *Geol. Soc. Am. Spec. Paper* **190**, 1–528.

Syme Gash, P. J. (1971). Dynamic mechanism for the formation of shatter cones. *Nature* **230**, 32–35.

Voight, B. (1966). Interpretation of in situ stress measurements. *Int. J. Roch Mech. Proc.* **3**, 332–348.

Voight, B. (1967). On photoelastic techniques, in situ stress and strain measurement and the field geologist. *J. Geol.* **75**, 46–58.

Voight, B. and Samuelson, B. (1969). On the application of finite-element techniques to problems concerning potential distribution and stress analysis in the earth sciences. *Pure Appl. Geophys.* **76**, 40–55.

Walsh, J. B. (1965). The effect of cracks on the uniaxial elastic compression of rocks. *J. Geophys. Res.* **70**, 399–411.

Rock products of faults and shear zones (Sessions 23 and 25)

Allen, A. R. (1979). Mechanism of frictional fusion in fault zones. *J. Struct. Geol.* **1**, 231–243.

Balk, R. (1952). Fabric of quartzites near thrust faults. *J. Geol.* **60**, 415–435.

Bell, T. H. and Etheridge, M. A. (1976). The deformation and recrystallisation of quartz in a mylonite zone, central Australia. *Tectonophysics* **32**, 235–267.

Bell, T. H. and Etheridge, M. A. (1973). The microstructure of mylonites and their descriptive terminology. *Lithos* **6**, 337–348.

Borradaile, G. J., Bayly, M. B. and Powell, C. McA. (1982). "Atlas of Deformation and Metamorphic Rock Fabrics", 1–551. Springer Verlag, New York.

Bouchez, J. L., Lister, G. S. and Nicolas, A. (1983). Fabric asymmetry and shear sense in movement zones. *Geol. Rundsch.* **72**, 401–419.

Boullier, A. M. and Guegen, Y. (1975). Origin of some mylonites by superplastic flow. *Contr. Mineral. Petrol.* **50**, 93–104.

Byerlee, J. D., Mjachkin, V., Summers, R. and Voevoda, O. (1978). Structures developed in fault gouge during stable sliding and stick-slip. *Tectonophysics* **44**, 161–171.

Chester, F. M., Friedman, M. and Logan, J. M. (1985). Foliated cataclasites. *Tectonophysics* **111**, 139–146.

Dietrich, D. and Honglin Song (1984). Calcite fabrics in a natural shear environment, the Helvetic nappes of western Switzerland. *J. Struct. Geol.* **6**, 19–32.

Edlington, J. W., Melton, K. N. and Cutler, C. P. (1976). Superplasticity. *Prog. Mater. Sci.* **14**, 63–100.

Engelder, T. (1974). Cataclasis and the generation of fault gouge. *Geol. Soc. Am. Bull.* **85**, 1515–1522.

Gilotti, J. A. and Kumpulainen, R. (1986). Strain softening induced ductile flow in the Sarv thrust sheet, Scandinavian Caledonides. *J. Struct. Geol.* **8**, 441–445.

Hadizadeh, J. and Rutter, E. H. (1983). The low temperature brittle-ductile transition in a quartzite and the occurrence of cataclastic flow in nature. *Geol. Rundsch.* **72**, 493–509.

House, W. M. and Gray, D. R. (1982). Cataclasites along the Saltville thrust, U.S.A. and their implications for thrust sheet emplacement. *J. Struct. Geol.* **4**, 257–269.

Hsü, K. J. (1985). A basement of melanges: A personal account of the circumstances leading to the breakthrough in Franciscan research. *Geol. Soc. Am. Spec.* **1**, 47–64.

Hsü, K. J. (1968). Principles of melanges and their bearing on the Franciscan-Knoxville paradox. *Geol. Soc. Am. Bull.* **79**, 1069–1074.

Jordan, P. (1987). The deformation behaviour of bimineralic limestone-halite aggregates. *Tectonophysics* **135**, 185–197.

Lapworth, C. (1885). The Highland controversy in British Geology: its causes, course and consequences. *Nature* **32**, 558–559.

Lister, G. S. and Snoke, A. W. (1984). S-C mylonites. *J. Struct. Geol.* **6**, 617–638.

McKenzie, D. and Brune, J. N. (1972). Melting on fault planes during large earthquakes. *Geophys. J.R. ast. Soc.* **29**, 65–78.

Mitra, G. (1978). Ductile deformation zones and mylonites: the mechanical process involved in the deformation of crystalline basement rocks. *Am. J. Sci.* **278**, 1057–1084.

Mitra, G. (1984). Brittle to ductile transition due to large strains along the White Rock thrust, Wind River Mountains, Wyoming. *J. Struct. Geol.* **6**, 51–61.

Philpotts, A. R. (1964). Origin of pseudotachylites. *Am. J. Sci.* **262**, 1008–1035.

Powell, D. and Treagus, J. F. (1970). Rotational fabrics in metamorphic minerals. *Min. Mag.* **27**, 801–814.

Rosenfeld, J. L. (1970). Rotated garnets in metamorphic rocks. *Geol. Soc. Am. Prof. Paper* **129**, 1–105.

Schmid, S. M., Boland, J. N. and Paterson, M. S. (1977). Superplastic flow in fine grained limestone. *Tectonophysics* **43**, 257–291.

Schmid, S. M., Casey, M. and Starkey, J. (1981). An illustration of the advantages of a complete texture described by the orientation distribution function (ODF) using quartz pole figure data. *Tectonophysics* **78**, 101–118.

Schmid, S. M., Casey, M. and Starkey, J. (1981). The microfabric of calcite tectonites from the Helvetic nappes (Swiss Alps). *In* "Thrust and Nappe Tectonics" (K. R. McClay and N. J. Price, eds.) *Geol. Soc. London Spec. Publ.* **9**, 151–158.

Schmid, S. M. and Casey, M. (1986). Complete fabric analysis of some commonly observed quartz c-axis patterns. *Am. Geophys. Union Monograph* **36**, 263–286.

Shand, S. J. (1916). The pseudotachylite of Parijs (Orange Free State) and its relation to "trapshotten gneiss" and "flinty crush rock". *J. Geol. Soc. London* **72**, 198–217.

Sibson, R. H. (1975). Generation of pseudotachylite by ancient seismic faulting. *Geophys. J. Royal Astr. Soc.* **43**, 775–794.

Spry, A. (1968). The origin and significance of snowball structure in garnet. *J. Petrol.* **4**, 211–222.

Stel, H. (1981). Crystal growth in cataclasites: diagnostic microstructure and implications. *Tectonophysics* **78**, 585–600.

White, S. (1979). Grain and sub-grain size variations across a Mylonite zone. *Contr. Mineral. Petrol.* **70**, 193–202.

White, S. (1977). Geological significance of recovery and recrystallisation processes in quartz. *Tectonophysics* **39**, 143–170.

White, S. (1982). Fault rocks of the Moine Thrust zone: a guide to their nomenclature. *Textures Microstructures* **4**, 211–221.

White, S. H., Burrows, S. E., Carreras, J., Shaw, N. D. and Humphreys, F. J. (1980). On mylonites in ductile shear zones. *J. Struct. Geol.* **2**, 175–188.

White, S. H., Evans, D. J. and Zhong, D. L. (1982). Fault rocks of the Moine Thrust zone: microstructures and textures of selected mylonites. *Textures Microstructures* **5**, 33–61.

Wiltshire, H. G. (1971). Pseudotachylite from the Vredefort Ring, S. Africa. *J. Geol.* **79**, 195–206.

Wojtal, S. and Mitra, G. (1986). Strain hardening and strain softening in fault zones from foreland thrusts. *Geol. Soc. Am. Bull.* **97**, 674–687.

Fluid inclusions (Session 25)

Fyfe, W. S., Price, N. J. and Thompson, A. B. (1978). Fluids in the Earth's crust. *Elsevier Pub. Co. Amsterdam*, 1–385.

Haas, J. L. (1976). Thermodynamic properties of the coexisting phases and thermochemical properties of the H_2O component in boiling NaCl solutions. *U.S. Geol. Surv. Bull.* **1421-B**, 1–71.

Knipe, R. J. and White, S. (1978). On the importance of bubbles in quartz deformation. *J. Geol. Soc. London* **135**, 313–316.

Mullis, J. (1976). Dans Wachstumsmilieu der Quarzkristalle im Val d'Illiez, Wallis, Schweiz. *Schweiz. Min. Petrol. Mitt.* **56**, 219–168.

Mullis, J. (1979). The system methane-water as a geologic thermometer and barometer from the external part of the Central Alps. *Bull Minéral.* **102**, 526–536.

Mullis, J. (1987). Fluid inclusion studies during very low grade metamorphism. *In* "Very Low Grade Metamorphism" (M. Frey, ed). Blackie, Glasgow.

Potter, R. W. (1977). pressure correlations for fluid inclusions, homogenisation temperatures based on the volumetric properties of the system $NaCl-H_2O$. *U.S. Geol. Surv. J. Res.* **5**, 603–607.

Poty, B. P., Stalder, H. A. and Weisbrod, A. M. (1974). Fluid inclusion studies in quartz from fissures of western and central Alps. *Schweiz. Min. Petrol. Mitt.* **54**, 717–752.

Roedder, E. (1962) and (1963). Studies of fluid inclusions 1. low temperature application of a dual purpose freezing and heating stage. 2. Freezing data and their interpretation. *Econ. Geol.* **57**, 1045–1061, **58**, 167–211.

Sorby, H. C. (1858). On the microscopic structures of crystals, indicating the origin of minerals and rocks. *J. Geol. Soc. London* **14**, 453–500.

Smith, F. G. (1953). Historical development of inclusion thermometry. *Unit. Toronto Press*, 1–149.

Shear zone geometry (Session 3 and 26)

Beach, A. (1975). The geometry of en-echelon vein arrays. *Tectonophysics* **28**, 245–263.

Beach, A. (1977). Vein arrays, hydraulic fractures and pressure solution structures in a deformed flysch sequence, S.W. England. *Tectonophysics* **40**, 201–225.

Berthé, D. and Brun, J. P. (1980). Evolution of folds during progressive shear in the South American shear zone, France. *J. Struct. Geol.* **2**, 127–133.

Brunel, M. (1986). Ductile thrusting in the Himalayas: shear sense criteria and stretching lineations. *Tectonics* **5**, 247–265.

Caby, R. (1968). Une zone de décrochements a l'échele de l'Afrique dans le Précambrien de l'Ahaggar occidental.

Géol. Soc. France Bull. **10**, 577–587.

Cobbold, P. R. (1977). Description and origin of banded deformation structures. *Canad. J. Earth Sci.* **14**, 2510–2532.

Coward, M. P. (1976). Archean deformation patterns in Southern Africa. *Phil. Trans. Roy. Soc. London* **283**, 313–332.

Coward, M. P. (1984). The strain and textural history of thin-skinned tectonic zones: examples from the Assynt region of the Moine thrust zone, NW Scotland. *J. Struct. Geol.* **6**, 89–100.

Coward, M. P. and Potts, G. J. (1983). Complex strain patterns developed at the frontal and lateral tips to shear zones. *J. Struct. Geol.* **5**, 383–399.

Dietrich, D. (1986). The orientation of conjugate sets of en-echelon veins in multilayered sequence, helvetic nappes of W. Switzerland. *Tect. Studies Gp. Meeting abstracts*, 17.

Fleuty, M. J. (1964). Tectonic slides. *Geol. Mag.* **101**, 452–456.

Grocott, J. (1977). The relationship between Precambrian shear belts and modern fault systems. *J. Geol. Soc. London* **133**, 257–262.

Grocott, J. and Watterson, J. (1980). Strain profile of a boundary within a large ductile shear zone. *J. Struct. Geol.* **2**, 111–118.

Harris, L. B. and Cobbold, P. R. (1985). Development of conjugate shear bands during bulk simple shearing. *J. Struct. Geol.* **7**, 37–44.

Ingles, J. (1983). Theoretical strain patterns in ductile zones simultaneously undergoing heterogeneous simple shear and bulk shortening. *J. Struct. Geol.* **5**, 369–381.

Knipe, R. J. and White, S. H. (1979). Deformation in low grade shear zones in the Old Red Sandstone, S.W. Wales. *J. Struct. Geol.* **1**, 53–66.

Laubscher, H. P. (1983). Detachment, shear, and compression in the central Alps. *Geol. Soc. Am. memoir* **158**, 191–211.

Laurent, P. and Burg, J. P. (1978). Strain analysis through a shear zone. *Tectonophysics* **47**, 15–42.

Lisle, R. J. (1984). Strain discontinuities within the Seve-Köli nappe complex, Scandinavian Caledonides. *J. Struct. Geol.* **6**, 101–110.

Passchier, C. W. (1984). The generation of ductile and brittle shear bands in a low-angle mylonite zone. *J. Struct. Geol.* **6**, 273–282.

Pavoni, N. (1961). Die nordanatolische Horizontalverschiebung. *Geol. Rundsch.* **51**, 122–139.

Platt, J. and Behrmann, J. H. (1986). Structures and fabrics in a crustal-scale shear zone, Betic Cordillera, SE Spain. *J. Struct. Geol.* **8**, 15–33.

Poirier, J. P. (1980). Shear localization and shear instability in materials in the ductile field. *J. Struct. Geol.* **2**, 135–142.

Pollard, D. D., Segall, P. and Delaney, P. T. (1982). Formation and interpretation of dilatant echelon cracks. *Geol. Soc. Am. Bull.* **93**, 1291–1303.

Rickard, M. J. and Rixon, L. K. (1983). Stress configuration in conjugate quartz vein arrays. *J. Struct. Geol.* **5**, 573–578.

Sanderson, D. (1973). Models of strain variation in nappes and thrust sheets, a review. *Tectonophysics* **137**, 289–302.

Sibson, R. H. (1980). Transient discontinuities in ductile shear zones. *J. Struct. Geol.* **2**, 165–171.

Simpson, C. (1983). Displacement and strain patterns from naturally occurring shear zone terminations. *J. Struct. Geol.* **5**, 597–506.

Tchalenko, J. S. (1970). Similarities between shear zones of different magnitudes. *Geol. Soc. Am. Bull.* **81**, 1625–1640.

Watts, M. J. and Williams, G. D. (1983). Strain geometry, microstructure and mineral chemistry in metagabbro shear zones: a study of softening mechanisms during progressive mylonitisation. *J. Struct. Geol.* **5**, 507–518.

Williams, G. D., Chapman, T. J. and Milton, N. J. (1984). Generation and modification of finite strain patterns by progressive thrust faulting in the Lakse fjord Nappe. *Tectonophysics* **107**, 177–186.

Joints (Session 27)

Bahat, D. and Engelder, T. (1984). Surface morphology on cross-fold joints of the Appalachian plateau, New York and Pennsylvannia. *Tectonophysics* **104**, 299–313.

Bevan, T. G. (1984). Tectonic evolution of the Isle of Wight: a Cenozoic stress history based on mesofractures. *Proc. Geol. Assoc. London* **96**, 227–235.

Carlisle, D. (1965). Sliding friction and overthrust faulting. *J. Geol.* **73**, 271–192.

Engelder, T. (1985). loading paths to joint propagation during a tectonic cycle: an example from the Appalachian Plateau, U.S.A. *J. Struct. Geol.* **7**, 459–476.

Firman, R. J. (1960). The relationship between joints and fault patterns in the Eskdale granite (Cumberland) and the adjacent Borrowdale volcanic series. *J. Geol. Soc. London* **116**, 317–347.

Gudmundson, A. (1983). Stress estimates from the length/width ratios of fractures. *J. Struct. Geol.* **5**, 623–626.

Hancock, P. L. (1984). Regional joint sets in the Arabian platform as indicators of intraplate processes. *Tectonics* **3**, 27–43.

Hancock, P. L. (1969). Jointing in the Jurassic limestones of the Cotswold Hills. *Proc. Geol. Assoc. London* **80**, 291–241.

Hodgson, R. A. (1961). Regional study of jointing in Comb Ridge-Navajo mountain area, Arizona and Utah. *Am. Assoc. Pet. Geol. Bull.* **45**, 1–38.

Holst, T. B. and Foot, G. R. (1981). Joint orientation in Devonian rocks of the northern portion of the lower peninsula of Michigan. *Geol. Soc. Am. Bull.* **92**, 85–93.

Kobold, F. (1977). Die Hebung der Alpen aus dem Vergleich des "Nivellement de précision" der Schweizerischen Geodätischen Kommission mit den Landesnivellementen der Eidgenössischen Landestopographie. *Vermessung. Photogrametrie, Kulturtechnik* **75**, 129–137.

Ladeira, F. L. and Price, N. J. (1981). Relationship between fracture spacing and bed thickness. *J. Struct. Geol.* **3**, 179–184.

La Pointe, P. R. and Hudson, J. A. (1985). Characterisation and interpretation of rock mass joint patterns. *Geol. Soc. Am. Spec. Paper* **199**, 1–37.

Meier, D. (1984). Zur Tektonik des schweizerischen Tafel—und Faltenjura. *Clausthaler Geowiss. Diss.* **14**, 1–75.

Murray, F. N. (1967). Jointing in sedimentary rocks along the Grand Hogback monocline, Colorado. *J. Geol.* **75**, 340–350.

Ruhland, M. (1973). Méthode d'étude de la fracturation naturelle des roches associée à divers modèles structuraux. *Sci. Géol. Bull. Strasbourg* **26**, 91–113.

Parker, J. M. (1942). Regional systematic jointing in slightly deformed sedimentary rocks. *Geol. Soc. Am. Bull.* **53**, 381–408.

Price, N. J. (1959). Mechanics of jointing in rocks. *Geol. Mag.* **96**, 149–167.

Roberts, J. C. (1961). Feather-fracture and the mechanics of rock jointing. *Am. J. Sci.* **259**, 481–492.

Reches, Z. (1976). Analysis of joints in two monoclines in Israel. *Geol. Soc. Am. Bull.* **87**, 1654–1662.

Segall, P. and Pollard, D. D. (1983). Joint formation in granitic rock of the Sierra Nevada. *Geol. Soc. Am. Bull.* **94**, 563–575.

Segall, P. (1984). Formation and growth of extensional fracture sets. *Geol. Soc. Am. Bull.* **95**, 454–462.

Spry, A. (1962). The origin of columnar jointing, particularly in basalt flows. *J. Geol. Soc. Australia* **8**, 191–216.

Syme-Gash, P. J. (1971). A study of surface features relating to brittle and semi-brittle fracture. *Tectonophysics* **12**, 349–391.

Terzaghi, R. D. (1965). Sources of errors in joint surveys. *Geotechnique* **15**, 287–304.

Woodworth, J. B. (1896). On the fracture systems of joints with remarks on certain great fractures. *Proc. Boston Soc. Nat. Hist.* **27**, 163–184.

INDEX

Page numbers in bold refer to citations where the term is defined or explained; page numbers in italic refer to figures.